ORGANIC REACTIONS AND THEIR NOMENCLATURE

ORGANIC REACTIONS AND THEIR NOMENCLATURE

Ramesh Chandra, Snigdha Singh, and Aarushi Singh

ARCLER PRESS

www.arclerpress.com

Organic Reactions and their nomenclature

Ramesh Chandra, Snigdha Singh, and Aarushi Singh

Arcler Press
2010 Winston Park Drive,
2nd Floor
Oakville, ON L6H 5R7
Canada
www.arclerpress.com
Tel: 001-289-291-7705
001-905-616-2116
Fax: 001-289-291-7601
Email: orders@arclereducation.com

ISBN: 978-1-77407-358-2 (Hardcover)

ABOUT THE AUTHORS

Prof. Ramesh Chandra is an outstanding scientist, revered teacher and an exceptionally successful administrator. He is currently heading Department of Chemistry, University of Delhi, where he is serving as Professor for the last more than 26 years and Founder Director of Dr. B. R. Ambedkar Center for Biomedical Research, University of Delhi, since March 1991. He has been Vice-Chancellor, Bundelkhand University, Jhansi for six years (1999-2005); Member, Planning Commission, Government of U.P, India as well as the President of the Indian Chemical Society (2004-2006). Professor Chandra started his research career at the University of Delhi, thereafter he went to The New York Hospital-Cornell University Medical Center and the Rockefeller University, New York; State University of New York at Stonybrook, USA as Assistant Research Professor. He conducted advanced research at the Harvard University Medical School-Massachusetts General Hospital, jointly at MIT, Cambridge, USA. Over the last 38 years, Professor Chandra has contributed largely in the field of Chemical Sciences and particularly in New Drug Discovery and Development as well as Drug Metabolism. He has to his credit several patents, published more than 300 original Scientific Research Papers/ Review Articles in International journals of repute and six of his internationally acclaimed scientific Books. Prof. Chandra is the recipient of several professional national/ international recognitions; these includes: Millennium Plaques of Honor (Life Time Achievement Award for Contribution in Science & Technology) by the Indian Science Congress Association (ISCA) for 2017-2018, Award of the Highest Honor of Soka University, Tokyo, Japan (2000); J William Fulbright Scholarship (1993); The Rockefeller Foundation USA-Biotechnology Career Award (1993); and several others.

Snigdha Singh completed her M.Tech. degree (Chemical Synthesis and Process Technologies) in 2016 from University of Delhi, India. After that she joined Prof. Ramesh Chandra group at Department of Chemistry, University of Delhi for her doctoral studies. She is involved in miscellaneous projects for development of novel hydroxyethylamine molecules as potent multistage Antimalarial. Her sincere efforts and excellent performance has culminated her into a keen researcher. Currently, she is working at University of Siena under the supervision of Prof. Maurizio Taddei on Synthesis of novel 8-hydroxyquinolines as Gli-1 Hedgehog Inhibitors. She has successfully optimized complex organic syntheses during her doctoral research. Her research efforts are directed towards the synthesis of bioactive heterocyclic molecules. She has already published 9 papers in international and national reputed Journals and also presented her work in many conferences.

Ms. Aarushi Singh completed her Masters degree in chemistry (2013) from University of Delhi, Delhi, India. Thereafter, she joined as senior research fellow at Indian Agricultural Research Institute, Pusa, Delhi, India. Then, she joined Lingaya's University, Faridabad, India as Assistant Professor in Chemistry.

TABLE OF CONTENTS

LIST OF FIGURES

LIST OF TABLES

LIST OF ABBREVIATIONS

1-Me-AZADO	1-methyl-2-azaadamantane N-oxyl
2D	two-dimensional
2IB acid	2-iodo benzoic acid
3D	three-dimensional
A	acceptors
ABNO	9-azabiciclo[3.3.1]nonane N-oxyl
AZADO	2-azaadamantane N-oxyl
BDBA	1,4-benzene boronic acid
BPA	bis(picolyl) amine
BPDA	biphenyldiboronic acid
BTBDBA	1,4-benzothiadiazole diboronic acid
CA	carbon atoms
CC	Couple Cluster
COFs	covalent organic frameworks
CTFs	covalent triazine-based frameworks
Cu	copper
D	donors
DBSA	p-dodecylbenzenesulfonic acid
DCC	dynamic covalent chemistry
DES	eutectic deep solvents
DFT-B	density-functional founded tight-binding
DIAD	diisopropyl azodicarboxylate
DKR	dynamic kinetic resolution
DMSO	dimethyl sulfoxide
DPB	4,40-diphenylbutadiynebis(boronic acid)

EGB	electro-generated base
Fe	Iron
GCMC	Grand Canonical Monte Carlo
HHTP	2,3,6,7,10,11hexahydroxy triphenylene
HQL	8-hydroxyquinoline
IBA	2-iodoso benzoic acid
IBX	o-iodoxybenzoic acid
ILs	ionic liquids
Ir	iridium
ketoABNO	9-azabiciclo[3.3.1]non-3-one N-oxyl
KetoABNO	9-azabiciclo[3.3.1]non-3-one N-oxyl
LJ	Lennard-Jones
MOFs	metal-organic frameworks
MP	Møller–Plesset
NBS	N-bromosuccinimide
NHPI	N-hydroxyphthalimide
Ni	Nickel
Nor-AZADO	2-azanoradamantane N-oxyl
NPs	nanoparticles
O2	oxygen
OH	hydroxyl group
PCC	pyridinium chlorochromate
PDBA	pyrene-2,7-diboronic acid
PES	potential energy surface
PINO	nitroxyl radical phthalimide-N-oxyl
Ru	ruthenium
RuH2(PPh3)3CO	Ruthenium-hydride precursor
STM	scanning tunneling microscopy
TBPM	tetra(4-dihydroxy boryl phenyl) methane
TBPS	tetra(4dihydroxyborylphenyl) silane
tBuSi(OH)3	tert-butylsilane triol
TEM	transmission electron microscopy

TEMPO	2,2,6,6tetramethylpiperidine-1-oxyl
THA	thymine-1-acetate
THB	1,2,4,5-tetra-hydroxybenzene
TOPO	trioctylphosphine
TriBOT	2,4,6-tris(benzyloxy)-1,3,5-triazine
V	vanadium
XRD	X-ray diffraction
ZIFs	zeolitic imidazolate frameworks

PREFACE

Organic chemistry literature contains a time-honored tradition of assigning name to many reactions after chemists or scientists who developed or discovered them. These so called "name reactions" are a substantial tool available to the chemists for assessing the feasibility of chemical reactions. Since the discovery of organic name reactions, these reactions have been implemented in organic synthesis in varying ways, and many reactions have been amended as well. New processes, reaction conditions, solvent types, and condensing species all have backed to their advancement and development. Moreover, the study of different organic mechanisms have resulted in a better understanding of many reactions. Consequently, the organic chemists and scientists of today have at their disposal tools which have been honed and refined. It seemed worthy, therefore, to gather some of these name reactions and make them accessible to the chemists in their present state.

An extensive or comprehensive coverage of the available literature for all the name reactions is beyond the scope of this book. Therefore, no effort has been made to compete with the renowned book "Organic Reactions," nor have the efforts been made to quote references exclusively on the basis of priority. However, the selection of topics has been made from the perspective of general interest, leading references and illustrative examples.

There are around 500 well-known reactions out of approximately a thousand name reactions presently available in literature. We provided a list of common name reactions in our book but only 7 major name reactions are discussed in the book. Unlike many books covering reactions, this book focusses on detailed analysis of mechanism and reactions taking place for a particular organic compound.

Fundamental knowledge of organic compounds and their nomenclature is important to understand different kinds of name reactions. Chapter 1 briefly encompasses the fundamentals of organic compounds and their nomenclature. Each of the following seven chapters mainly focusses on a single name reaction in detail. Chapter 2 discusses the phenomenon of catalytic oxidation of alcohol under different catalytic conditions. Chapter 3 introduces us with the fundamentals of covalent organic frameworks. Design, synthesis and functionality of covalent organic frameworks is discussed in the chapter. Chapter 4 focusses on the reactions involving the synthesis of carboxylic acid esters. Both electrophilic and nucleophilic reactions are discussed in the chapter.

Fundamentals of carboxylic acids, including their chemical and physical properties, are discussed in Chapter 5. Moreover, the chapter also contain details about the nomenclature and use of different compounds containing carboxylic acid. Aldehydes are important species in organic chemistry which are used to manufacture a wide range of compounds. Chapter 6 contains the fundamentals of aldehydes along with different reactions involving aldehydes. Chapter 7 discusses the historical aspects of Clemmensen reduction mechanism followed by the description of complete Clemmensen reduction mechanism. Carbonyl group is considered an essential entity in organic chemistry. There is an extensive variety of compounds containing carbonyl group. Chapter 8 contains information about carbonyl group, their presence in aldehydes and ketones and reactions of compounds containing carbonyl group.

This book contains detailed information about key organic name reactions. Therefore, the book is a ready reference for students, researchers and chemists working in a chemical industry.

1

Organic Compounds and Their Nomenclature

CONTENTS

1.1. INTRODUCTION

The branch of chemistry involving the study of carbon as well as its compounds is known as Organic chemistry. Presently, carbon is famous to produce apparently infinite number of compounds. We use organic compounds excessively in our daily lives in agriculture, medicine, and general life.

In principle (Oparin, 1923), the beginning of organic chemistry might be with the big bang when the components of nitrogen, ammonia, CO_2, and methane combined to produce amino acids. This experiment has already been verified in the laboratory (Miller, 1950). Egyptians and Romans made use of organic chemicals in ancient times to produce medicines, dyes, and poisons from natural sources; however, the chemical composition of the substances was still not known (Brooke, 1968).

In the late 16th century, organic compounds in their pure state, were isolated from nature (Cassebaum & Kauffman, 1976). After that, analytical methods were established to determine elemental composition (Lavoisier, 1784).

It was believed by numerous scientists that organic chemicals that are found in nature have a special "vital force" which direct their natural synthesis, and hence, it would not be possible to achieve a laboratory creation of the chemicals (Ramberg, 2000). Luckily, after some decades, it was discovered by Frederich Wöhler (1828) that urea, which is a natural component in urine, can be manufactured in the laboratory through heating of ammonium cyanate. This discovery nullified the initial idea that the natural "vital force" is necessary for synthesis of organic compounds, and therefore, cemented the way for multiple chemists to synthesize organic compounds artificially.

By the mid of the 19th century, several advancements had been made into the discovery, analysis as well as synthesis of numerous new organic compounds. Understanding of the structures of organic chemistry took place with a theory of bonding known as valence theory (Baskaran et al., 1987; Mayer, 1984; Palmer & Schloss, 2010).

In the 19th century, organic chemistry developed into an exciting and productive science. Since then, numerous new reaction mechanisms, production methods, analytical techniques, as well as structural theories have been developed. By the end of the century, a great deal of the knowledge of organic chemistry has been extended to the learning of biological systems like DNA and proteins. A great deal of information is published monthly in books, journals, and electronic media both about biological and organic chemistry.

The huge information available presently means that for fresh students of organic chemistry, a great deal of learning is required. It is important that students learn about organic reactions, mechanism, their synthesis and analysis, and as well as their biological function. Even though, the study of organic chemistry is complex, but at the same time, it is very interesting.

1.2. THE CHEMICAL BOND

1.2.1. Atomic Theory

In the early 19th century, the atomic theory of electrons began and was accepted near 1926 after Schroedinger and Heisenberg discovered mathematical solutions to the electronic energy levels present in atoms, this field of study is now known as quantum mechanics.

Electrons are present in energy levels surrounding the nucleus of the atom. As they get distant from the nucleus, the energy of these levels starts increasing. The energy levels are known as shells, and inside these shells, there are other energy levels present known as orbitals or subshells containing up to two electrons. The calculations from atomic theory provide the below mentioned results for electron energy and orbitals. The most important are the results for the shell 1 and shell 2 for bonding in organic chemistry.

Table 1.1: Electronic Shells and Orbitals Configuration

Shell	Total Possible Electrons	Orbitals			
		s	p	d	f
1	2	1			
2	8	2	3		
3	18	3	3	5	
4	32	1	3	5	7

Energy level 1 have up to 2 electrons in a spherical orbital known as 1s orbital.

- **Energy level 2** have up to eight electrons; two electrons are in an 2s-orbital and the other two are in each of three orbitals labeled as 2p-orbitals. The shape of p-orbitals is like a barbell and are aligned along the x axes, y axes, and z axes. They are, therefore, known as the px, py, and pz orbitals.
- **Energy level 3** have up to eighteen electrons, two electrons are present in a 3s orbital, six electrons are present in the three 3p orbitals, whereas ten electrons are present in the five 3d orbitals.
- **Energy level 4** have up to thirty-two electrons, two electrons are there in a 4s-orbital, six electrons are present in the three 4p-orbitals, ten electrons are present in the five 4 d-orbitals, whereas, fourteen electrons are there in the seven 4f-orbitals.

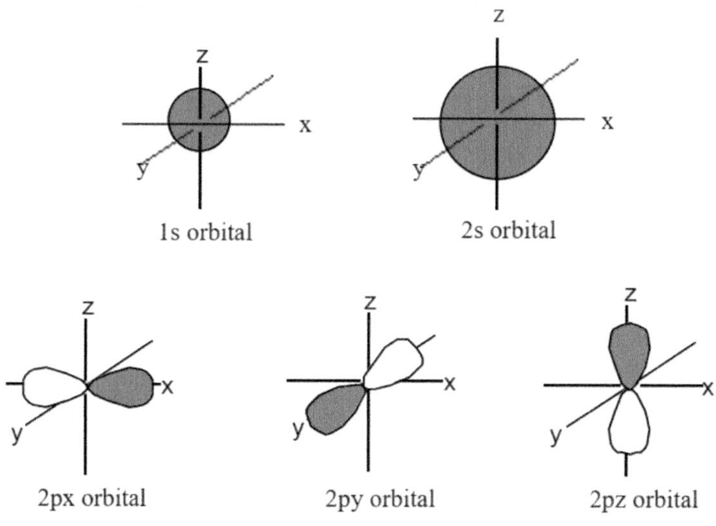

Figure 1.1: Configuration of electronic orbitals.

[Source: http://www.siue.edu/~tpatric/Ch%2001%20Intro%20H%20T%20I.pdf]

Electrons at first, fills the lower energy levels till each of the electrons is used (Aufbau principle). The number of electrons in an element is equal to its atomic number. For the elements present in the first and second row, configurations are comparatively simple.

Element	(Atomic number)	Electron Configuration
Hydrogen	(1)	1s1 (1st shell, s orbital, one electron)
Helium	(2)	1s2
Lithium	(3)	1s2, 2s1
Beryllium	(4)	1s2, 2s2
Boron	(5)	1s2, 2s2, 2p1
Carbon	(6)	1s2, 2s2, 2p2
Nitrogen	(7)	1s2, 2s2, 2p3
Oxygen	(8)	1s2, 2s2, 2p4
Fluorine	(9)	1s2, 2s2, 2p5
Neon	(10)	1s2, 2s2, 2p6 (inert, completely filled)

1.2.2. Electronegativity

The capability of an atom to attract electrons to itself is known as electronegativity and usually increases as we move from left side to the right side across the periodic table.

Least Most

Electronegative element Li < Be < B < C < N < O < F electronegative element

Increase in electronegativity is also observed as we move from the bottom place to the top of a column in the periodic table.

Least Most

Electronegative element I < Br < Cl < F electronegative element

Elements which can easily lose electron and resultantly acquires a positive charge are known as **electropositive** elements. Alkali metals are normally electropositive elements.

1.2.3. Bonding

Atoms can get bonded with each other. The type of bond which is formed is dictated by their electronic structure. There are two main two types of bonds which are formed and are known as **ionic bond** and **covalent bond**.

1.2.3.1. Ionic Bond

Ionic bonding is significant among atoms having massively different electronegativity. The bond is formed as a result of one atom giving up an electron whereas, another atom accepts that electron. Resultantly, both atoms acquire a stable Nobel gas configuration.

In the lithium fluoride, the electron present in $2s^1$ orbital of lithium is shifted to the $2p^5$ orbital of fluorine. The lithium atom releases an electron to produce the positively charged lithium cation with $1s^2$, $2s^0$ configuration, whereas, the fluorine atom accepts an electron to produce a fluoride anion having configuration of $1s^2$, $2s^2$, $2p^6$. Hence, the outer most energy levels of both ions are fully filled. Electrostatic attraction of the negative and positive and ions is responsible for holding the ions together.

$$\text{Li} + \text{F} \longrightarrow \text{Li}^{+} + \text{F}^{-}$$

Li	F	Li$^+$	F$^-$
$1s^2$	$1s^2$	$1s^2$	$1s^2$
$2s^1$	$2s^2$ $2p^5$	$2s^0$	$2s^2$ $2p^6$

1.2.3.2. Covalent Bond

A **covalent bond** is produced as a result of sharing of two electrons by two atoms. A hydrogen atom containing the $1s^1$ electron lines with another hydrogen atom having $1s^1$ configuration. The two atoms then create a covalent bond having two electrons through of sharing their electrons.

Figure 1.2: Covalent bonding of hydrogen.

[Source: http://www.siue.edu/~tpatric/Ch%2001%20Intro%20H%20T%20I.pdf]

In case of hydrogen fluoride, the hydrogen 1s electron is united with a $2p^5$ electron present in fluorine ($1s^2$, $2s^2$, $2p^5$), and the resultant molecule is held together through a covalent bond. Here, the electronegativity of fluorine atom is much more as compared to the hydrogen atom and the electrons present in the bond are likely to stay near the fluorine atom.

This is known as a **polar covalent bond,** and the atoms contain a tiny partial charge represented by the Greek δ symbol.

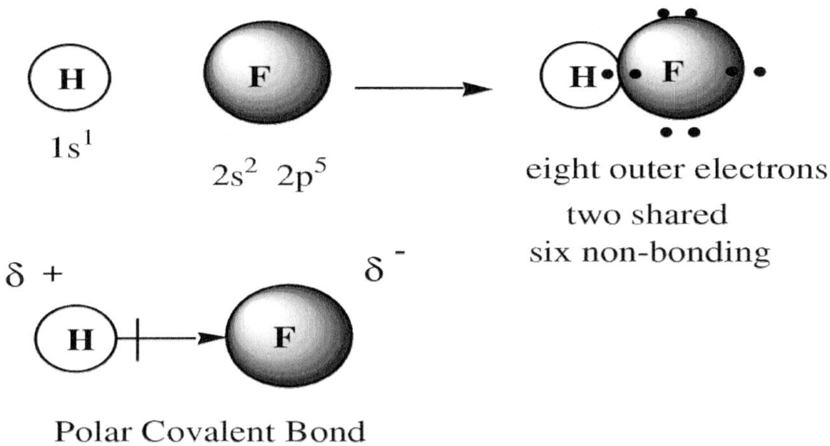

Figure 1.3: Poor covalent bond between fluorine and hydrogen.

[Source: http://www.siue.edu/~tpatric/Ch%2001%20Intro%20H%20T%20I.pdf]

1.3. BONDING IN CARBON COMPOUNDS

One unique property of carbon is that it has the ability to create bonds with itself and hence permits a great number of organic chemicals having numerous diverse properties. Carbon can create a single, double as well as triple bonds with itself as well as with other atoms (Reed & Weinhold, 1985; Bauer et al., 1987). This multiple bond ability of carbon permits carbon compounds to have a variability of shapes. Carbon creates four bonds in all carbon compounds. The types of bonds employed by the carbon atom are identified as sigma () bonds and pi (л) bonds. Different combinations of sigma and pi bonds results in creation of carbon single bonds, double bonds as well as triple bonds (Saunders & Hyne, 1958; Zerger et al., 1974).

1.3.1. Carbon-Hydrogen Single Bond: The Sigma () Bond

Till now, majority of the bonds formed in carbon compounds are covalent bonds found generally in the carbon-hydrogen single bond.

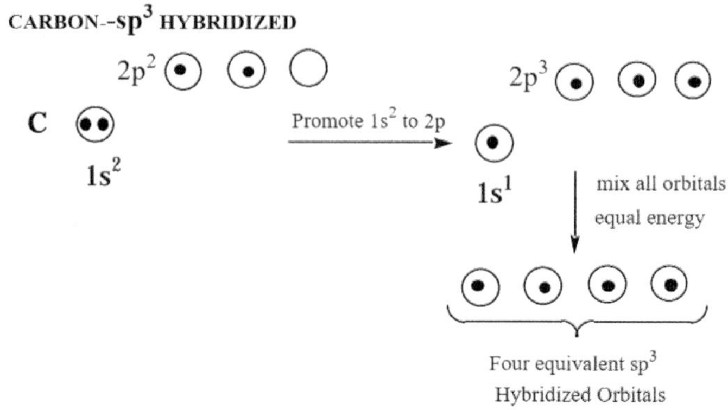

Figure 1.4: Sp3 hybridization of C orbitals.

[Source: http://www.siue.edu/~tpatric/Ch%2001%20Intro%20H%20T%20I.pdf]

In carbon one of the electrons of the $2s^2$ orbital is shifted to the 3rd 2p0 orbital. The s and 3 p orbitals hybridize to create four new orbitals having equal energy known as sp^3 hybrid orbitals. The electrons present in the four sp^3 hybridized orbitals create bond through overlap with the 1s1 hydrogen orbital. The resultant single covalent bond is known as a **sigma ()** bond. After that, the sp^3 bonds arrange themselves as distant from each other as they can. The shape of methane molecule, CH_4, is tetrahedral having bond angles of 109.5° (Haddon et al., 1980; Fink et al., 1983).

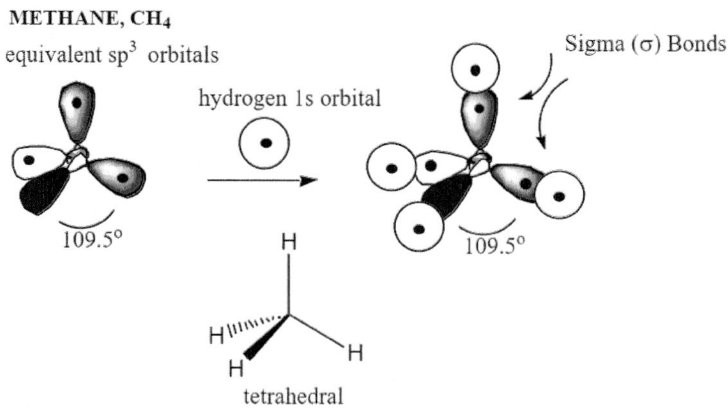

Figure 1.5: Bonding of sp3 hybridized C orbitals to produce methane.

[Source: http://www.siue.edu/~tpatric/Ch%2001%20Intro%20H%20T%20I.pdf]

The unique ability of carbon which discriminates it from the other elements permits the formation of numerous different organic compounds. Hence, adding of another carbon atom to CH_4 leads to creation of ethane which contains covalent sigma bonds to the hydrogen atoms whereas, a covalent sigma bond among the carbon atoms. Adding of any more carbon atoms results in several more compounds (Allen & Roberts, 1981; Alcarazo et al., 2009).

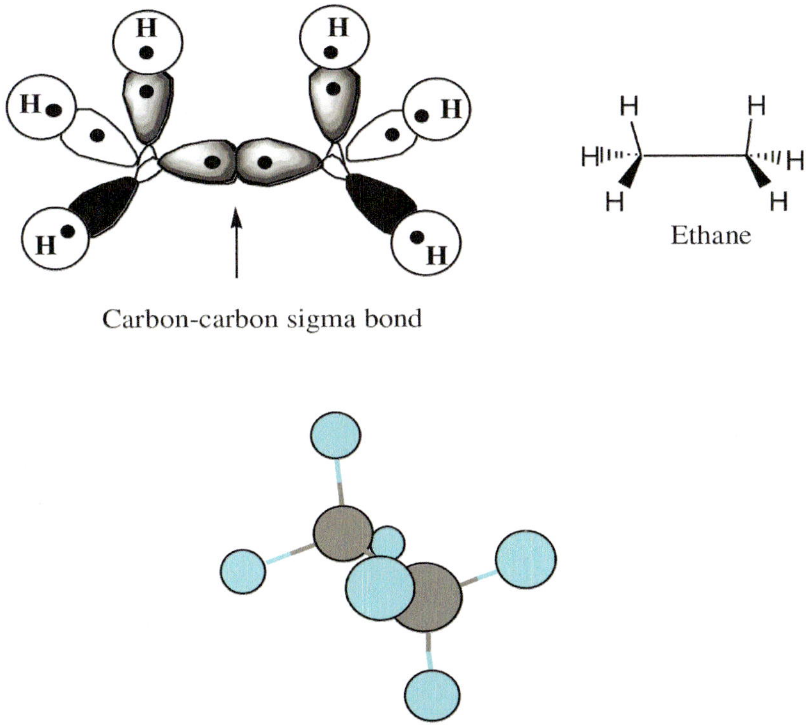

Figure 1.6: Bonding of sp3 hybridized orbitals with H and sigma bond between two carbons.

[Source: http://www.siue.edu/~tpatric/Ch%2001%20Intro%20H%20T%20I.pdf]

1.3.2. Carbon-Carbon Double Bond: The Pi (л) Bond

Carbon produces an extensive variety of compounds containing carbon bonded through a double bond to another carbon. These compounds are categorized as alkenes (older name is olefins). The blow mentioned orbital model describes the carbon-carbon double bond. The configuration of carbon electron illustrates one s electron being shifted to a p orbital. However, now just three orbitals are mixed which are one s orbital and two p orbital, that are known as sp2 hybrid orbitals and are used for creating single bonds (sigma bonds). There is one electron in the p orbital (Tonner & Frenking, 2007; 2008).

The grouping of two of the sp^2 hybridized C atom results in two C atoms being combined through overlap of sp^2 orbitals to create a C-C single bond, and the side-to-side overlay of the p orbitals to create another bond called a **pi** (л) bond. In the ethane molecule presented below, a total of five sigma bonds as well as one pi bond is present. As a consequence of the bonding in an ethane molecule, the molecule is planar having bond angles of 120° and a C=C bond length which is lengthier than the C-H bond length (Melaimi et al., 2009; Klein et al., 2010).

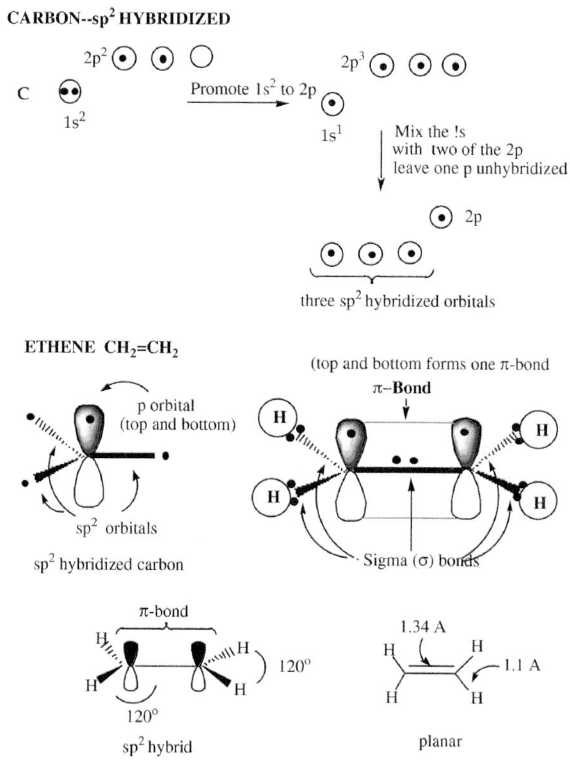

Figure 1.7: Creation of sp^2 hybridized orbitals and pi bonds.

[Source: http://www.siue.edu/~tpatric/Ch%2001%20Intro%20H%20T%20I.pdf]

When carbon creates a bond with an electronegative atom like nitrogen, oxygen, sulfur or else a halogen, the resultant bond is a polar covalent bond having electrons of the bond living closer to the electronegative atom.

1.3.3. The Carbon-Carbon Triple Bond

Another kind of bond formed by carbon with itself is the triple bond observed in a class of compounds known as alkynes. After 2s electron promotes to a 2p orbital, one s orbital is mixed with one p orbital to yield two hybrid sp orbitals. The residual two p orbitals are employed to create p bonds. Hence, the C is bound to hydrogen by a sigma bond from one of the sp hybrid orbital, to the other atom of C through a sigma bond from one of the sp hybrid orbitals, while the two C atoms are bound by two π bonds from side-to-side overlay of the two p orbitals. The two sp hybrid orbitals are 180° apart and therefore ethyne molecule is linear with the hydrogen atoms 180° apart (Alcarazo, 2011; Frenkin & Tonner, 2011).

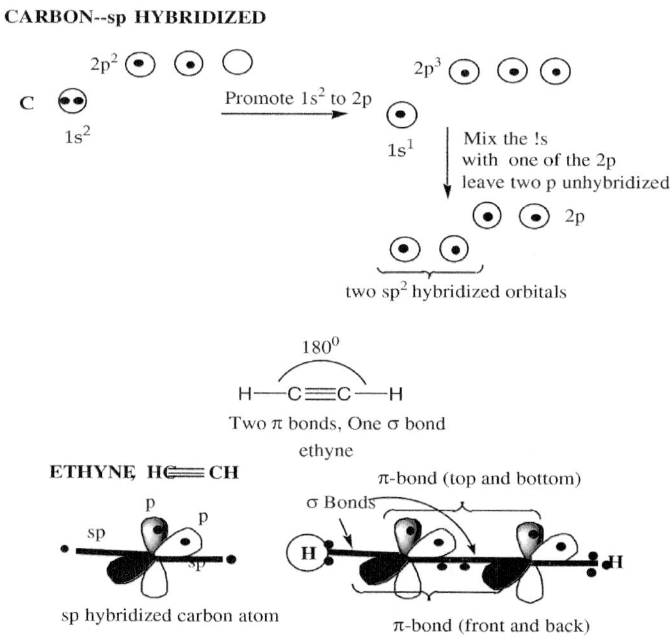

Figure 1.8: Formation of sp orbitals and 2 pi bonds.

[Source: http://www.siue.edu/~tpatric/Ch%2001%20Intro%20H%20T%20I.pdf]

1.3.4. Polar Covalent Bonds in Carbon

Carbon creates all single, double as well as triple bonds with other elements. The atoms which are generally involved in the bonding are nitrogen, oxygen, sulfur as well as the halogens. As compared to Carbon, these elements are much more electronegative and therefore attract the electrons to themselves. Thus, the bonds are polar covalent bonds. Bonds containing a separation of charge have a dipole moment which is a property contributing to the whole polarity of the molecule.

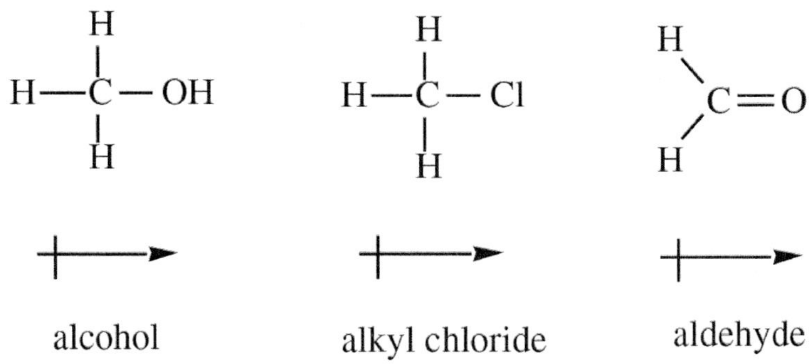

Figure 1.9: Polar covalent bonding shown in carbon compounds.

[Source: http://www.siue.edu/~tpatric/Ch%2001%20Intro%20H%20T%20I.pdf]

1.3.5. Hydrogen Bonds and Bond Polarity

The bonds such as N-H, O-H, and F-H are highly polar covalent bonds as the electronegative lures electrons away from the H atom. In each case, the atom of hydrogen has a partial positive charge.

$$N\delta^{-}\text{-}H^{\delta+}\ O^{\delta-}\text{-}H^{\delta+}\ F^{\delta-}\text{-}H^{\delta+}$$

As a consequence of molecules having highly polarized bonds with the H atom partially +ve in nature, the hydrogen atom is attracted more towards the basic site in other molecules like the non-bonding electrons on nitrogen and oxygen. This attraction is known as hydrogen bonding and is beneficial for explaining high melting points and high boiling points of equally low mass molecules. Hence, hydrogen bonding observed in water clarifies why the compound having just three atoms boils comparatively high as compared to other molecules having comparable mass. Additional energy is necessary for breaking the hydrogen bonds during the boiling procedure.

Hydrogen bonding is very significant in organic compounds for describing the solubility and boiling characteristics of acids and alcohols.

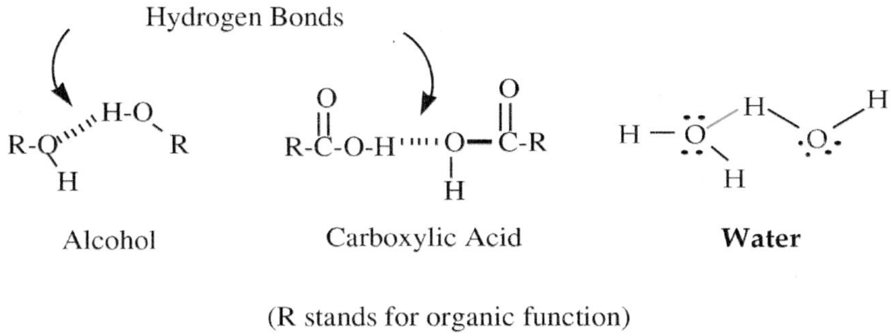

Figure 1.10: Hydrogen bonding in various compounds.

[Source: http://www.siue.edu/~tpatric/Ch%2001%20Intro%20H%20T%20I.pdf]

1.4. ORGANIC STRUCTURES

The study of organic chemistry encompasses understanding the shapes and energy of simple organic structures. The study of organic chemistry at advance stages also involves studying complicated three-dimensional structures and lastly the study of reactions of organic molecules. We make use of theories of bonding as well as the facts about basic or acidic properties of the molecules in order to understand the complete picture of what is taking place (Roeges & Baas, 1994; Crews et al., 2010). However, the overall purpose is to improve mankind by developing new and better procedures which can be applied in agriculture, medicine, and biomedical science to make life much better. Hence, we should be able to study complicated molecules as well (Langley & Hulliger, 1999; Nitschke, 2007).

While writing an alkane like pentane which contain five carbon atoms, we can write it by merely showing carbon atoms, however, it does not show carbon with 4 bonds. More appropriately it requires all of its hydrogen atoms, nonetheless writing all the hydrogen atoms for compounds can get burdensome. Therefore, we write condensed structures as can be seen below (Zerkowski et al., 1990; Ishii et al., 1999).

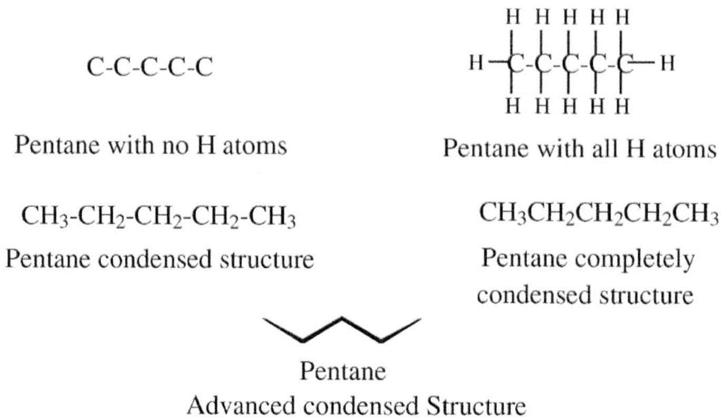

Figure 1.11: Several structures of pentane.

[Source: http://www.siue.edu/~tpatric/Ch%2001%20Intro%20H%20T%20I.pdf]

Carbon compounds also occur in rings, known as cyclic compounds. Below mentioned is the compound known as cyclohexane written in many different ways.

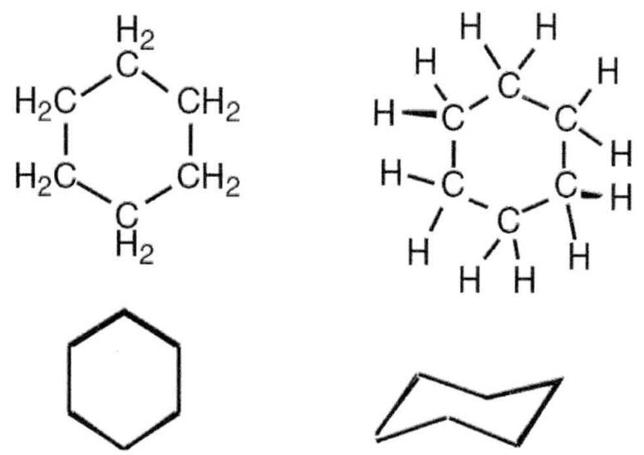

Figure 1.12: Shorthand made structures of organic compounds.

[Source: http://www.siue.edu/~tpatric/Ch%2001%20Intro%20H%20T%20I.pdf]

It is important to realize even in the most condensed structures that all the hydrogen atoms and all the carbon atoms are present even if the H's and the C's are not written. These shorthand approaches of writing structures are quiet useful in fast writing of equations, and therefore understanding them is significant (Chen et al., 2000; Rodríguez et al., 2012).

The structures of organic chemicals can be very interesting and very complicated at the same time. At times, their structure is so complex that only condensed structures are the reasonable presentation. Nonetheless as always, the carbon atom contains four bonds. Different advanced organic structures are represented below (PeteráClayton & John, 1990; Angelos et al., 1993).

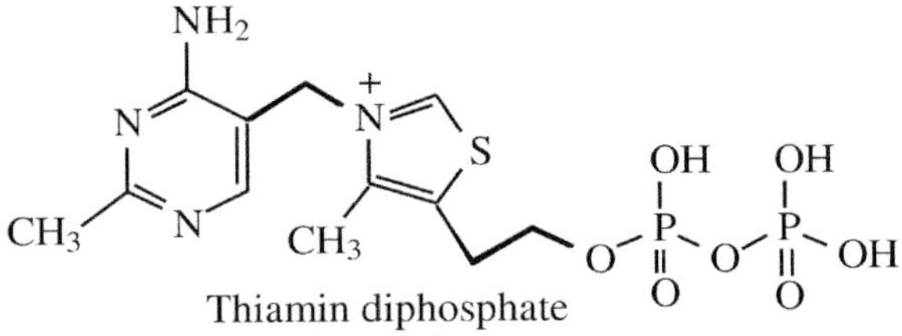

Thiamin diphosphate
Vitamin B_1
(grains, liver, pork)

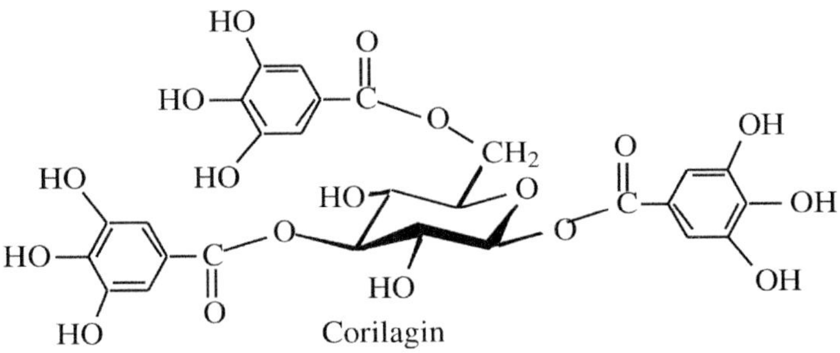

Corilagin
a molecule in the tannin family
found in cranberries
prevents bacteria from attaching to the kidneys

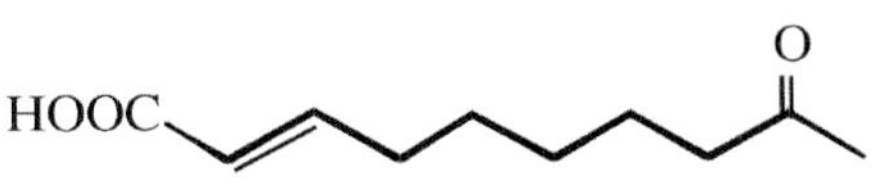

sex pheromone of bees

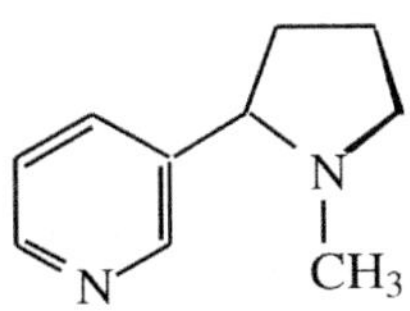

Nicotine
Addictive component
of tobacco

Figure 1.13: Advanced organic structures.

[Source: http://www.siue.edu/~tpatric/Ch%2001%20Intro%20H%20T%20I.pdf]

1.5. CLASSIFICATION OF ORGANIC COMPOUNDS

Organic compounds are frequently classified as per the type of functional groups present in them. The on-sight recognition of functional groups is necessary for studying organic chemistry.

Table 1.2: Organic Functional Groups

Alkane	CH_3CH_3	Alkene	$CH_2=CH_2$
Alkyne	$HC \equiv CH$	Alkyl Halide	R-halogen
Aromatic		Alcohol	R-OH
Phenol	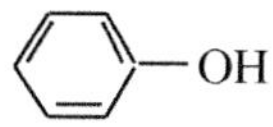	Ether	R-O-R
Aldehyde	RCH=O	Ketone	R-C(=O)-R
Acid	RCOOH	Ester	RCOOR
Anhydride	$(RCO)_2O$	Amide	$RCONH_2$
Nitrile	RCN	Amine	$R-NH_2$

1.6. ORGANIC NOMENCLATURE

Due to the vast variety of names of compounds, confusion can take place in organic chemistry; common names, systematic names, and trade names are predominant. For instance, a formula of compound, C_6H_6O is widely known as phenol, phenic acid, carbolic acid, phenylic acid, phenyl hydroxide, oxobenzene, and hydroxybenzene (Tukey et al., 1954; Hagenbuch & Meier, 2004)

In order to abolish the multiplying of numerous names for a single compound, scientists have derived a systematic IUPAC naming system to exclusively name the several million different organic compounds according to their structure (Son et al., 1954; Goodson, 1980).

Generally, compounds are named and classified by considering the following factors:

- the quantity and types of atoms present,
- the types of bond present in the molecule, and
- The geometry of the molecule.

Before setting the rules to name a compound, it is important to review few aspects of formulae. The molecular formula of any compound does not provide any clear information about the structure of that compound. For example, the formula C_2H_6O does not provide any information about the arrangement of various atoms. Actually two different compounds are represented by this same formula but have entirely different structures and differ vastly in their properties (Young, 1974; Bonchev, 1983).

ethanol   dimethyl ether

Figure 1.14: Structural formulae of dimethyl ether and ethanol.

[Source: https://people.ok.ubc.ca/pshipley/teaching/chem203/worksheets/nomenclature.pdf]

We can differentiate among these two compounds by writing their formulae in a structural manner.

ethanol	CH_3-CH_2-OH	or	CH_3CH_2OH	or	C_2H_5OH
dimethyl ether	$CH_3—O—CH_3$	or	CH_3OCH_3		

While writing these formulae, the atoms written after a carbon designate the groups or elements, attached to that carbon atom. We should remember that all carbon atoms in organic compounds essentially be involved in four bonds.

$CH_3CHClCCl_3$

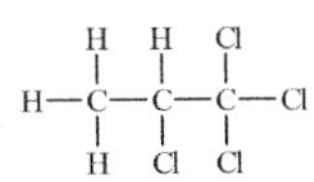

$CH_3CH(CH_3)CH_2CH{=}CH_2$

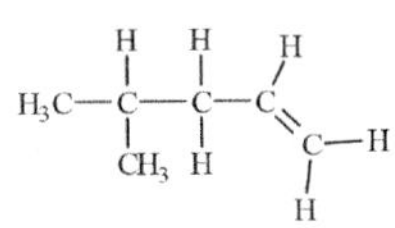

$C_6H_5CH(NH_2)COOH$

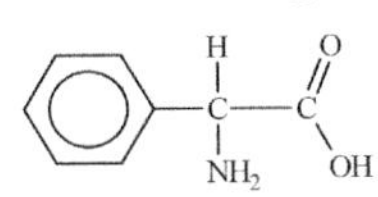

Figure 1.15: Structure formulae of several organic compounds.

[Source: https://people.ok.ubc.ca/pshipley/teaching/chem203/worksheets/nomenclature.pdf]

In the above example, many times, the bond angles are drawn as right angles and the molecules seem planar. It must be noted that we are drawing two-dimensional representations of three-dimensional molecules and in reality, the bond angles are hardly ever 90° (Favre et al., 1999).

propane: C_3H_8 or $CH_3CH_2CH_3$ or 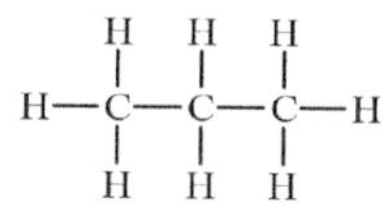 or 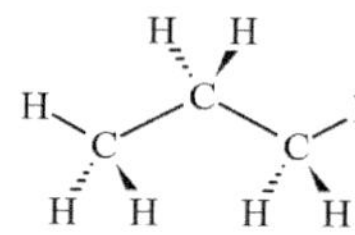

In the above representation, dashed lines denotes bonds behind the plane of the page whereas, solid lines represents bonds coming out from the page. All the remaining lines are in the plane of the page whereas, all bond angles are of 109°.

1.7. NAME REACTIONS

Name reactions honor the discoveries of revolutionary chemical reactions or modifications of former known transformations in a manner that numerous scientists have their names attached to a phenomenon, an effect, a constant or to an equation, etc. The names of reaction names can also just describe the type of reaction, frequently by using the initials or else referring to structural characteristics (Mundy et al., 2005; Polshettiwar & Varma, 2008).

For example, carbon-carbon bond formation is very significant field in chemical synthesis, and numerous name reactions exist describing such transformations. The development of a method for making use of organo-magnesium compounds by Victor Grignard resulted in completely new addition reactions which extremely extended the scope of organic synthesis. Grignard was not the first person to use such reagents; however, he slightly simplified the procedure through generation of highly reactive reagent in situ (Cablewski et al., 1994; Negishi, 2007). Grignard reagents are most frequently employed in additions to carbonyl compounds which provide alcohols or some other products in great yields, and this procedure is presently referred to as the Grignard reaction. On the other hand, several C-C bond-creating reactions are endorsed by palladium catalysis, which led to more effective usage of reagents and more freely accessible conditions (Liu & Lai, 1981; Dzhemilev et al., 2004). One example is biaryl moieties synthesis which are substructures occurring commonly in compounds of interest in field of medicinal chemistry (England et al., 2001; Dzhemilev, 2008).

Besides using the names of chemists, we also have revolutionary reactions which came to be known through abbreviations of a descriptive name, for instance INOC (intramolecular nitrile oxide cyclization) or "RCM" (ring-closing metathesis). We rarely use the name of the chemist who established RCM (Robert Grubbs) for referring to the reaction, nevertheless, as an alternative, his contribution is acknowledged through application of his name to the ruthenium-catalysts used (Yamada et al., 1981). Therefore, we speak of the "2nd generation Grubbs catalyst" or "Grubbs catalyst." Besides using names like "RCM," few regularly used reactions are named after their structural features of the product. Few examples include the "aldol reaction" or the "pinacol rearrangement." Majority of the common name reactions are written below:

- Acetaldehyde
- Actualization
- Acylation
- Alcohol oxidation
- Aldol condensation
- Allylic rearrangement
- Anthraquinone
- Baeyer-Drewson indigo synthesis
- Baeyer–Villiger oxidation
- Baeyer's reagent
- Baker–Venkataraman rearrangement
- Balz–Schiemann reaction
- Bamberger rearrangement
- Bamford–Stevens reaction
- Barfoed's test
- Beckmann rearrangement
- Benzoin condensation
- Birch reduction
- Bischler-Möhlau indole synthesis
- Bischler–Napieralski reaction
- Biuret test
- Blaise ketone synthesis

- Branching (polymer chemistry)
- Cannizzaro reaction
- Carbohydrate acetalisation
- Carbonyl reduction
- Carboxybenzyl
- Carboxylic acid
- Citrate synthase
- Claisen condensation
- Claisen rearrangement
- Clemmensen reduction
- Covalent organic framework
- Darzens reaction
- Dieckmann condensation
- Emil Knoevenagel
- Ester
- Knoevenagel condensation
- Mannich reaction
- Pechmann condensation
- Perkin reaction
- Phenanthrene
- Phenol formaldehyde resin
- Povarov reaction
- Self-condensation
- Thorpe reaction
- Wolffenstein-Böters reaction

1.8. NOMENCLATURE OF ORGANIC COMPOUNDS

As previously stated, compounds are classified according to their structure and are named in view of that. The simple most classification is that of the hydrocarbons, compounds of hydrogen and carbon. Furthermore, hydrocarbons are identified as being aromatic or aliphatic.

The aromatic hydrocarbons contain one or more than one benzene rings; whereas, aliphatics might be alkenes, alkanes, or alkynes. It is significant for students to get a good understanding of the convention that are used in naming the alkanes, as the naming of other classes is just an extension of alkane nomenclature.

1.8.1. Alkanes

Alkenes contain only hydrogen and carbon. Alkanes are:

- *Saturated*, that is, contain just single bonds.
- Straight chain alkanes are named as per the number of carbon atoms present.
- Normal alkanes creates a series, a homologous series having formula C_nH_{2n+2} where n is an integer (Colby et al., 1977; Ambrose & Tsonopoulos, 1995; Atkinson, 1997).

Table 1.3: Nomenclature of Carbon Compounds (Jones & Feher, 1983)

Name	Number of Carbon atoms (n)
Methane	1
Ethane	2
Propane	3
Butane	4
Pentane	5
Hexane	6
Heptane	7
Octane	8
Nonane	9
Decane	10

- It is important to memorize these names.
- The structure of butane, C_4H_{10}, is

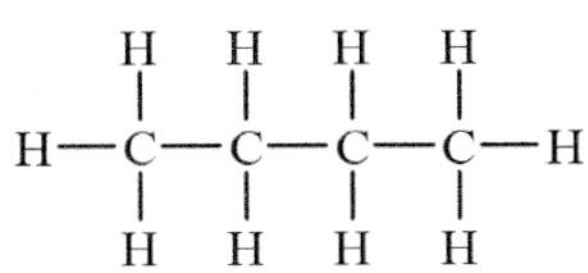

- Complications take place when branching occurs. The below mentioned compound also has the formula C_4H_{10}.

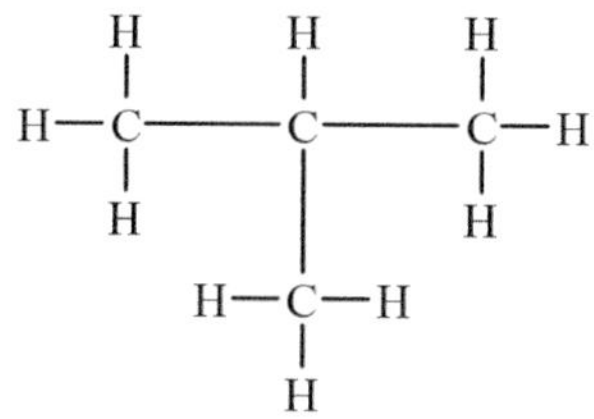

- Compounds of the same formula are known as isomers.
- Structural isomers have the same formula nonetheless different groupings branch from the major chain of carbon.

The below mentioned few structures only represent the carbon atoms, this method of drawing structures and marks is not acceptable. The only purpose of showing it here is that you can clearly see the differences (Webster, 1977; Sun, 1998; Shah et al., 2005). The ideal method of drawing structures is shown below.

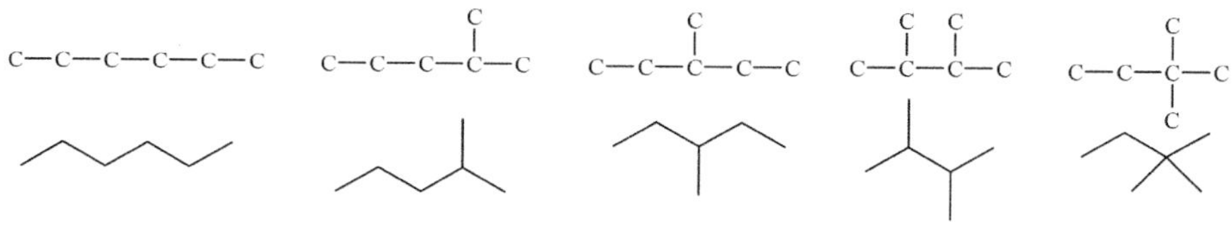

Figure 1.16: Shorthand isomeric structures of hexane.

[Source: https://people.ok.ubc.ca/pshipley/teaching/chem203/worksheets/nomenclature.pdf]

- Each of these compounds must be identified by a unique name, we can't call all of them to be hexane.
- Hexane is applied to the straight chain structure.
- Name the lengthiest continuous carbon chain present in the molecule as the parent name.
- Recognize the side groups that are attached to this chain and place them prior to the parent name in alphabetical manner.

alkane -------→ alkyl group CH_4 (methane) -------→ CH_3— (methyl) C_2H_6 (ethane) -------→ CH_3CH_2— or C_2H_5— (ethyl)

- If more than one groups of the same kind are attached to the major chain, only list the groups once by employing the suitable numerical prefix such as di, tri, tetra, penta, etc. to designate how many times that particular side group appears.

- Allocate a number to each of the side groups to specify where the group is attached to the major chain. The numbering of the major chain should be started from whichever end of the major chain will provide the lowest set of numbers (Qin et al., 1992).

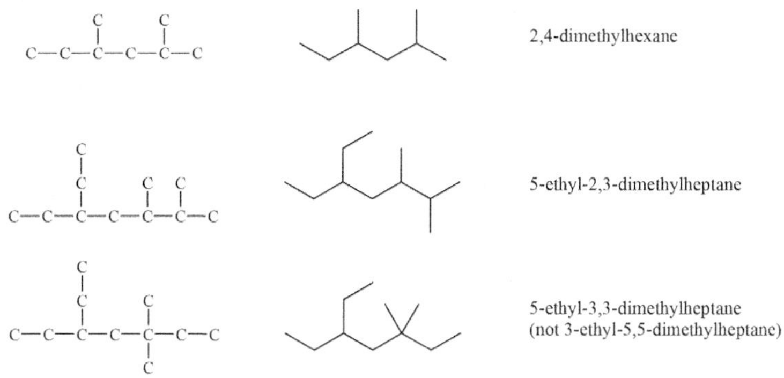

Figure 1.17: Various isomers of an organic compound having same chemical formula.

[Source: https://people.ok.ubc.ca/pshipley/teaching/chem203/worksheets/nomenclature.pdf]

a) hyphens **should** separate letters and numbers

b) commas **should** separate numbers

c) the di, tri, tetra etc. are not involved in the alphabetizing procedure.

d) n, s, and t are not involved in the alphabetizing procedure, however, iso is.

e) the prefix 'cyclo' is employed in case of cyclic alkanes.

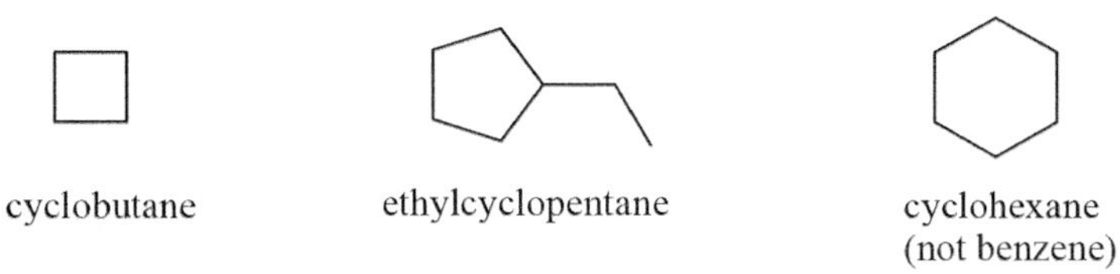

Figure 1.18: Cyclic organic compounds.

[Source: https://people.ok.ubc.ca/pshipley/teaching/chem203/worksheets/nomenclature.pdf]

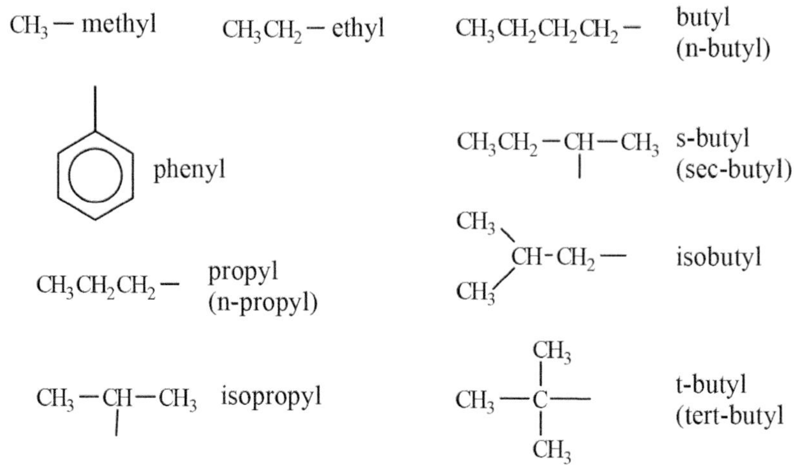

Figure 1.19: Common organic side groups.

[Source: https://people.ok.ubc.ca/pshipley/teaching/chem203/worksheets/nomenclature.pdf]

1.8.2. Alkenes

Alkenes are such hydrocarbons containing at minimum one carbon-carbon double bond (C=C) (Denis et al., 1974; Chan, 1977). The method for identifying alkene include below mentioned procedure (Millward et al., 2000; Blakemore, 2002):

- Select the longest continuous carbon chain as the parent structure which contains the carbon double bond. Substitute 'ane' with 'ene.'
- After that, number this chain from the end giving the carbon atom starting the double bond the lowest number. Attach the number with this name.
- Treat side-groups as did for alkanes.
- Dienes have two double bonds, whereas, trienes have three, etc.

$CH_2{=}CH\text{-}CH_2{-}CH_3$ 1-butene $CH_3{-}CH{=}CH{-}CH_3$ 2-butene

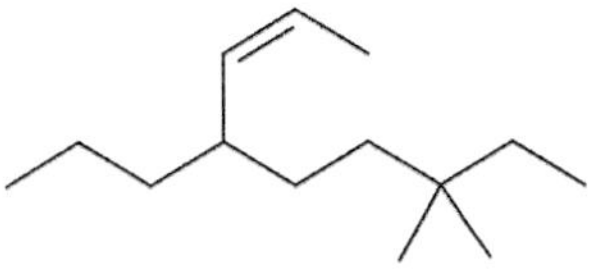

7,7-dimethyl-4-propylnon-2-ene

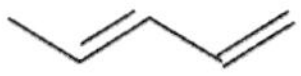

1,3-pentadiene

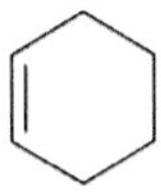

cyclohexene

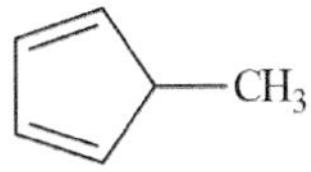

5-methylcyclopent-1,3-diene

Figure 1.20: Few common alkene structures.

[Source: https://people.ok.ubc.ca/pshipley/teaching/chem203/worksheets/nomenclature.pdf]

Consider the compounds (Tromp et al., 2002; Marciniec & Pietraszuk, 2003):

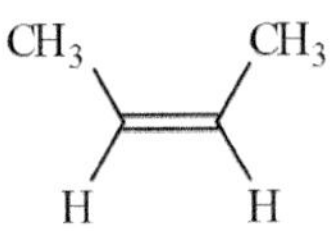

A

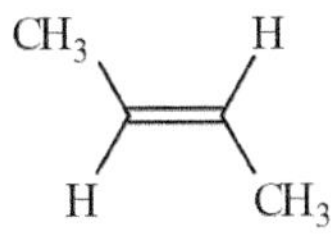

B

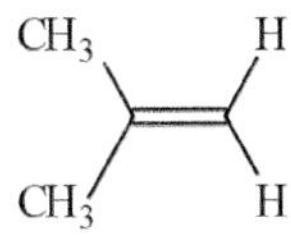

C

Figure 1.21: Alkene isomers.

[Source: https://people.ok.ubc.ca/pshipley/teaching/chem203/worksheets/nomenclature.pdf]

Both A and B are configurational isomers. The two groups that are attached to the C atom on the left side of the double bond in both A and B, have the similar orientation in space. Considering the two groups being attached to the C atom on the right side of the double bond, we observe that A has somewhat different arrangement in space than B. we cannot super impose A and B on each other. In order to name these two isomers, we first consider the left side of the double bond and prioritize the two groups on the basis of their atomic number. After that, the two groups present on the right side of the double bond are arranged (Asaro et al., 2000; Anderson et al., 2006). Earlier A was called *cis* and B *trans* and this type of nomenclature is still sometimes used for simple compounds (Kang et al., 2002).

In the image above, C is a different structural isomer; both A and B are the same structural isomer nonetheless they are different configurational isomers. The name of these three compounds are: A: (Z)-2-butene, B: (E)-2-butene, C: 2-methyl-1-propene.

1.8.3. Alkynes

Alkynes are hydrocarbons containing at minimum of one carbon-carbon triple bond (Rostovtsev et al., 2002; Tornøe et al., 2002; Alonso et al., 2004)

- The numbering and nomenclature of alkynes is the same as alkenes, apart from replacing 'ene' with 'yne.'
- Alkynes do not display configurational isomerism owing to the linear nature of the carbon triple bond.

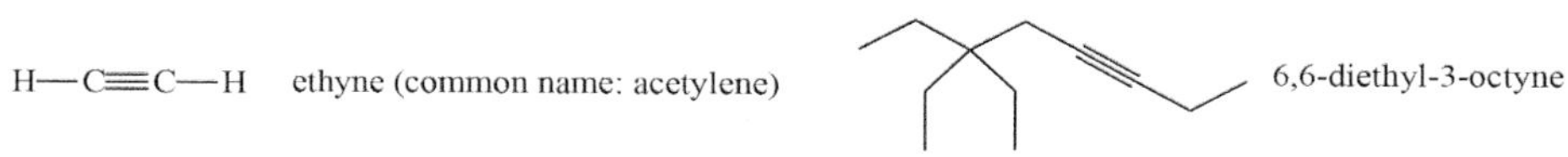

Figure 1.22: Structure and nomenclature of alkynes.

[Source: https://people.ok.ubc.ca/pshipley/teaching/chem203/worksheets/nomenclature.pdf]

1.8.4. Aromatic Hydrocarbons

Aromatics are based on the existence of benzene-alike rings. Constituents are treated like in cyclic alkanes (Snieckus, 1990; Wang & Frenklach, 1997).

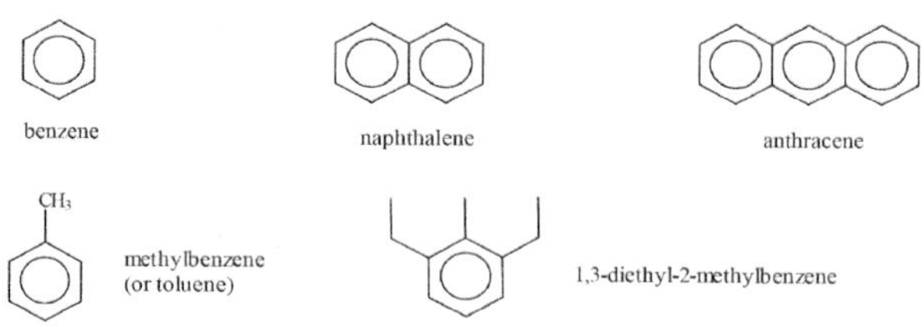

Figure 1.23: Aromatic hydrocarbons structure (Beak & Snieckus, 1982).

[Source: https://people.ok.ubc.ca/pshipley/teaching/chem203/worksheets/nomenclature.pdf]

1.8.5. Alkyl Halides

These are significant compounds in organic synthesis reactions. The halogens such as F, Br, Cl, I are regarded as fluoro, bromo, chloro, and iodo side groups for the purpose of nomenclature.

2-chloro-2-fluoro-3,3-dimethylhexane

1,1-diiodo-2-methylprop-1-ene

Figure 1.24: Structure and nomenclature of alkyl halides.

[Source: https://people.ok.ubc.ca/pshipley/teaching/chem203/worksheets/nomenclature.pdf]

1.8.6. Functional Groups

Having talked about alkanes, alkenes and alkynes now we will consider other classes of organic compounds, particularly the common classes which contain nitrogen and oxygen. A brief summary of such functional groups is tabularized below.

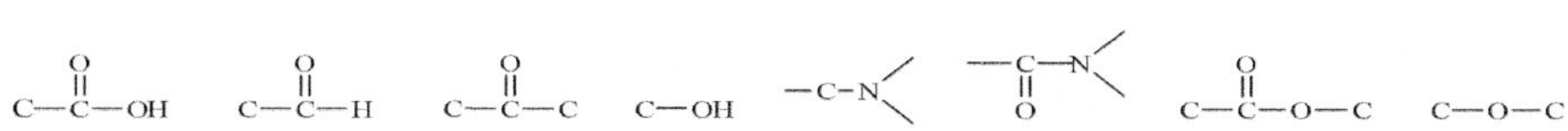

carboxylic acid aldehyde ketone alcohol amine amide ester ether

We, therefore, classify a compound by the presence of one or more of such functional groups.

1.8.7. Alcohols

Alcohols consist of the hydroxyl group (-OH), which is bonded to a C atom.

- Select the longest continuous carbon chain containing the carbon attached to the alcohol group as the parent structure. Substitute ‘ane’ with ‘anol.’
- Number the chain from the end in such a way that the -OH is attached to the C with the lowest number. Begin the name with this number.

- Treat side groups as earlier.

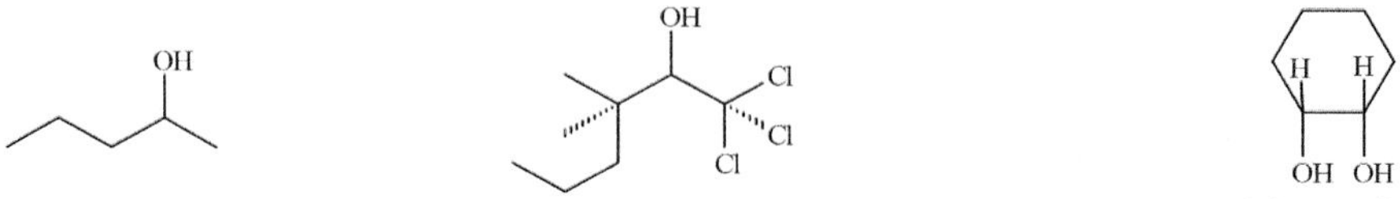

pentan-2-ol   1,1,1-trichloro-3,3-dimethylhexan-2-ol   (Z)-1,2-cyclohexanediol

Figure 1.25: Structure and nomenclature of alcohols.

[Source: https://people.ok.ubc.ca/pshipley/teaching/chem203/worksheets/nomenclature.pdf]

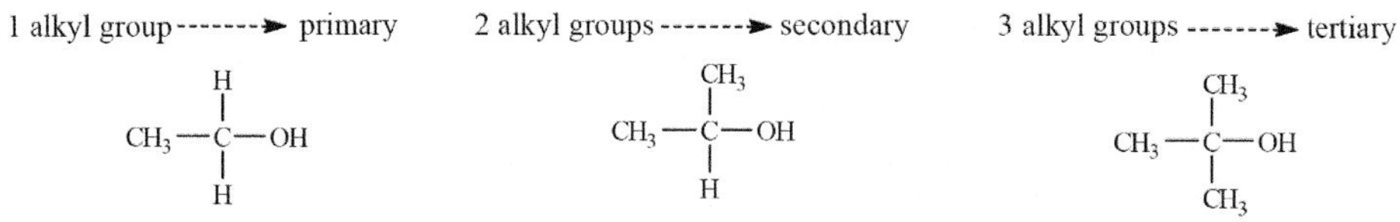

Figure 1.26: Primary, secondary and tertiary alkyl groups.

[Source: https://people.ok.ubc.ca/pshipley/teaching/chem203/worksheets/nomenclature.pdf]

1.8.8. Ethers

Ethers are composed of an oxygen bridge (R-O-R'). We treat the shortest (R-O-) group as an alkoxy side group on the lengthiest chain of carbon (Pedersen et al., 1988; Darnerud et al., 2001).

$CH_3—O—CH_2CH_3$

1-methoxyethane
(common name ethyl methyl ether)

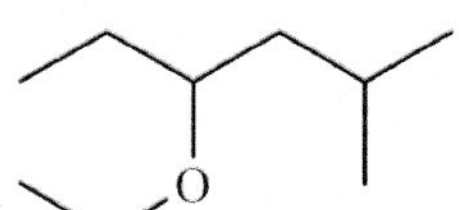

4-ethoxy-2-methylhexane

Figure 1.27: Structure and nomenclature of ethers (Richman & Atkins, 1974; Hites, 2004).

[Source: https://people.ok.ubc.ca/pshipley/teaching/chem203/worksheets/nomenclature.pdf]

1.8.9. Aldehydes

Aldehydes have a carbonyl group attached at the end of a carbon chain

- Select the longest carbon chain comprising the terminal aldehyde group as the parent structure. Substitute 'ane' with 'anal'
- Number the chain in such a manner that the aldehyde carbon is atom number one.
- Treat side groups as earlier.

The number '1' for the 'al' group does not need to be included in the name.

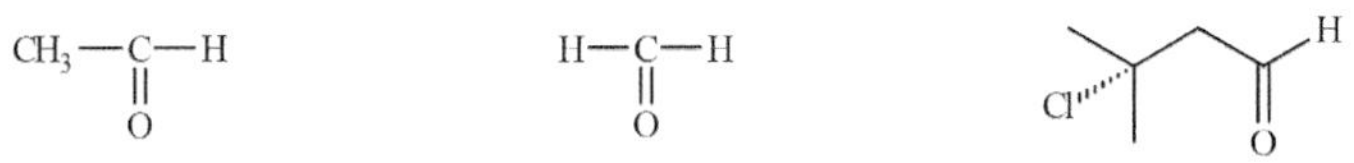

ethanal   methanal (formaldehyde)   3-chloro-3-methylbutanal

Figure 1.28: Structure and nomenclature of aldehydes.

[Source: https://people.ok.ubc.ca/pshipley/teaching/chem203/worksheets/nomenclature.pdf]

1.8.10. Ketones

Ketones have a non-terminal carbonyl group (Evans et al., 1988; Ohkuma et al., 1995):

- Select the longest continuous carbon chain comprising the carbonyl group carbon atom as the parent structure. Substitute 'ane' with 'anone.'
- Number the chain in a manner that the carbonyl carbon has the lowest number. Precede the name with this number.
- Treat side groups as earlier.

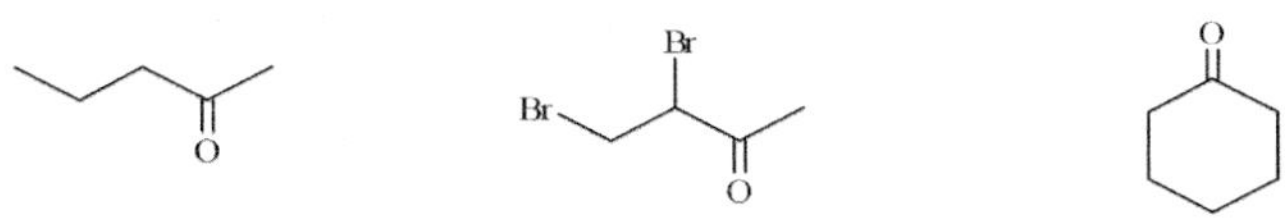

2-pentanone   3,4-dibromobutan-2-one   cyclohexanone

Figure 1.29: Structure and nomenclature of ketones (Dess & Martin, 1983).

[Source: https://people.ok.ubc.ca/pshipley/teaching/chem203/worksheets/nomenclature.pdf]

1.8.11. Organic Acids

Organic acids are composed of the carboxylic acid group

- Select the longest continuous carbon chain containing the carboxylic acid group as the parent structure. Substitute 'ane' with 'anoic acid'
- Number the chain in a manner that the carboxylic acid carbon is atom number one. Precede the name with this number.
- 3 Treat side groups as earlier.

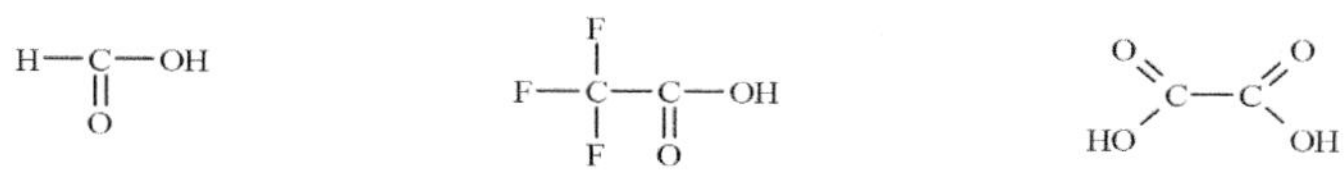

methanoic acid (formic acid) 2,2,2-trifluoroethanoic acid 1,2-ethanedioic acid (oxalic acid)

Figure 1.30: Structure and nomenclature of organic acids.

[Source: https://people.ok.ubc.ca/pshipley/teaching/chem203/worksheets/nomenclature.pdf]

1.8.12. Esters

Esters can be viewed as a derivative of carboxylic acid. Esters contain R and R' being alkyl or aryl (benzene-like) groups (Metcalfe & Schmitz, 1961; Morrison & Smith, 1964). For naming of esters, following points are kept in mind:

- Identify the carboxylic acid part of the molecule and name it as the parent carboxylic acid. Substitute 'anoic acid' with 'oate.' Name the remaining molecule (alcohol part) as a radical, located in front of the parent name.
- Carbon atom number '1' is known as the ester carbonyl.

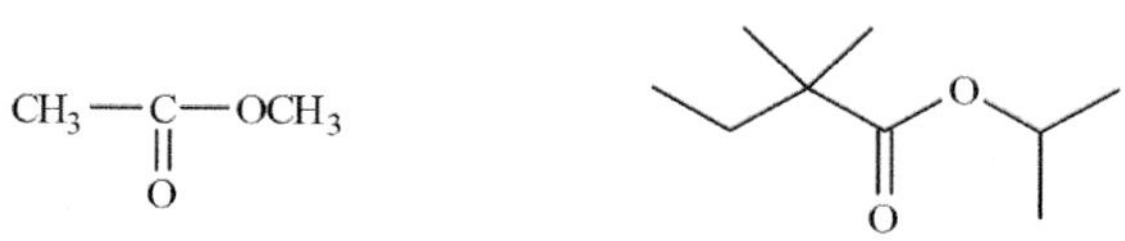

methyl ethanoate isopropyl 2,2-dimethylbutanoate

Figure 1.31: Structure and nomenclature of esters (Castagna et al., 1982).

[Source: https://people.ok.ubc.ca/pshipley/teaching/chem203/worksheets/nomenclature.pdf]

1.8.13. Amines

Amines are composed of nitrogen attached to sp^3carbon atoms (Jeanmaire & Van Duyne, 1977; Fukuyama et al., 1995).

Primary Amines

The naming of simple primary amines is done either through:

- addition of the suffix amine to the name of the alkyl group which is bonded to the nitrogen atom of the amine. We writ the name of the name of the amine as a single word.
- or else by replacement of the last e of the IUPAC name of the parent alkane with *amine.*

2-methyl-2 ethanamine 2-butanamine cyclohexanamine

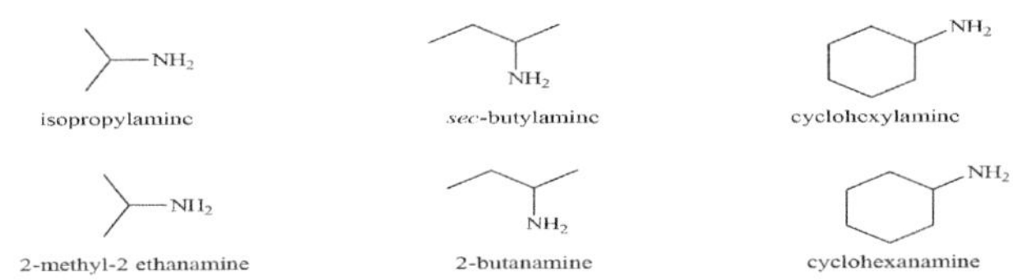

Figure 1.32: Structure and nomenclature of amines (Falck et al., 1962).

[Source: https://people.ok.ubc.ca/pshipley/teaching/chem203/worksheets/nomenclature.pdf]

Secondary and Tertiary Amines

- We name symmetrical secondary amines and tertiary amines by addition of the prefix *di* or *tri* to the name of the alkyl group.

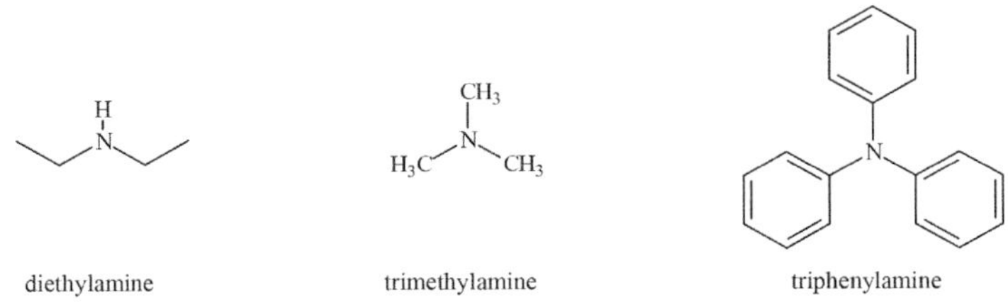

Figure 1.33: Structure and nomenclature of symmetrically replaced secondary and tertiary amines.

[Source: https://people.ok.ubc.ca/pshipley/teaching/chem203/worksheets/nomenclature.pdf]

- Whereas, we name unsymmetrically substituted secondary amines and tertiary amines as N-substituted primary amines. We chose the largest of the alkyl substituents as the parent chain. Writing letter N designates that the alkyl groups are attached to the N atom and not to a C atom of the parent alkyl chain.

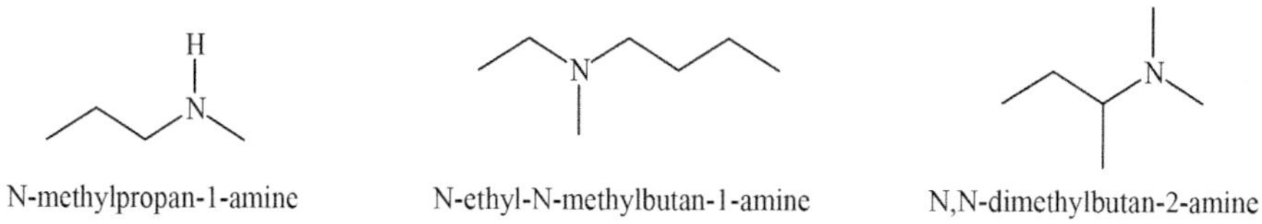

Figure 1.34: Structure and nomenclature of unsymmetrically substituted secondary and tertiary amines.

[Source: https://people.ok.ubc.ca/pshipley/teaching/chem203/worksheets/nomenclature.pdf]

- The structure of $-NH_2$ group is more complicated and is known as an amino group and treated as any other substituents.

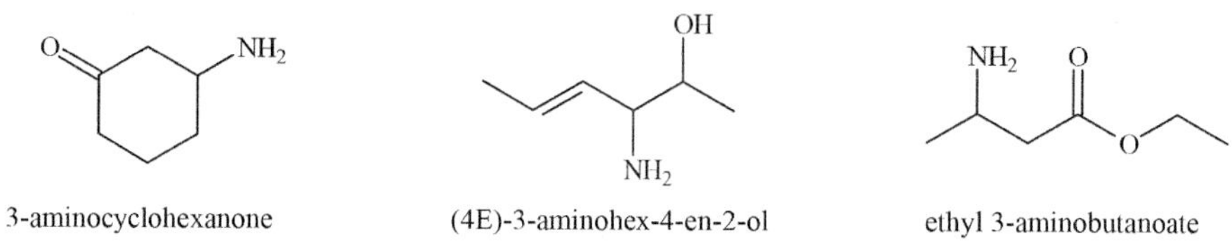

Figure 1.35: Structure and nomenclature of complex amino groups.

[Source: https://people.ok.ubc.ca/pshipley/teaching/chem203/worksheets/nomenclature.pdf]

1.8.14. Amides

Amides have an amide group with a single bonded nitrogen and a double bonded carbon. Amides are named similar to esters, treated as a derivative of carboxylic acid. We name the amine part of the molecule as a radical positioned in front of the parent name. There is not any space among the alkyl group attached to the N and the parent name (Gutowsky & Holm, 1956; Cravatt et al., 1996).

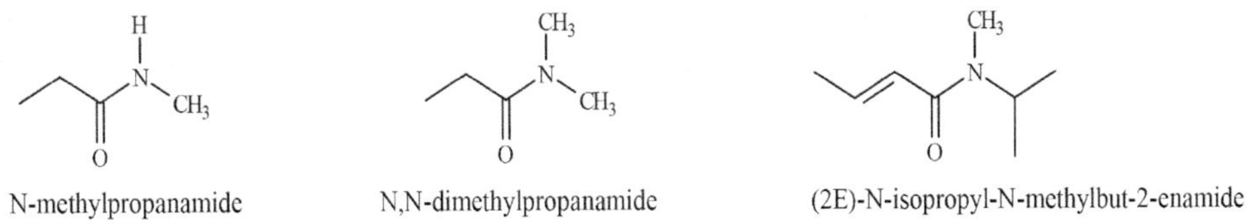

Figure 1.36: Structure and nomenclature of amides (Jorgensen & Swenson, 1985).

[Source: https://people.ok.ubc.ca/pshipley/teaching/chem203/worksheets/nomenclature.pdf]

1.8.15. Aromatics

The numerous functional groups which we have encountered might also be present in aromatic compounds. In several instances, we still use common names.

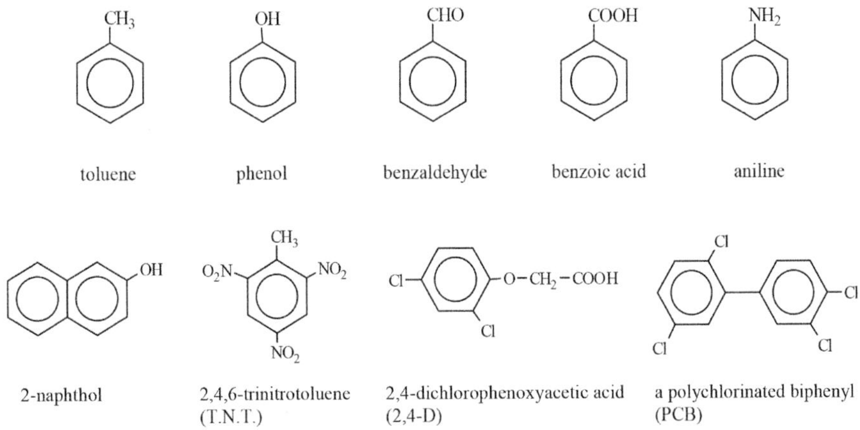

Figure 1.37: Structure and nomenclature of aromatic compounds.

[Source: https://people.ok.ubc.ca/pshipley/teaching/chem203/worksheets/nomenclature.pdf]

REFERENCES

1. Alcarazo, M. (2011). On the metallic nature of carbon in allenes and heterocumulenes. *Dalton Transactions*, *40*(9), 1839–1845.
2. Alcarazo, M., Lehmann, C. W., Anoop, A., Thiel, W., & Fürstner, A. (2009). Coordination chemistry at carbon. *Nature Chemistry, 1*(4), 295.
3. Allen, M., & Roberts, J. D. (1981). Effects of protonation and hydrogen bonding on carbon-13 chemical shifts of compounds containing the group. *Canadian Journal of Chemistry*, *59*(2), 451–458.
4. Alonso, F., Beletskaya, I. P., & Yus, M. (2004). Transition-metal-catalyzed addition of heteroatom-hydrogen bonds to alkynes. *Chemical reviews, 104*(6), 3079–3160.
5. Ambrose, D., & Tsonopoulos, C. (1995). Vapor-liquid critical properties of elements and compounds. 2. Normal alkanes. *Journal of Chemical and Engineering Data*, *40*(3), 531–546.
6. Anderson, D. R., Hickstein, D. D., O'Leary, D. J., & Grubbs, R. H. (2006). Model Compounds of Ruthenium−Alkene Intermediates in Olefin Metathesis Reactions. *Journal of the American Chemical Society, 128*(26), 8386–8387.
7. Angelos, S. A., Lankin, D. C., Meyers, J. A., & Raney, J. K. (1993). The structural identification of a methyl analog of methaqualone via 2-dimensional NMR techniques. *Journal of Forensic Science*, *38*(2), 455–465.
8. Asaro, F., Lenarda, M., Pellizer, G., & Storaro, L. (2000). A multinuclear NMR study of [Pt0 (PPh3) 2 (alkene)] compounds containing asymmetric olefins. *Spectrochimica Acta Part A: Molecular and Biomolecular Spectroscopy*, *56*(11), 2167–2175.
9. Atkinson, R. (1997). Gas-phase tropospheric chemistry of volatile organic compounds: 1. Alkanes and alkenes. *Journal of Physical and Chemical Reference Data*, *26*(2), 215–290.
10. Bach, A., & Oparin, A. (1923). About fermentation in germinating plant seeds. *Biochem. Zeitschr*, *134*, 183–189.
11. Baskaran, G., Zou, Z., & Anderson, P. W. (1987). The resonating valence bond state and high-Tc superconductivity—a mean field theory. *Solid state communications*, *63*(11), 973–976.
12. Bauer, W., Winchester, W. R., & Schleyer, P. V. R. (1987). Monomeric organolithium compounds in tetrahydrofuran: tert-butyllithium, sec-

butyllithium, supermesityllithium, mesityllithium, and phenyllithium. Carbon-lithium coupling constants and the nature of carbon-lithium bonding. *Organometallics*, *6*(11), 2371–2379.

13. Beak, P., & Snieckus, V. (1982). Directed lithiation of aromatic tertiary amides: an evolving synthetic methodology for polysubstituted aromatics. *Accounts of chemical research*, *15*(10), 306–312.
14. Blakemore, P. R. (2002). The modified Julia olefination: alkene synthesis via the condensation of metallated heteroarylalkylsulfones with carbonyl compounds. *Journal of the Chemical Society, Perkin Transactions 1*, (23), 2563–2585.
15. Bonchev, D. (1983). Principles of a Novel nomenclature of Organic Compounds. *Pure and Applied Chemistry*, *55*(2), 221–228.
16. Brooke, J. H. (1968). Wöhler's urea, and its Vital Force?—A verdict from the Chemists. *Ambix*, *15*(2), 84–114.
17. Cablewski, T., Faux, A. F., & Strauss, C. R. (1994). Development and application of a continuous microwave reactor for organic synthesis. *The Journal of Organic Chemistry, 59*(12), 3408–3412.
18. Cassebaum, H., & Kauffman, G. B. (1976). The analytical concept of a chemical element in the work of Bergman and Scheele. *Annals of Science*, *33*(5), 447–456.
19. Castagna, M., Takai, Y., Kaibuchi, K., Sano, K., Kikkawa, U., & Nishizuka, Y. (1982). Direct activation of calcium-activated, phospholipid-dependent protein kinase by tumor-promoting phorbol esters. *Journal of Biological Chemistry*, *257*(13), 7847–7851.
20. Chan, T. H. (1977). Alkene synthesis via. beta.-functionalized organosilicon compounds. *Accounts of Chemical Research*, *10*(12), 442–448.
21. Chen, A. E., O'Leary, D. J., Miura, S. S., & Anet, F. A. (2000). From the teaching laboratory: A half-cage isodrin derivative for demonstrating the nuclear Overhauser effect. *Concepts in Magnetic Resonance*, *12*(1), 1–5.
22. Colby, J., Stirling, D. I., & Dalton, H. (1977). The soluble methane mono-oxygenase of Methylococcus capsulatus (Bath). Its ability to oxygenate n-alkanes, n-alkenes, ethers, and alicyclic, aromatic and heterocyclic compounds. *Biochemical Journal*, *165*(2), 395–402.
23. Cravatt, B. F., Giang, D. K., Mayfield, S. P., Boger, D. L., Lerner, R. A., & Gilula, N. B. (1996). Molecular characterization of an enzyme

that degrades neuromodulatory fatty-acid amides. *Nature, 384*(6604), 83.

24. Crews, P., Rodríguez, J., & Jaspars, M. (2010). *Organic structure analysis* (Vol. 636, pp. 1–30). New York: Oxford university press.
25. Darnerud, P. O., Eriksen, G. S., Jóhannesson, T., Larsen, P. B., & Viluksela, M. (2001). Polybrominated diphenyl ethers: occurrence, dietary exposure, and toxicology. *Environmental health perspectives, 109*(suppl 1), 49–68.
26. Denis, J. S., Oliver, J. P., Dolzine, T. W., & Smart, J. B. (1974). Intramolecular metal double bond interactions: V. 1H NMR investigation of group II metal alkene compounds. *Journal of Organometallic Chemistry, 71*(3), 315–323.
27. Dess, D. B., & Martin, J. C. (1983). Readily accessible 12-I-5 oxidant for the conversion of primary and secondary alcohols to aldehydes and ketones. *The Journal of Organic Chemistry, 48*(22), 4155–4156.
28. Dzhemilev, U. M. (2008). Catalytic replacement of transition metal atoms in metallacarbocycles by the atoms of nontransition metals. *Mendeleev Communications, 18*(1), 1–5.
29. Dzhemilev, U. M., Ibragimov, A. G., Gilyazev, R. R., & Khafizova, L. O. (2004). Zirconium-catalyzed preparation of aluminacyclopentanes and synthesis of five-membered carbo-and heterocycles. *Tetrahedron, 60*(6), 1281–1286.
30. England, D. B., Kuss, T. D., Keddy, R. G., & Kerr, M. A. (2001). Cyclopentannulation of 3-alkylindoles: a synthesis of a tetracyclic subunit of the kopsane alkaloids. *The Journal of Organic Chemistry, 66*(13), 4704–4709.
31. Evans, D. A., Chapman, K. T., & Carreira, E. M. (1988). Directed reduction of. beta.-hydroxy ketones employing tetramethylammonium triacetoxyborohydride. *Journal of the American Chemical Society, 110*(11), 3560–3578.
32. Falck, B., Hillarp, N. Å., Thieme, G., & Torp, A. A. (1962). Fluorescence of catechol amines and related compounds condensed with formaldehyde. *Journal of histochemistry & cytochemistry, 10*(3), 348–354.
33. Favre, H. A., Hellwich, K. H., Moss, G. P., Powell, W. H., & Traynham, J. G. (1999). Corrections to'A Guide to IUPAC Nomenclature of Organic Compounds.' *Pure and Applied Chemistry, 71*(7), 1327–1330.

34. Fink, J. T. J. A. P., Müller-Heinzerling, T., Pflüger, J., Bubenzer, A., Koidl, P., & Crecelius, G. (1983). Structure and bonding of hydrocarbon plasma generated carbon films: an electron energy loss study. *Solid state communications*, *47*(9), 687–691.

35. Frenking, G., & Tonner, R. (2011). Carbodicarbenes—divalent carbon (0) compounds exhibiting carbon–carbon donor–acceptor bonds. *Wiley Interdisciplinary Reviews: Computational Molecular Science*, *1*(6), 869–878.

36. Fukuyama, T., Jow, C. K., & Cheung, M. (1995). 2-and 4-Nitrobenzenesulfonamides: Exceptionally versatile means for preparation of secondary amines and protection of amines. *Tetrahedron Letters*, *36*(36), 6373–6374.

37. Goodson, A. L. (1980). Graph-based chemical nomenclature. 1. Historical background and discussion. *Journal of Chemical Information and Computer Sciences*, *20*(3), 167–172.

38. Gutowsky, H. S., & Holm, C. H. (1956). Rate processes and nuclear magnetic resonance spectra. II. Hindered internal rotation of amides. *The Journal of Chemical Physics*, *25*(6), 1228–1234.

39. Haddon, R. C., Wasserman, S. R., Wudl, F., & Williams, G. R. J. (1980). Molecular orbital study of sulfur-nitrogen and sulfur-carbon conjugation: mode of bonding in (SN) x and related compounds. *Journal of the American Chemical Society*, *102*(22), 6687–6693.

40. Hagenbuch, B., & Meier, P. J. (2004). Organic anion transporting polypeptides of the OATP/SLC21 family: phylogenetic classification as OATP/SLCO superfamily, new nomenclature and molecular/functional properties. *Pflügers Archiv*, *447*(5), 653–665.

41. Hites, R. A. (2004). Polybrominated diphenyl ethers in the environment and in people: a meta-analysis of concentrations. *Environmental science & technology*, *38*(4), 945–956.

42. Ishii, H., Sugiyama, K., Ito, E., & Seki, K. (1999). Energy level alignment and interfacial electronic structures at organic/metal and organic/organic interfaces. *Advanced Materials, 11*(8), 605–625.

43. Jeanmaire, D. L., & Van Duyne, R. P. (1977). Surface Raman spectroelectrochemistry: Part I. Heterocyclic, aromatic, and aliphatic amines adsorbed on the anodized silver electrode. *Journal of electroanalytical chemistry and interfacial electrochemistry*, *84*(1), 1–20.

44. Jones, W. D., & Feher, F. J. (1983). Alkane carbon-hydrogen bond activation by homogeneous rhodium (I) compounds. *Organometallics*, *2*(4), 562–563.

45. Jorgensen, W. L., & Swenson, C. J. (1985). Optimized intermolecular potential functions for amides and peptides. Structure and properties of liquid amides. *Journal of the American Chemical Society*, *107*(3), 569–578.

46. Kang, M., Sen, A., Zakharov, L., & Rheingold, A. L. (2002). Diametrically opposite trends in alkene insertion in late and early transition metal compounds: relevance to transition-metal-catalyzed polymerization of polar vinyl monomers. *Journal of the American Chemical Society*, *124*(41), 12080–12081.

47. Klein, S., Tonner, R., & Frenking, G. (2010). Carbodicarbenes and related divalent carbon (0) compounds. *Chemistry – A European Journal, 16*(33), 10160–10170.

48. Langley, P. J., & Hulliger, J. (1999). Nanoporous and mesoporous organic structures: new openings for materials research. *Chemical Society Reviews*, *28*(5), 279–291.

49. Lavoisier, A. Memoire on the Formation of Acid Calls Fixed Air or Chalky Acid, which I now designate under the name of Coal Acid. *Works by Lavoisier*, 1, 422.

50. Liu, H. J., & Lai, H. K. (1981). Regioselectivity of the Cyclization of Dithiol Esters. An Alternative Route to Cyclic γ-Substituted β-Keto Thiolesters. *Synthetic Communications*, *11*(1), 65–69.

51. Marciniec, B., & Pietraszuk, C. (2003). Synthesis of unsaturated organosilicon compounds via alkene metathesis and metathesis polymerization. *Current Organic Chemistry*, *7*(8), 691–735.

52. Mayer, I. (1984). Bond order and valence: Relations to Mulliken's population analysis. *International Journal of Quantum Chemistry*, *26*(1), 151–154.

53. Melaimi, M., Parameswaran, P., Donnadieu, B., Frenking, G., & Bertrand, G. (2009). Synthesis and ligand properties of a persistent, all-carbon four-membered-ring allene. *Angewandte Chemie International Edition, 48*(26), 4792–4795.

54. Metcalfe, L. D., & Schmitz, A. A. (1961). The rapid preparation of fatty acid esters for gas chromatographic analysis. *Analytical Chemistry*, *33*(3), 363–364.

55. Miller, J. A., & Miller, E. C. (1953). The carcinogenic aminoazo dyes. In *Advances in cancer research* (Vol. 1, pp. 339–396). Academic Press.
56. Millward, D. B., Sammis, G., & Waymouth, R. M. (2000). Ring-opening reactions of oxabicyclic alkene compounds: Enantioselective hydride and ethyl additions catalyzed by group 4 metals. *The Journal of Organic Chemistry, 65*(13), 3902–3909.
57. Morrison, W. R., & Smith, L. M. (1964). Preparation of fatty acid methyl esters and dimethylacetals from lipids with boron fluoride–methanol. *Journal of lipid research*, *5*(4), 600–608.
58. Mundy, B. P., Ellerd, M. G., & Favaloro, F. G. (2005). *Name reactions and reagents in organic synthesis* (vol. 1, pp. 820–821). Hoboken, NJ: Wiley.
59. Negishi, E. I. (2007). Transition metal-catalyzed organometallic reactions that have revolutionized organic synthesis. *Bulletin of the Chemical Society of Japan*, *80*(2), 233–257.
60. Nitschke, J. R. (2007). Construction, substitution, and sorting of metallo-organic structures via subcomponent self-assembly. *Accounts of chemical research*, *40*(2), 103–112.
61. Ohkuma, T., Ooka, H., Hashiguchi, S., Ikariya, T., & Noyori, R. (1995). Practical enantioselective hydrogenation of aromatic ketones. *Journal of the American Chemical Society*, *117*(9), 2675–2676.
62. Palmer, S. E., & Schloss, K. B. (2010). An ecological valence theory of human color preference. *Proceedings of the National Academy of Sciences*, *107*(19), 8877–8882.
63. Pedersen, C. J. (1988). The discovery of crown ethers (Noble Lecture). *Angewandte Chemie International Edition in English*, *27*(8), 1021–1027.
64. PeteráClayton, J., & John, H. C. (1990). Isolation, structure elucidation, and synthesis of novel penicillin degradation products: thietan-2-ones. *Journal of the Chemical Society, Perkin Transactions 2*, (9), 1559–1566.
65. Polshettiwar, V., & Varma, R. S. (2008). Microwave-assisted organic synthesis and transformations using benign reaction media. *Accounts of Chemical Research*, *41*(5), 629–639.
66. Qin, A., Hoffman, D. E., & Munk, P. (1992). Excess volumes of mixtures of alkanes with carbonyl compounds. *Journal of Chemical and Engineering Data*, *37*(1), 55–61.

67. Ramberg, P. J. (2000). The death of vitalism and the birth of organic chemistry: Wohler's urea synthesis and the disciplinary identity of organic chemistry. *Ambix*, *47*(3), 170–195.

68. Reed, A. E., & Weinhold, F. (1985). A theoretical model of bonding in hyperlithiated carbon compounds. *Journal of the American Chemical Society*, *107*(7), 1919–1921.

69. Richman, J. E., & Atkins, T. J. (1974). Nitrogen analogs of crown ethers. *Journal of the American Chemical Society*, *96*(7), 2268–2270.

70. Rodríguez, J., Crews, P., & Jaspars, M. (2012). Contemporary strategies in natural products structure elucidation. *Handbook of marine natural products*, 1, 423–517.

71. Roeges, N. P., & Baas, J. M. A. (1994). *A Guide to the Complete Interpretation of Infrared Spectra of Organic Structures* (p. 282ff). Chichester etc.: Wiley.

72. Rostovtsev, V. V., Green, L. G., Fokin, V. V., & Sharpless, K. B. (2002). A stepwise huisgen cycloaddition process: copper (I)-catalyzed regioselective "ligation" of azides and terminal alkynes. *Angewandte Chemie International Edition*, *41*(14), 2596–2599.

73. Saunders, M., & Hyne, J. B. (1958). Study of hydrogen bonding in systems of hydroxylic compounds in carbon tetrachloride through the use of NMR. *The Journal of Chemical Physics*, *29*(6), 1319–1323.

74. Shah, S. D., Ogunyoku, T. A., Miller, J. W., & Cocker, D. R. (2005). On-road emission rates of PAH and n-alkane compounds from heavy-duty diesel vehicles. *Environmental science & technology*, *39*(14), 5276–5284.

75. Snieckus, V. (1990). Directed ortho metalation. Tertiary amide and O-carbamate directors in synthetic strategies for polysubstituted aromatics. *Chemical Reviews*, *90*(6), 879–933.

76. Son, L. L. S., Holmstedt, B., & Tjus, E. (1954). Some considerations regarding the nomenclature of organic phosphorus compounds. *Acta chem. scand*, *8*(9), 1–10.

77. Sun, H. (1998). COMPASS: an ab initio force-field optimized for condensed-phase applications overview with details on alkane and benzene compounds. *The Journal of Physical Chemistry B*, *102*(38), 7338–7364.

78. Tonner, R., & Frenking, G. (2007). C (NHC) 2: Divalent Carbon (0) Compounds with N-Heterocyclic Carbene Ligands—Theoretical

Evidence for a Class of Molecules with Promising Chemical Properties. *Angewandte Chemie International Edition*, *46*(45), 8695–8698.

79. Tonner, R., & Frenking, G. (2008). Divalent Carbon (0) Chemistry, Part 2: protonation and complexes with main group and transition metal Lewis acids. *Chemistry–A European Journal*, *14*(11), 3273–3289.
80. Tornøe, C. W., Christensen, C., & Meldal, M. (2002). Peptidotriazoles on solid phase:[1, 2, 3]-triazoles by regiospecific copper (I)-catalyzed 1, 3-dipolar cycloadditions of terminal alkynes to azides. *The Journal of Organic Chemistry*, *67*(9), 3057–3064.
81. Tromp, D. S., Duin, M. A., Kluwer, A. M., & Elsevier, C. J. (2002). Synthesis of new (σ2-N, N′-diazadiene)(η2-alkene) platinum (0) compounds. *Inorganica Chimica Acta*, *327*(1), 90–97.
82. Tukey, H. B., Went, F. W., Muir, R. M., & Van Overbreek, J. (1954). Nomenclature of chemical plant regulators: report by a committee of the American Society of Plant Physiologists. *Plant Physiology*, *29*(3), 307.
83. Wang, H., & Frenklach, M. (1997). A detailed kinetic modeling study of aromatics formation in laminar premixed acetylene and ethylene flames. *Combustion and Flame*, *110*(1–2), 173–221.
84. Webster, D. E. (1977). Activation of Alkanes by Transition Metal Compounds. In *Advances in Organometallic Chemistry*(Vol. 15, pp. 147–188). Academic Press.
85. Wöhler, F. (1828). Ueber künstliche bildung des harnstoffs. *Annalen der Physik*, *87*(2), 253–256.
86. Yamada, Y., Kimura, M., & Hosaka, K. (1981). A direction controlled Dieckmann type cyclization of half-thiol diesters. *Tetrahedron Letters*, *22*(14), 1353–1354.
87. Young, J. A. (1974). Revised nomenclature for highly fluorinated organic compounds. *Journal of Chemical Documentation*, *14*(2), 98–100.
88. Zerger, R., Rhine, W., & Stucky, G. (1974). Stereochemistry of polynuclear compounds of the main group elements. Bonding and the effect of metal-hydrogen-carbon interactions in the molecular structure of cyclohexyllithium, a hexameric organolithium compound. *Journal of the American Chemical Society*, *96*(19), 6048–6055.
89. Zerkowski, J. A., Seto, C. T., Wierda, D. A., & Whitesides, G. M. (1990). The design of organic structures in the solid state: hydrogen-

bonded molecular" tapes." *Journal of the American Chemical Society, 112*(24), 9025–9026.

2

Catalytic Alcohol Oxidation

CONTENTS

2.1. INTRODUCTION

Oxidation of the alcohols to carbonyl-containing compounds or the full oxidations of alcohols are among the fundamental reactions in the organic chemistry and are of concern for the growth of environment friendly processes, production of the new energy sources and new materials (Sheldon et al., 2000; 2002; Allen et al., 2013). Due to their crucial role in the industrial fields and anticipated further uses, these reactions carry on to attract the great attention, revealing new oxidants, catalysts, substrates, with odd features and uses (Mallat & Baiker, 2004; Tojo & Fernández, 2006).

Regarding oxidants, the stoichiometric oxidations with transition sulfoxides or metal compounds are still in common utilization, in spite of the formation of a large number of unwanted products (Bianchini & Shen, 2009; Sarathy et al., 2009). The most utilized oxidants comprise of small organic molecules based reagents, e.g., Swern, Moffatt, Dess-Martin periodinane, SO_3/pyridine, Corey-Kim oxidants, some of them are expensive and moisture-sensitive, or systems which are metal-based (such as Collins, Jones, Oppenauer reagents, pyridinium dichromate, pyridinium chlorochromate (PCC), barium permanganate, ruthenium tetroxide, manganese dioxide, silver carbonate). A latest environmental sustainability and compatibility method lead towards aerobic oxidations with the transition-metal catalysts and hydrogen peroxide or di-oxygen or as oxidants (Sigman & Jensen, 2006; Dobereiner & Crabtree, 2009). The utilization of the molecular oxygen as the stoichiometric re-oxidant in amalgamation with the catalytic metal has practical benefits because of the favorable economics linked with O_2 and the creation of environment-friendly by-products.

Advances in the growth of new approaches, oxidation agents, catalysts and the uses have been surveyed regularly, and the field stays to be the most widely and actively examined areas of present organic synthesis (Liang & Zhao, 2012; Ryland & Stahl, 2014). Newly advanced green oxidations of the alcohols normally involve selective and active recyclable catalysts which ideally must work with air, di-oxygen, or other inexpensive oxidants, not leaving behind contaminated or useless by-products (Hage & Feringa, 2010; Murahashi & Komiya, 2010).

However, in spite of some significant advances, only some of the known approaches are capable of providing the practical and economic oxidation towards the specific industrially vital transformation (Ishii et al., 2004; Arends & Sheldon, 2010). Several of the discovered catalytic systems undergo high reagent cost, employment of unsafe metals or oxidants, instability, harsh reaction conditions, functional group incompatibility, operational complexity, or production of un-process able wastes. Therefore, there is a rising demand for the latest catalytic systems which could overcome these challenges (Bäckvall, 2011; Trincado et al., 2014).

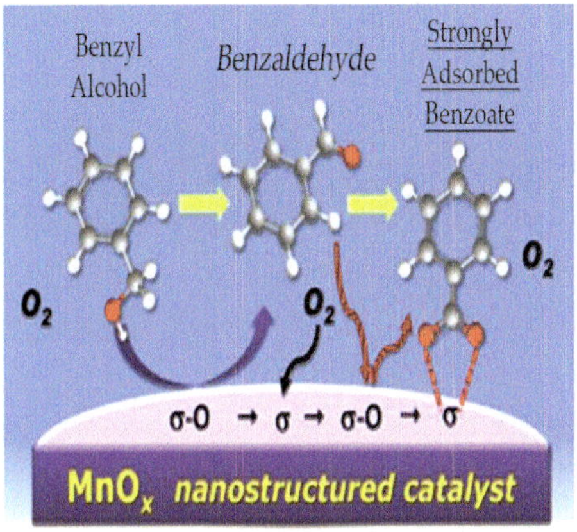

Figure 2.1: Catalytic oxidation of the alcohol on MnO_x nanostructured catalyst.

[Source: https://www.sciencedirect.com/science/article/pii/S0926337315000454]

Moreover, other viewpoints for the oxidation of alcohol have been tested, comprising of the hydrogen transfer and production, atom-efficient transformations, oxidation of the natural substrates, like cellulose, sequential and cascade reactions, etc. (Sheldon & Arends, 2004; Mizuno, 2009; Da Silva et al., 2011).

2.2. AEROBIC AND PER-OXIDATIVE OXIDATIONS

Aerobic and per-oxidative oxidations of the alcohols, specifically of the benzylic alcohols, are usual model reactions because of their significance and generality; cheap O_2, H_2O_2 or TBHP (tert-butyl hydroperoxide) oxidants, and easy procedures are normally involved. In this section, description of some fascinating catalytic systems introduced lately into the area of oxidation of alcohol is given (Roberts & Whittall, 2007; Griffith, 2011). Furthermore, a glimpse at new oxidants and substrates which could effectively be utilized in the near future and brings the variations in terms of selectivity, efficiency, economy, and sustainability of the procedures, is also described (Zakzeski et al., 2010; Wigington et al., 2012).

2.2.1. Metal Catalysts

Historically, Pd, Pt, Ru, Rh, and Ir complexes were amongst the most efficient catalysts for the alcohol oxidation. The series was extended to 3d metals, e.g., Cu/TEMPO and V systems which have been invented again and developed, and includes the representatives of several groups and sub-groups of transition and non-transition metals. As compared to the noble metals, the lately introduced catalysts centered on rich 3d and associated metals usually function by redox mechanisms which normally involve one-electron processes (Zhan & Thompson, 2004; Schultz & Sigman, 2006). During the last period, the exploration for effective and new catalysts for the alcohol oxidation has primarily concentrated on finding inexpensive and more efficient metal-ligand combinations, accomplishing enantio and region selective reactions, and clarifying mechanistic information of action of recognized catalytic systems. Additionally, recent reports displayed that there is a constant interest regarding the structural particulars of the catalysts. For example, the part of nuclearity in the multinuclear Cu (II) complexes were lately discussed (Besson & Gallezot, 2000; Figiel et al., 2009; 2010).

When utilizing di-oxygen, the challenge to overcome worries the point that it is the four-electron oxidant, whereas the aerobic oxidation of the alcohols to the carbonyl compounds comprises of two electrons. Despite that, oxygen species reduced to some extent are normally more reactive than the O_2 itself. Therefore, the introduction of the special oxidation buffer agents which can stabilize the particular energetic necessities of the substrates by means of the potentials of the oxidant is a significant task, and intricate, organo-catalysts or metallorganic, can play the part of such agents. Normally, the efficient catalytic systems include the organic components, as the ligand in the coordination compound, but at times simple salts, like Mn(II) acetate, can also prove to be effective catalysts (Räisänen et al., 2014; Zhang et al., 2014).

Organic ligands in the complexes can have different roles: amend steric and electronic properties, offer the needed solubility or preparation of the central metal ions or safeguard them from reduction or over-oxidation. Therefore, the systematic examination of the catalytic action of the palladium complexes with available pyridine-comprising ligands discovered the circumstances where the precipitation of Palladium black doesn't occur

(Grßinne, 2010; John et al., 2013). Likewise, tertiary phosphine oxides can be utilized as the ligands for Palladium catalysts.[31] Recently, the significance of the tri-nuclear Pd_3O_2 intermediate $[(LPd^{II})_3(\mu^3\text{–}O)_2]^{2+}$ (L¼2,9-dimethylphenanthroline) in Palladium-assisted catalysis was exposed. This tri-nuclear compound is the creation of the oxygen activation by decreased palladium species and is the vital intermediate (Gowrisankar et al., 2013; Ingram et al., 2014).

The introduction of the latest ligands is a vital task in the growth of the new catalytic systems. For example, few of the polymers can be utilized as macro-ligands to the host metal ions. The polymer ligands can't only offer the catalyst re-usage, but also balance the essential metal ions and avert their aggregation (Giachi et al., 2012). This method is relatively attractive and merges the benefits of heterogeneous and homogeneous catalysts. It includes coordination of the Pd(II) by 4-pyridine methylene-end-capped poly(caprolactone) and poly(l-lactide) (Figure 2.2). The polymer-anchored catalyst is soluble under the practical catalytic conditions, but on the addition of the methanol or n-pentane, the polymer-anchored catalysts precipitate and therefore can be separated easily from the mixture of reaction. These catalysts are efficient in the oxidation of various secondary and primary alcohols with O_2.

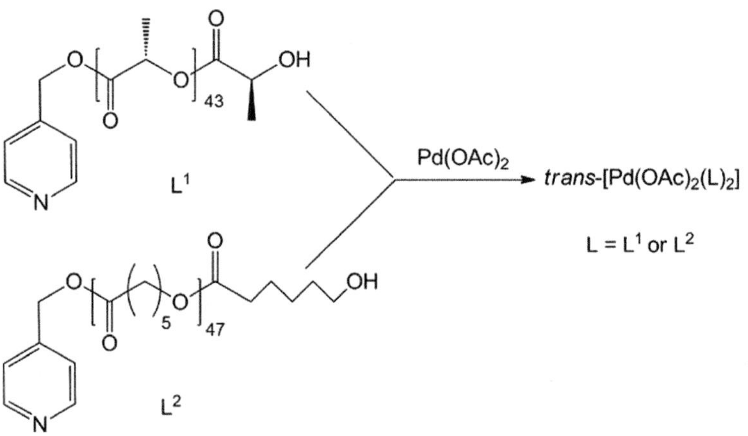

Figure 2.2: Synthesis of palladium (II) complexes with poly (caprolactone) and poly(l-lactide) macro ligands.

[Source: https://www.researchgate.net/publication/283763126_Catalytic_Oxidation_of_Alcohols_Recent_Advances]

In a linked work, the design of enzyme-stimulated star block copolymers with bifurcated topologies and protein like quaternary or tertiary structures was executed (Mugemana et al., 2014). These polymers include polydentate, hydrophilic, and super-hydrophobic metal-binding positions and self-accumulate in water. The main feature of these star block copolymers is the incorporation of per-fluorocarbons and, because of that, their suspensions in water can invite and pre-concentrate O_2 in the neighborhood of the energetic metal site. Addition of TEMPO and Cu (II) leads to an efficient catalytic system for the oxidation of the alcohols to aldehydes in the water.

The series of tetra-dentate pyridyl-imine ended Schiff-bases, bis(pyridyl-imine) ended siloxane and other associated polymers can be utilized as ligands to the host Cu (II) ions (Hu & Kerton, 2012). These $CuBr_2$/TEMPO/polyL systems (polyL is an abbreviation for polydimethylsiloxane) are efficient for the aerobic oxidations of secondary and primary alcohols under the aqueous conditions. Chiral N, O-ligands, e.g., cheap L-proline, can be utilized to prepare Cu catalysts which are particularly efficient for the oxidation of the sterically stuck, heterocyclic or allylic alcohols like 1-(2-furfuryl)ethanol, 1-(3-pyridyl)ethanol, 2-thienyl, 3-pyridyl methanol and 2-furyl (Kopylovich et al., 2011a; 2011b; 2012; 2013). Associated mono- and di-Cu (II) amino-polyalcohols were prepared easily by self-assembly. (Kirillov et al., 2011; Zhang et al., 2013). The selectivity factors for the oxidative transformations were discussed and measured, assisting free-radical mechanisms. Cu complexes with the hydrazone ligands have been studied extensively during the last era as an oxidation catalyst and their family grows continuously (Mahmudov et al., 2010; 2014). For instance, an easy to produce and to handle tri-nuclear di-hydrazone Cu (II) complex $[Cu_3(L)(\mu^2{-}Cl)_2(H_2O)_6]$ can be utilized as a reusable catalyst for the choosy oxidation of the wide range of alcohols, not deactivated by the N/S-heteroatom-including substrates.

Cu-containing MOFs (metal-organic frameworks) based on the 5-(4pyridyl) tetrazole building blocks, prepared easily in situ by the 1,3-dipolar cycloaddition amongst azide and 4-cyanopyridine in the existence of Cu (II) chloride, were applied successfully as pre-catalysts for low power microwave-assisted per-oxidative oxidation of the secondary alcohols producing conforming ketones with produces up to 86% and TOFs (turnover frequencies) equal to 430 h^1 after one hertz, in the deficiency of added solvent or additive. Zr (IV)-centered strong metal-organic frameworks with open Cr- and Fe-mono-catecholate metal locations on the configuration of the organic connectors were made by the post-synthetic metal interchange, the tactic which permits good regulator over the number of the metal-binding locations, and can be utilized as the facile and effective way to get MOFs which can't be synthesized directly under solvothermal conditions. The Cr-metallated metal-organic frameworks are versatile, efficient and recyclable heterogeneous catalysts (Borthakur et al., 2014; Nasani et al., 2014).

Biomimetic Copper (II) and Iron(II) complexes with bis- and tris-pyridyl imino and amino thioether ligands and unoccupied coordination sites (Figure. 2.3) are active as catalysts precursors for the halogen- and solvent-free microwave-assisted oxidation of the 1-phenyl ethanol by tert-butyl hydroperoxide, in the existence of pyridazine or some other N-centered additives. Maximum turnover frequency of 5220 hertz was accomplished just after five minutes of reaction time. The same authors described the catalytic activity of associated iron, copper, and vanadium (V) systems with varied-N, S pyridine thioether ligands (Fernandes et al., 2011a; 2011b; Fei et al., 2014).

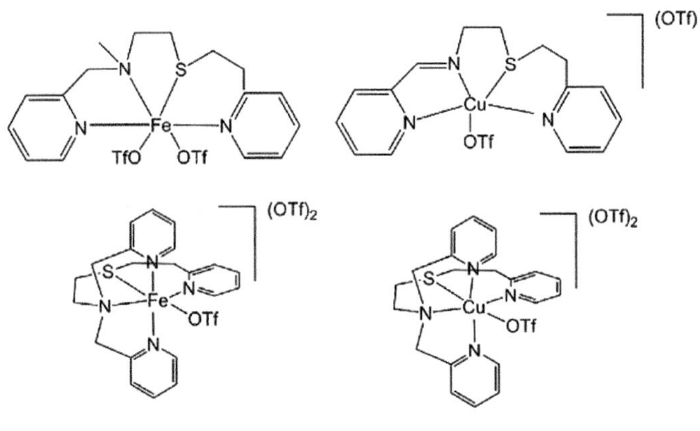

Figure 2.3: Iron (Fe) and copper (Cu) complexes with bis- & tris-pyridyl imino and amino thioether ligands.

[Source: https://www.researchgate.net/publication/283763126_Catalytic_Oxidation_of_Alcohols_Recent_Advances]

The Fe and Cu complexes demonstrated to be beneficial catalysts in several microwave-supported alcohol oxidations with tert-butyl hydroperoxide, at 80°C. Therefore, S-containing ligands can be utilized to make efficient catalyst precursors (Al-Hunaiti et al., 2010; Join et al., 2011). Another easy and green to prepare Fe-centered catalyst, $[Fe(BPA)_2]$ $(OTf)_2$, with available BPA{ bis(picolyl)amine} ligand (Figure. 2.4), chemo-selectively oxidizes the range of the secondary alcohols in the existence of major ones into the conforming hydroxy ketones within 15 minutes at the room temperature with three mol percentage catalyst loading and the H_2O_2 as oxidant (Lenze & Bauer, 2013).

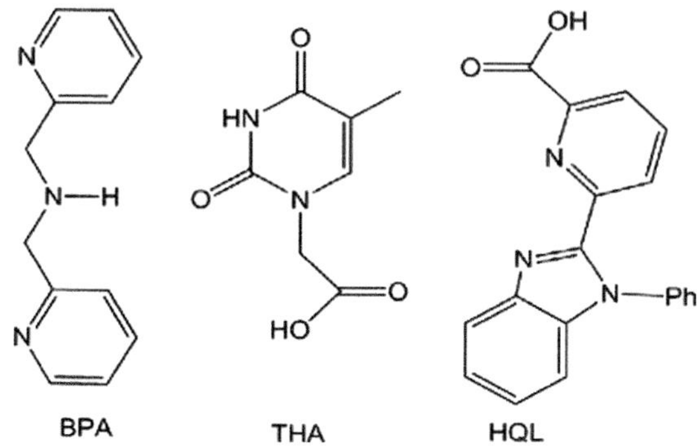

Figure 2.4: BPA (bis (picolyl) amine), THA (thymine-1-acetate), and HQL (8-hydroxyquinoline).

[Source: https://www.researchgate.net/publication/283763126_Catalytic_Oxidation_of_Alcohols_Recent_Advances]

The complexes can also be produced in situ and work correspondingly to the one performed. In situ produced FeCl complexes with 6-(N-phenyl benzimidazole)-2-pyridine carboxylic acid or thymine-acetate ligands (Figure. 2.4) has been synthesized recently and demonstrated to be a choosy and suitable catalyst for the oxidation of allylic and benzylic alcohols. They can be useful for sensitive compounds like geraniol, perillyl alcohol, or carveol, whereas diols could be oxidized in decent yields without oxidative cleavage of the products. Mechanistic investigations disclose that the thymine-acetate owns the organo-catalytic activity for alcohols oxidation (Son & Toste, 2010; Hanson et al., 2011).

Aerobic alcohol oxidation with V (vanadium) catalysts carries on to broaden their substrate possibility and applications (Ohde & Limberg, 2010; Sedai et al., 2011). The lately studied process of the intramolecular oxidation of the benzyl alcohol ligands in the 8-hydroxyquinoline(L) V(vanadium) complexes of the kind [LV(O) (OR)] bear a resemblance to those suggested for particular metalloenzyme-catalyzed oxidation and includes intermolecular deprotonation and rare ligand interchange at the benzylic location (Figure. 2.5).

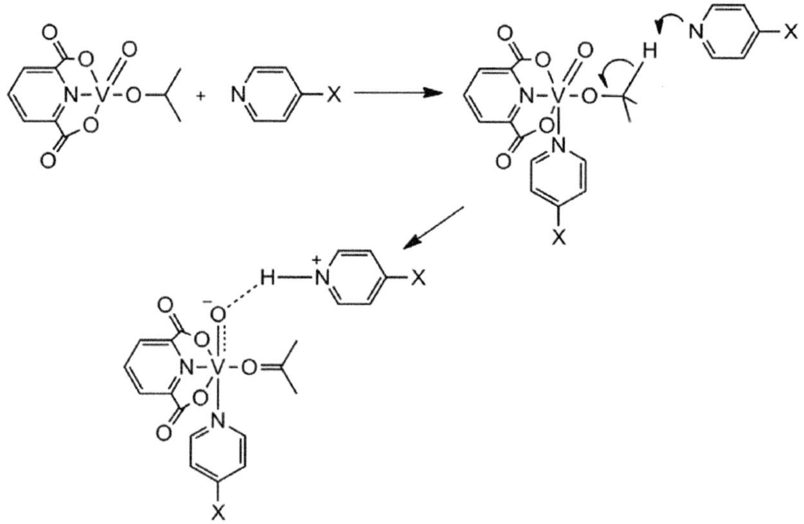

Figure 2.5: Main steps in base-supported dehydrogenation for the [LV(O)(OR)] complexes.

[Source: https://www.researchgate.net/publication/283763126_Catalytic_Oxidation_of_Alcohols_Recent_Advances]

This biomimetic pathway varies from the formerly recognized hydride-relocation and radical procedures for the transition metal-mediated alcohols oxidation. As an outcome, new ways to improve the selectivity and activity of vanadium (V) catalysts were proposed. They contain the regulation of the external coordination sphere and the use of ligands with suitably located pendant bases to assist as proton shuttles.

Related V (vanadium) complexes demonstrate activity towards the oxidative decay of the pinacol with Cd-C bond cleavage and the aerobic oxidation of lignin model compounds and 4-methoxy benzyl alcohol (Zhang et al., 2012; Dönges et al., 2014). Other oxide-vanadium complexes with cis-2,6-bis-(methanolate)-piperidine ligand of the kind shown in Figure. 2.6 were made practical as catalysts to transform prochiral alkenols into the 2-(tetrahydropyran-2-yl)-2-propanols, 2-(tetrahydrofuran-2-yl)-2-propanols, epoxides, and oxepan-3-ols and on oxidative alkenol cyclization with tert-butyl hydroperoxide as oxidant (Figure. 2.6). These catalysts are unstable stable and owns enhanced chemo-selectivity, e.g., epoxidation of the geraniol takes place enantio-selectively. It was finalized that the vanadium tert-butyl peroxy complex creation is the main step to trigger peroxides.

Silver NHC (N-heterocyclic carbene) catalysts (Figure. 2.7) can be made practical not only for selective oxidation of the alcohols to carboxylic acid or aldehydes but also for the added tandem one-pot combination of imines (Han et al., 2014).

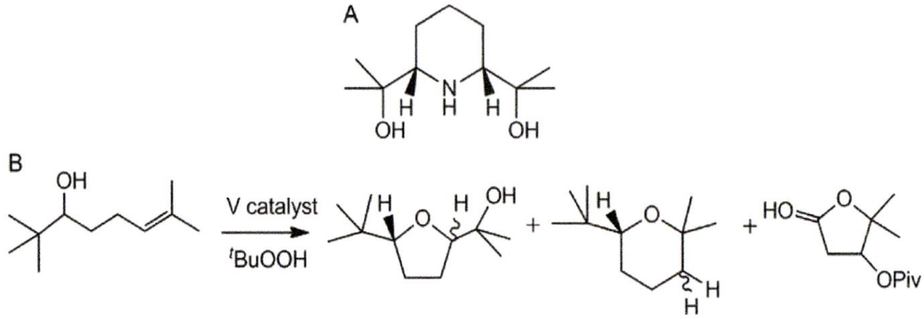

Figure 2.6: (A) Example of the cis-2,6-bis-(methanolate)-piperidine ligand (B) and oxidation of the alkenols by tert-butyl hydroperoxide, catalyzed by piperidine-derived V (vanadium) complexes.

[Source: https://www.researchgate.net/publication/283763126_Catalytic_Oxidation_of_Alcohols_Recent_Advances]

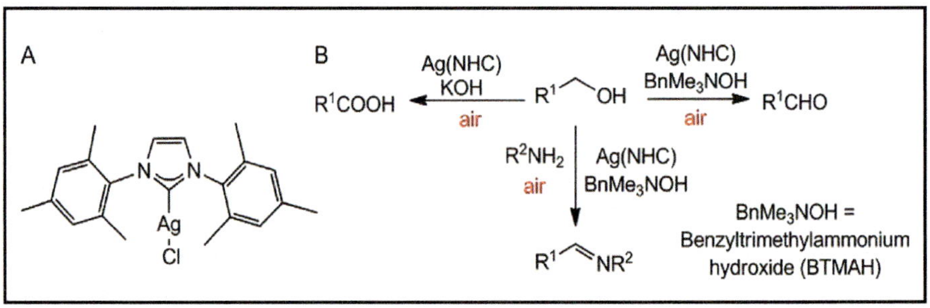

Figure 2.7: (A) Example of the Ag(NHC) catalyst (B) and oxidation of the alcohols to carboxylic acid or aldehydes and tandem one-pot combination of the imines catalyzed by them.

[Source: https://pubs.acs.org/doi/abs/10.1021/ol501353q]

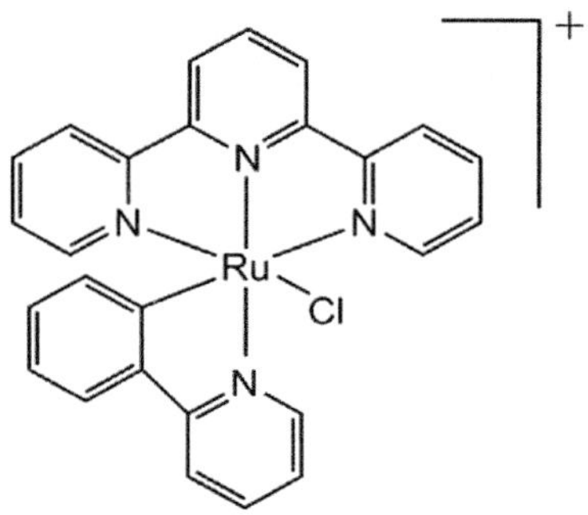

Figure 2.8: Example of the ruthenium (III) catalyst for aerobic oxidative dehydrogenation of the benzyl alcohols.

[Source: https://www.sciencedirect.com/science/article/pii/S0040403912007575]

The complexes of Rhodium porphyrin have been applied successfully as the catalysts for choosy oxidation of the functionalized alcohols, as they bear the range of functional groups, like methoxy, thiofuran moieties, and C] C (Liu et al., 2010). The recommended catalyst is tough and doesn't damage under the studied circumstances. The complex of porphyrin rhodium (III) methoxide was recognized as the main intermediate in the recommended mechanism. The complex of cyclo-metallated bearing the phenyl pyridine [RuCl(ppy)(tpy)][PF_6] (Figure. 2.8) is an instance of the ruthenium(III) catalyst utilized for aerobic oxidative dehydrogenation of the benzyl alcohols to the benzaldehydes. The complex was applied as a catalyst for a one-pot combination of benzonitriles from the benzyl alcohol with NH_3 (ammonia) (Taketoshi, et al., 2012; Wang et al., 2012).

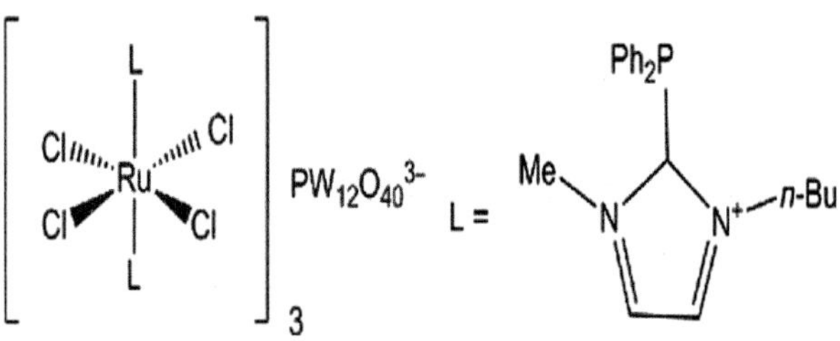

Figure 2.9: An ionic catalyst comprising of the Ru(III) complex cation and the α-Keggin-kind phosphotungstate anion.

[Source: https://www.sciencedirect.com/science/article/pii/S1566736712003299]

Another method to produce fresh catalysts is the amalgamation of diverse metals and even diverse kinds of complexes in a system, e.g., the ionic compound comprising of a Ru(III) complex cation and the α-Keggin-kind phosphotungstate anion (Figure. 2.9). This compound is strong as the phosphor-tungstate anion, although displaying the tiny contribution to the initiation of benzyl alcohol, importantly stabilizes the structure. Conversely, because of the ionic nature and high polarity of the complex, ILs (ionic liquids) can be utilized as solvents. The catalyst was made functional as an effective catalyst for the aerobic alcohols oxidations, free of nitroxyl-radical and base, and can be reprocessed without noteworthy loss of activity (Borthakur et al., 2013). Other heteronuclear complexes are efficient for the solvent-free per-oxidative oxidation of secondary and primary alcohols.

Finally, it is important that the oxidation of alcohol reaction can be utilized for a direct one-pot combination of the coordination compounds. The application of this technique permits to produce in situ ketones, aldehydes, and some other carbonylic derivatives, which aren't accessible commercially, are unsteady or can't be made and separated by the conventional methods. As an outcome, new fascinating compounds can be made with the ligands which aren't achievable by typical synthetic methods (Sanmartín-Matalobos et al., 2014).

2.2.2. Organic Radicals, Organocatalysts, and Other Additives

Despite their high motion, catalytic systems that engage transition metals display a number of drawbacks. For example, substrates with the chelating capability can fix to metal and obstruct the reaction. Furthermore, the occurrence of the metal can demonstrate some environmental influence. Thus, fresh transition-metal free schemes have been examined. One of such methods mimics the Anelli–Montanari procedure and engages an oxoammonium salt which transmits substrate oxidation whereas NO_2 restores the salt (Cao et al., 2014). It is probable to function aerobic NO_x structures under halide free circumstances; but, the contribution of halides) as energetic co-oxidants can possibly broaden the opportunity of such reactions. Additionally, an effective transition-metal free catalytic scheme mediated by NBS (N-bromosuccinimide) for the aerobic oxidation of numerous aromatic alcohols, under the mild environments, was reported recently (Tong et al., 2014). For example, benzyl alcohol is generally oxidized to benzaldehyde having 99% translation by the 2,3-dichloro-5,6-dicyano-1,4-benzoquinone–$NaNO_2$– N-bromosuccinimide system under the 0.3 MPa of oxygen (O_2) for 2 hours at 90 °C.

The NH_4NO_3 /H^+ /TEMPO catalytic system was stated as effective, under minor aerobic circumstances, for chemoselective oxidation of the complete variety of alcohols, counting those having oxidizable heteroatoms alkyl-, cycloalkyl-, (S, N, O), and allyl-type replaced substrates (Ma et al., 2014; Prebil et al., 2014). Very newly, tetra-n-butylammonium bromide was effectively applied as an easy but effective organocatalyst for peroxidative with TBHP oxidation of the diversity of functionalized allylic/benzylic alcohols under minor conditions. It displays outstanding selectivity for the secondary benzylic alcohols over the aliphatic alcohols. The analog tetra-n-butylammonium iodide was engaged for alfa-oxyacylation of the ketones by the benzylic alcohols prominent to alfa-acyloxyketones, with tert-butyl hydroperoxide, affording reasonable to good produces (Guo et al., 2014).

1,2-Di(1-naphthyl)-1,2-ethanediamine effectively catalyzes the oxidation of the alcohols by utilizing tert-butyl hydroperoxide as oxidant. Secondary benzyl alcohols are usually oxidized in measurable produces, and the catalyst shows the high motion towards the delayed cycloaliphatic secondary alcohols (Al-Hunaiti et al., 2014). Quinine-derived CH_4N_2O has been recognized as the highly effective organocatalyst for enantioselective oxidation of the 1,2-diols utilizing bromination substances as the oxidants, at room temperature, to produce the wide variety of α-hydroxy ketones in good production and exceptional enantioselectivity(Ciriminna & Pagliaro, 2009; Rong et al., 2013).

Nitroxyl radicals (Figure.2.10), like ABNO {9-azabiciclo[3.3.1]nonane N-oxyl}, ketoABNO {9-azabiciclo[3.3.1]non-3-one N-oxyl}, AZADO (2-azaadamantane N-oxyl), 1-Me-AZADO (1-methyl-2-azaadamantane N-oxyl) and Nor-AZADO (2-azanoradamantane N-oxyl), and particularly TEMPO, are broadly utilized agents for the aerobic oxidations of alcohols to conforming carbonyl compounds because of their high selectivity and efficiency. TEMPO is made functional in industrial processes utilizing Anelli–Montanari or aerobic type conditions (Figure. 2.11): an oxoammonium salt of TEMPO transmits the alcohol oxidations in the existence of a surplus of the hypochlorite. On the contrary, some applications of the nitroxyl radicals as the catalysts for peroxidative oxidations of alcohols were stated, in spite of its enhancing influence on the transformation of alcohols to the particular ketones (Sabbatini et al., 2014).

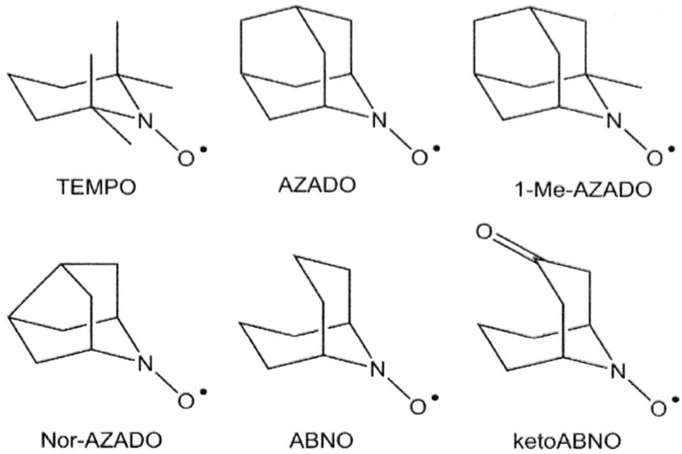

Figure 2.10: (From left to the right) structures of nitroxyl radicals TEMPO (2,2,6,6tetramethylpiperidine-1-oxyl), AZADO (2-azaadamantane N-oxyl), 1-Me-AZADO (1-methyl-2-azaadamantane N-oxyl), NorAZADO (2-azanoradamantane N-oxyl), ABNO (9-azabiciclo[3.3.1]nonane N-oxyl), and KetoABNO (9-azabiciclo[3.3.1]non-3-one N-oxyl).

[Source: https://www.researchgate.net/publication/283763126_Catalytic_Oxidation_of_Alcohols_Recent_Advances]

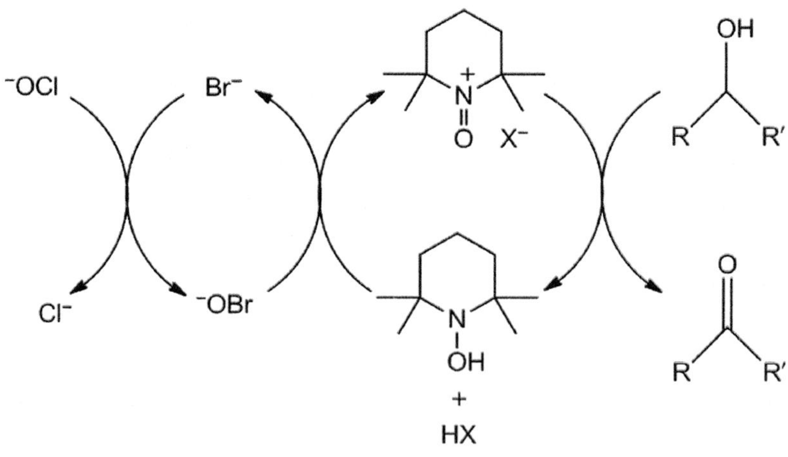

Figure 2.11: Anelli Montanari's oxidation of the alcohols.

[Source: https://pubs.acs.org/doi/10.1021/jo00388a038]

Compared to the TEMPO, the less obstructed ABNO and AZADO radicals exhibit considerably improved reactivity towards the wide variety of alcohols, comprising of structurally obstructed secondary alcohols which TEMPO somehow fails to effectively oxidize because of the steric congestion close to its active center (Sheldon & Arends, 2006; Hayashi et al., 2011).

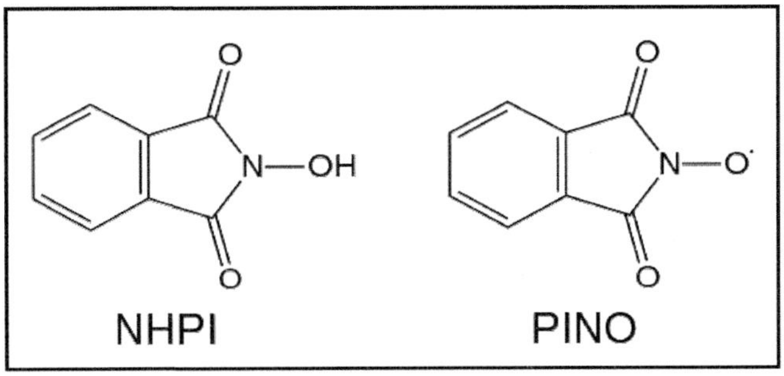

Figure 2.12: NHPI (N-hydroxy phthalimide) and PINO(phthalimide-N-oxyl nitroxyl) radical.

[Source: https://www.researchgate.net/publication/283763126_Catalytic_Oxidation_of_Alcohols_Recent_Advances]

Furthermore, the sterically unobstructed Nor-AZADO is catalytically more active than ABNO, AZADO, TEMPO, and 1-Me-AZADO in the aerobic oxidations of alcohols to the conforming carbonyl compounds.

Others associated with additives of TEMPO have been utilized, e.g., NHPI (N-hydroxyphthalimide) which produces the PINO (nitroxyl radical phthalimide-N-oxyl), (Figure. 2.12). The PINO analogs and radicals have been used for the range of aerobic oxidations reactions. PINO has been demonstrated to be more reactive as compared to TEMPO, but it isn't stable and is normally made from N-hydroxyphthalimide in situ with the help of an initiator.

Normally, the firm nitroxyl radicals can't alone directly catalyze the oxidations of alcohols with peroxide or dioxygen, so they depend on the support of several cocatalysts which play a vital role in triggering the oxidation agent. The most utilized cocatalysts are the first-row transition metals complexes where copper compounds with several N-donor ligands justify for the major ones (Gartshore & Lupton, 2010). In many examples, this combination assists as some type of models to associate the catalytic properties of Cu compounds. For instance, the performances of 2 asymmetric dinuclear (with $\{Cu_2(\mu{-}O)_2N_2O_2\}$ core) and tetranuclear (with $\{Cu_4(\mu{-}O)_2(\mu^3{-}O)_2N_4O_4\}$ core) and Cu(II) complexes were related in the catalytic TEMPO facilitated aerobic oxidations of benzylic alcohols. Despite their similarity, the complexes execute differently: tetranuclear Cu(II) complex is quite highly active generating yields up to 99% and TONs of around 770, whereas the dinuclear complex isn't so effective under the similar conditions.

However, the nitroxyl radicals are ineffective for the oxidations of secondary and aliphatic alcohols and need the surplus base for the oxidations to proceed. One of the major reason for the bad performance towards the secondary alcohols is supposed to be the steric interference and the participation of the bimolecular reaction amongst the TEMPO and Cu-alkoxide on the last step. This can be overwhelmed by swapping from the TEMPO to less sterically interfered nitroxyl radicals AZADO or ABNO. Therefore, Copper(I)/ABNO systems efficiently oxidize the wide variety of secondary alcohols, containing substrates with immense groups near to the alcohol groups, and still be efficient for primary alcohols. With the AZADO systems, substrates including secondary and primary alcohols can be oxidized in quite good to excellent produces (Steves & Stahl, 2013; Sasano et al., 2014).

Compared to the Copper/TEMPO amalgamation, Iron(III)/TEMPO systems eagerly catalyze the aerobic oxidations of secondary alcohols and don't usually need any source or less sterically interfered ligands (Wang et al., 2005). The finest activity was noticed with weakly organizing solvents unlike Copper/TEMPO systems, which generally are more vigorous in acetonitrile. The substrate scope and performance of Iron–TEMPO catalyst systems is enhanced by the existence of NaCl.

Lately, functionalized TEMPO was halted with the IL on silica covered magnetic NPs (nanoparticles).[71d] They catalyze aerobic oxidations of the variety of alcohols and catalysts can be isolated with an outer magnet and reprocessed 10 times without a substantial loss of activity.

The first instance of vanadium (V) as the only metallic component in the TEMPO-catalyzed aerobic alcohols oxidation in the acetonitrile was reported recently (Du et al., 2010). Though, the catalyst didn't perform well with the secondary aliphatic alcohols.

The level of complexity of stable radicals might increase in the forthcoming as it is clear that the conduct of catalysts can usually be enhanced by tuning both electronic and steric influences of these radicals (Shibuya et al., 2011). Another successful method is the utilization of the heterogenized radicals.

Therefore, the free-radical leaky matching polymer [Cu(DPIO)$_2$(SiF$_6$)] [DPIO¼4,7-bis(4-pyridyl)-1,1,3,3-tetramethylisoindolin-2-yloxyl] (Figure. 2.13), owns 1-D channels with merged nitroxyl catalytic sites.

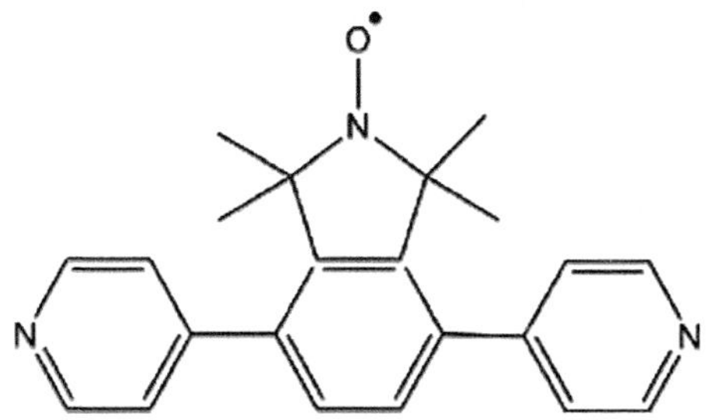

Figure 2.13: DPIO {4,7-Bis(4-pyridyl)-1,1,3,3-tetramethyl isoindoline-2-yloxyl}.

[Source: https://www.researchgate.net/publication/283763126_Catalytic_Oxidation_of_Alcohols_Recent_Advances]

When air or dioxygen is utilized as an oxidant, the polymer performs as an effective, recyclable, and extensively appropriate catalyst for the selective oxidation of numerous alcohols to the conforming ketones or aldehydes (Zhu et al., 2012).

Fullerene has been engaged effectively to anchor TEMPO components and utilized as the organocatalyst for the aerobic oxidations of secondary and primary alcohols in the existence of TBN (tert-butyl nitrite) as cocatalyst. The reaction demonstrated general applicability to numerous alcohols, and the catalyst was easily recovered and could be reprocessed for at least 7 cycles with no significant loss in the catalytic activity (Beejapur et al., 2014). In the rare instance of Iron/TEMPO catalyst schemes, $NaNO_2$ or $Fe(NO_3)_3$ weren't utilized; rather, the $FeCl_36H_2O$/TEMPO amalgamation was coupled with the silica support (Wang et al., 2012). In the linked study, TEMPO was covalently bound to the silica and united with $FeCl_36H_2O/NaNO_2$ (Wang et al., 2013).

2.2.3. Potential Substrates and the Oxidation Agents

Developing or finding new preliminary materials for the per-oxidative and aerobic oxidation of alcohols is a vital task, specifically by broadening the range of the substrates which are of importance in chemical synthesis. For example, as aldehydes are made by the oxidation of alcohol, whereas the aldehydes can participate by themselves in further reactions, amino-oxidation of the alcohols to amide synthesis and nitriles can be accomplished directly utilizing Mn catalyst for both of the conversions in one-pot (Fig, 2.14) (Ishida et al., 2012).

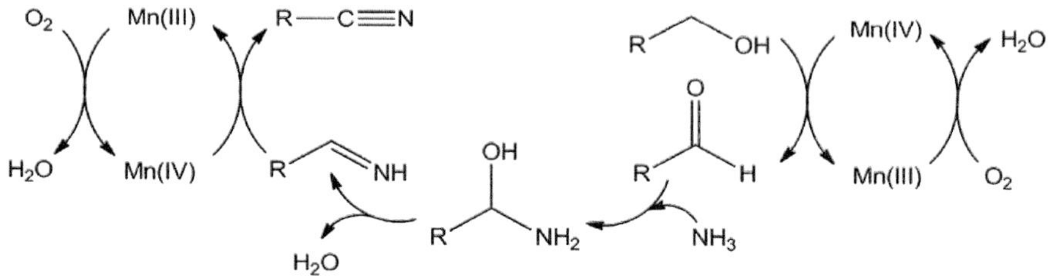

Figure 2.14: Anticipated reaction path for the ammoxidation of the alcohols over MnO_2.

[Source: https://www.sciencedirect.com/science/article/abs/pii/S0926860X12001391]

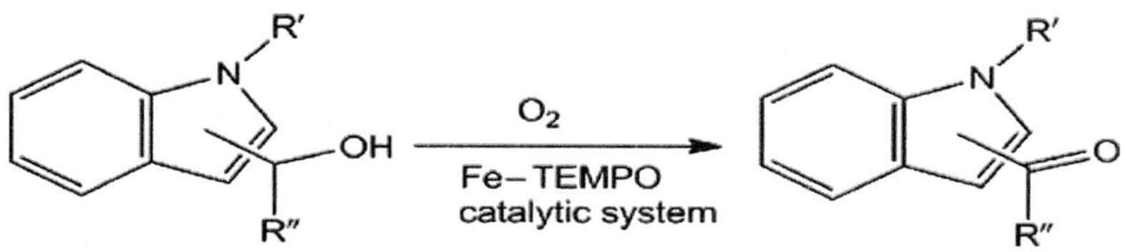

Figure 2.15: Oxidation of the indole carbinols utilizing the Iron(NO3)39H2O/TEMPO catalytic system.

[Source: https://pubs.rsc.org/en/content/articlelanding/2013/ob/c3ob40226f]

Indole derivatives having carbonyl units (Figure. 2.15) are found in the natural products, own adaptable bioactivity and are significant intermediates in numerous organic syntheses. The introduction of carbonyl moieties to the indoles enhances their reactivity considerably and can be accomplished by the oxidations of indole carbinols with Iron$(NO_3)_3 9H_2O$/NaCl/TEMPO/ at atmospheric pressure of di-oxygen and room temperature, utilizing toluene as the solvent (Liu & Ma, 2013). NaCl is the vital booster for the oxidation, decreasing the time of reaction from 15 to 2 hours; though, the part of Cl isn't quite clear.

Lately, there has been a rising interest in new procedures to derive fuels and chemicals from renewable carbon feedstocks. Therefore, the study of homogeneous copper and oxo-vanadium catalysts towards aerobic oxidation of the lignin model compound is of great prospective (Sedai et al., 2013). In this conversion, the vanadium catalyst gives mainly ketone products, whereas the system of copper catalyst gives the products of Cd-C bond cleavage reactions, therefore reflecting diverse mechanisms of oxidation.

Glycerol is one more substrate of significance in the biomass transformation, and glycerol can be converted to the added value di-hydroxyacetone with the assisted palladium catalyst (Painter et al., 2010). The related method is the added value amendment of the nature-derived materials. For example, pullulan, the polysaccharide removed from the fermentation medium of Aureobasidium pullulans bacteria, it can be made to function by the institution of several groups, like the carboxylic ones, which considerably affects its hydrophilicity, etc. (Spatareanu et al., 2014). Vanadium was discovered to be vigorous in the direct conversion of 4-pentanol to the 3-acyloxy-γ-butyrolactones, utilizing TBHP as the oxidant (Figure. 2.16) (Amberg et al., 2014). In this fascinating oxidative transformation, stereocenters of tetrahydrofuran moiety maintain their configuration. Allylic alcohols can also be oxidized to the stereo-defined α,β-unsaturated ketones/aldehydes with the retaining of the C-C double-bond configuration, utilizing Iron$(NO_3)_3 9H_2O$/NaCl/TEMPO, at room temperature and under atmospheric pressure of the oxygen (Liu & Ma, 2013a; 2013b). The similar catalyst was discovered to be efficient in the transformation of homopropargylic alcohols to their conforming homopropargylic ketones. Iron catalysts are efficient in the oxidative arrangement assembly of 3-(2-oxoethyl) indolin-two-ones from alcohols and N-aryl acrylamides via particular 1,2-difunctionalization of the C-C double-bond in N-aryl acrylamides (Figure. 2.17) (Ouyang et al., 2014).

Until very freshly, it was quite tough to make carbonyl compounds from the alcohols with robust electron-withdrawing groups near the RCHOH moiety. For example, 2,2,2-trifluoroethanol is frequently utilized as the solvent in oxidation reactions as it has been reflected inactive to the oxidation. Though, aerobic oxidation of inactive perfluoro-replaced alcohols to their conforming carbonyl byproducts has been achieved recently utilizing Pt(II) complexes with the dipyrido[3,2-a:20,30-c]-phenazine ligands (Figure. 2.18). Therefore, in the existence of O_2 and H_2SO_4 oxidant, trifluoroethanol was oxidized successfully to tri-fluoroethyl trifluoroacetate with >98% selectivity (Ben-David et al., 2013).

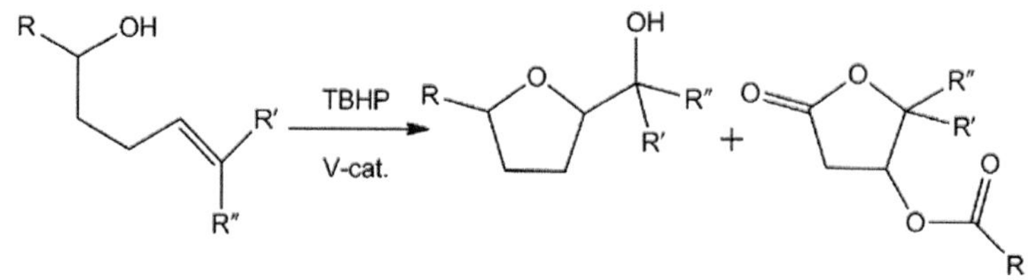

Figure 2.16: Oxidation of the 4-pentanol to the 3-acyloxy-γ-butyrolactones.

[Source: https://www.sciencedirect.com/science/article/pii/S0040402014006711]

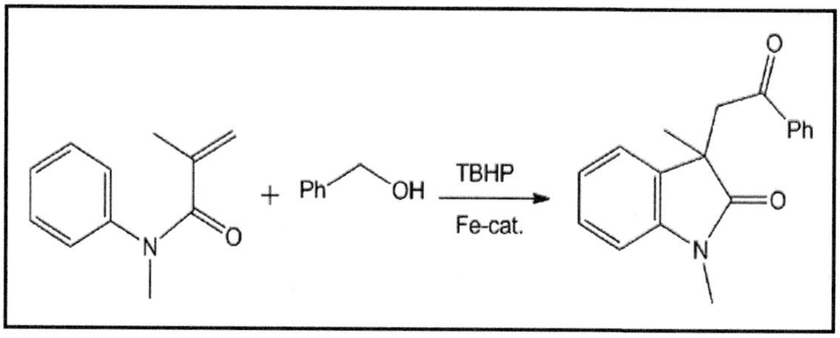

Figure 2.17: Synthesis of the 3-(2-oxoethyl)indolin-2-ones from the N-aryl acrylamides & alcohols.

[Source: https://onlinelibrary.wiley.com/doi/abs/10.1002/ejoc.201400043]

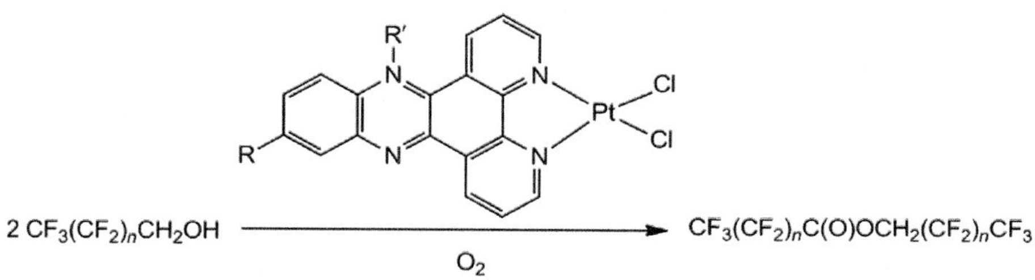

Figure 2.18: Aerobic oxidation of the perfluoro-replaced alcohols.

[Source: https://pubs.rsc.org/en/content/articlelanding/2013/CC/c2cc37759d]

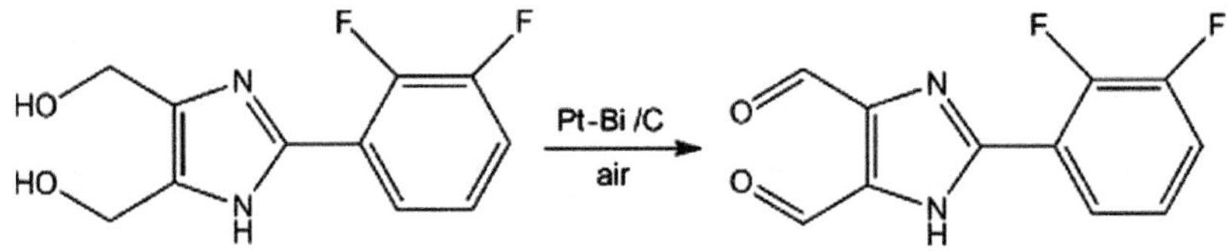

Figure 2.19: Oxidation of the imidazole diol to produce imidazole-4,5-di carbaldehyde.

[Source: https://www.ncbi.nlm.nih.gov/pubmed/24171666]

The bimetallic Pt/C/Bi catalyst was utilized for the oxidation of the imidazole diol to produce imidazole-4,5-dicarbaldehyde, the main step in the production of pro-drugs of the hepatitis C virus replicas' inhibitors (Figure. 2.19) (Bowman et al., 2013).

Introduction of fresh oxidative agents is a significant task as the classical oxidants, like pyridinium chromates, are usually not ecological, yield high amounts of the toxic agents and normally aren't selective enough. Oxidations with the eco-friendly peroxides, air, and oxygen, and, despite being striking in theory and on the small scale, come across a number of problems when being scaled. Actually, oxidations in combustible organic solvents will almost eradicate such oxidants because of the threat of irritation and other threats linked to the high oxygen pressures. In numerous examples, these attractive oxidants don't work or the reaction goes via an unwanted way, e.g., over-oxidation or production of the number of byproducts. Therefore, the search for fresh efficient oxidants is growing continuously.

Instead of mild and selective oxidizing agents, specifically hypervalent halogens or their precursors have been extensively familiarized into the oxidation of alcohols during the last few years (Duschek & Kirsch, 2011). IBX, less dangerous IBA (2-iodosobenzoic acid) and even the accessible 2IBAcid (2-iodobenzoic acid) (Figure. 2.20) in the existence of Oxone, an environmentally tolerable reagent, as the co-oxidant, were demonstrated to be efficient for the oxidation of several secondary and primary alcohols in eco-and user-friendly solvent mixtures (Thottumkara et al., 2005). In this procedure, the condensed form of IBX, specifically 2-iodoso benzoic acid is utilized in catalytic quantity and is re-oxidized with Oxone, in the aqueous media, therefore providing the striking green protocol.

Conversely, the use of IBX as an oxidant can give some unpredictable outcomes, e.g., unanticipated de-homologation of principal alcohols to 1-C shorter carboxylic acids (Figure. 2.21) (Xu et al., 2014). In this situation, the grouping of molecular iodine and IBX affords the different kind of energetic hypervalent iodine (I) species, which was separated and demonstrated to be critical for the reaction. It must be noted that normally the de-homologation includes complicated multistep processes, while in the defined oxidative approach it proceeds efficiently in one step. Another beneficial direct oxidation of the secondary alcohols was executed utilizing performic acid within 15 minutes (Li et al., 2013).

The adjustment of IBX with the fluorous tag (IBXF, Figure. 2.22) permits its recovery as the insoluble fluorous 2-iodoso benzoic acid can be isolation from the mixture of reaction by filtration, and be reutilized without substantial loss of the activity (Miura et al., 2012). Moreover, IBXF can be generated easily in situ from inexpensive and readily accessible Oxone. Other adjustments of the oxidizing components are being introduced. For example, the combination of $NaIO_4$ and NH_2OHHCl was proposed recently for the mild and selective oxidation of the alcohols to conforming carbonyl compounds at the room temperature (Majee et al., 2012). Concerning specific substrates, the broad range of the β-hydroxy ketones was

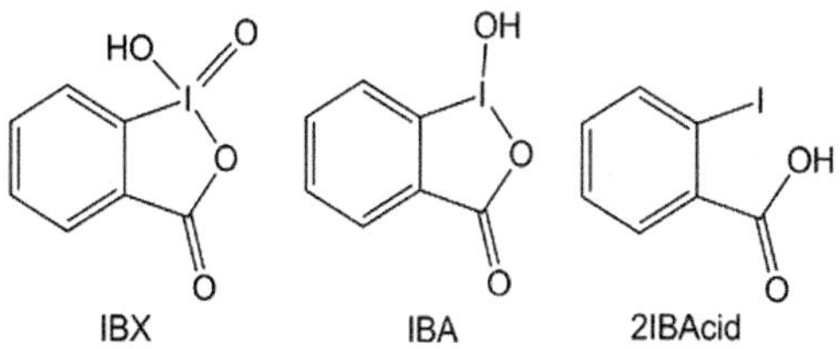

Figure 2.20: IBX (o-Iodoxybenzoic acid), IBA (2-iodoso benzoic acid), and 2IBAcid (2-iodo benzoic acid).

[Source: https://www.researchgate.net/publication/283763126_Catalytic_Oxidation_of_Alcohols_Recent_Advances]

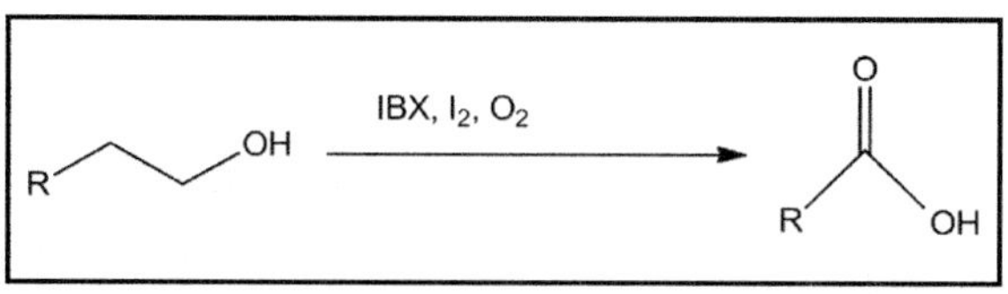

Figure 2.21: Oxidative dehomologation of the principal alcohols to 1-C shorter carboxylic acids.

[Source: https://pubs.rsc.org/en/content/articlelanding/2014/cc/c3cc49160a]

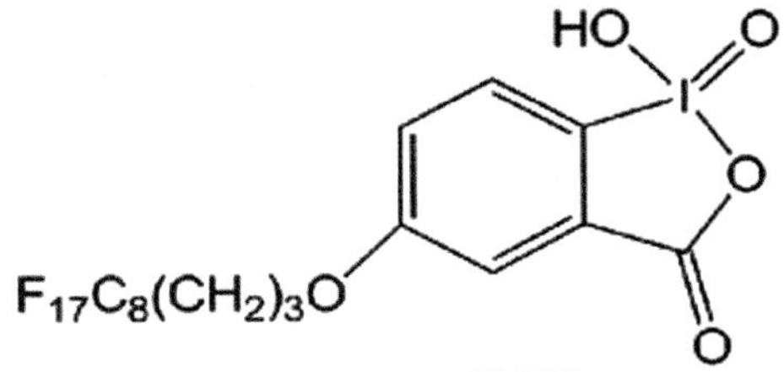

IBXF

Figure 2.22: Altered IBX with the fluorous alternative for its quick recovery.

[Source: https://www.researchgate.net/publication/283763126_Catalytic_Oxidation_of_Alcohols_Recent_Advances]

Oxidized selectively in quantitative produces to β-diketones with the IBX being the higher level oxidant for this conversion (Bartlett & Beaudry, 2011).

The ability to execute oxidations without producing species injurious for impending intermediates of further conversion is vital to execute multistep synthesis, as the domino reactions. In this regard, the utilization of aryl halides as available, cheap and stable oxidants is the powerful option because of the production of inactive, de-halogenated aryl byproducts in the anaerobic conditions. Commercially available Ni and Pd complexes with the NHC ligands were determined to be vigorous in the temperature-controlled domino oxidation (Berini et al., 2009). Tert-Butylnitrite (t-Buono) was introduced recently as the easy-removable and convenient oxidant for the environment-friendly conversion of secondary and primary benzylic alcohols to aldehydes and ketones, which can be isolated readily by evaporation of the mixtures of reaction as t-Buono decomposes producing only instable side products (Hamasaki et al., 2012). The oxidation needs neither organic catalysts nor metal-based reagents and apparently includes the nitrosyl exchange and successive thermal decomposition of the benzylic nitrites. t-Buono can possibly be recovered, as various alkyl nitrites are identified to be produced industrially from the conforming alcohols and gaseous nitrogen oxide under the O_2 atmosphere. The related method was recognized for the high-yield transformation of several chiral and achiral alcohols to the carbonyl compounds utilizing AZADO or TEMPO in combination with $LiBF_{4\ or}$ BF_3OEt_2 as the pre-catalysts and t-Buono as an oxidant (Holan & Jahn, 2013). The NO^+/NO pair were utilized for the mild anaerobic nitroxide re-oxidation, which permitted the oxidation of pure substrates without any racemization.

$K_3[Fe(CN)_6]$ was made functional as the secondary oxidant in the chemo-selective osmium-catalyzed oxidation of allylic, benzylic, and propargylic alcohols (Fernandes & Bethi, 2014). The unusual oxidation agent, DIAD (diisopropyl azodicarboxylate), can be utilized as an efficient terminal oxidant; in this situation, 1,2-diols were usually oxidized to diketones or hydroxyl ketones which is dependent on the quantity of DIAD utilized (Hayashi et al., 2012). Diaziridinone (Figure. 2.23) as oxidant permits the reaction to

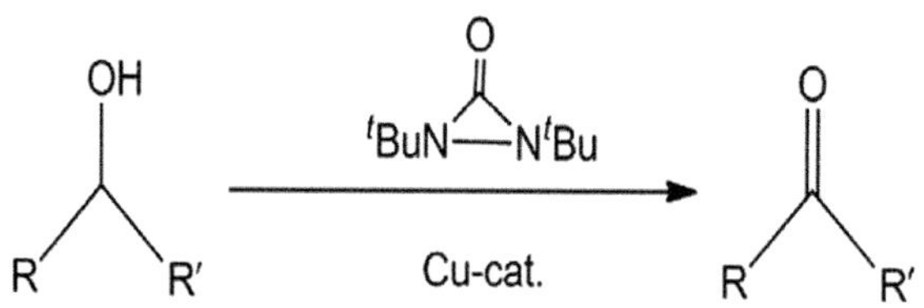

Figure 2.23: Diaziridinone as the oxidant for the base- or acid-sensitive substrates.

[Source: https://www.researchgate.net/publication/283763126_Catalytic_Oxidation_of_Alcohols_Recent_Advances]

be executed in neutral circumstances and makes it well-suited with base- or acid-sensitive substrates (Zhu et al., 2013). As an outcome, several cyclic and acyclic secondary benzylic alcohols with alkynyl, alkenyl, thioether, amide, silyl ether, carbamate, ester, heterocyclic moieties, and ketal were efficiently oxidized to the ketones in 73 to 99% produce, whereas no racemization of the stereocenters took place during the oxidation.

The definite way of evading the poisonous chemicals is the utilization of electric current as an oxidizing agent. Therefore, an electrochemical procedure for choosy oxidation of the 1,2-diols to conforming α-hydroxy ketones in the water utilizing $[Me_2SnCl_2]$ catalyst, platinum, and KBr electrodes has been presented. The Br^+ ions, produced at the anode, are the oxidants, whereas the OH ions, electro-produced at the cathode, support the role of the base. Conversely, in this synthetic approach, the poisonousness of the organotin catalysts is the drawback. This problem was presented in another study, where the methyl-boronic acid $[MeB(OH)_2]$ was utilized as a safe alternative. In this situation, selective oxidation of the 1,2-diols to conforming α-hydroxy ketones in the aqueous medium most possibly takes place through the creation of boronate esters (Willium et al., 2013; 2014).

2.3. ACCEPTOR-LESS DEHYDROGENATIVE OXIDATIONS

The per-oxidative and aerobic oxidations, described above, despite being attractive in several aspects, can yield a substantial amount of byproducts. Other issues worry about the over-oxidation and hazard of explosion because of the existence of oxygen and the organic reactants. In this regard, the acceptor- and oxidant-free dehydrogenation (Figure. 2.24) is a substitute environmentally friendly way for the transformation of alcohols into ketones or aldehydes, as gaseous hydrogen can be easily isolated from the mixture of reaction. Furthermore, this approach is much more striking from the atom effectiveness viewpoint and can offer a promising way for H_2 storage and synthesis (Nielsen et al., 2011).

Therefore, some improvements in the acceptor or oxidant-free oxidations centered on catalytic dehydrogenation of the alcohols complemented by the development of hydrogen gas are described here. Some new attractive procedures were recently introduced in terms of substrates and efficiency, and examples are given below.

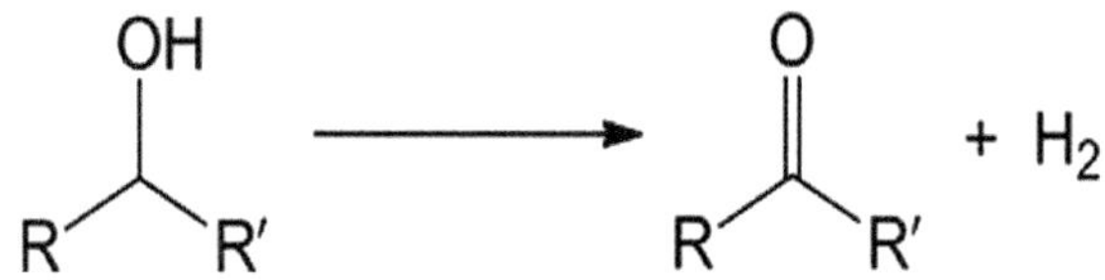

Figure 2.24: Acceptor- and oxidant-free dehydrogenation of the alcohols.

[Source: https://www.researchgate.net/publication/283763126_Catalytic_Oxidation_of_Alcohols_Recent_Advances]

Until now, the homogeneous catalytic methods for dehydrogenative oxidation of the alcohols are primarily based on iridium, ruthenium, and, in a lesser amount, other metal-organic compounds and transition metal complexes. For example, water-soluble Cp*Ir complexes with bi pyridonate or α-hydroxy pyridine ligands have demonstrated catalytic activity in the dehydrogenative oxidation of the broad range of secondary and primary alcohols and reversible hydrogenation–dehydrogenation amongst acetone and 2-propanol (Royer et al., 2010; Baratta et al., 2011). The catalysts of Ir are recyclable and can speed up the choosy dehydrogenation of important biologically and complex molecules (Figure. 2.25).

Production of the H_2 gas can be accomplished with the catalysts of Ir from 2-propanol, therefore providing the prototype for the hydrogen storage system, centered on the exchange amongst acetone and 2-propanol (Figure. 2.26). The reversible conversion with hydrogen storage and development was repeated many times without any loss of catalytic activity. The ligand-promoted process of dehydrogenation was suggested, performing the ligand as the proton acceptor in the activation step and the proton donor in the dehydrogenation step, therefore playing the dual role in the cooperative catalysis.

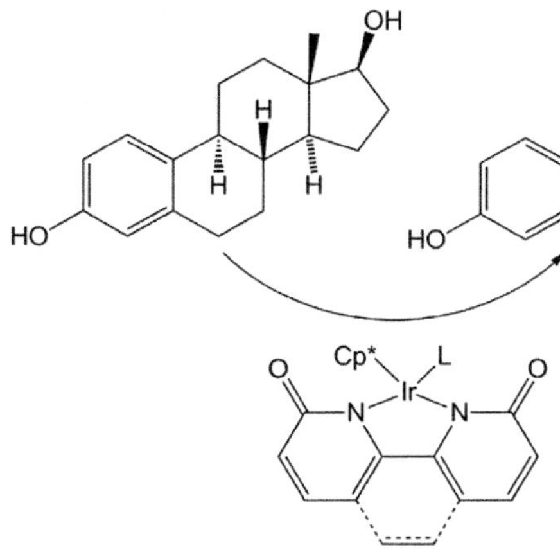

Figure 2.25: Careful dehydrogenation of β-estradiol to the estrone.

[Source: https://pubs.acs.org/doi/10.1021/om100901b]

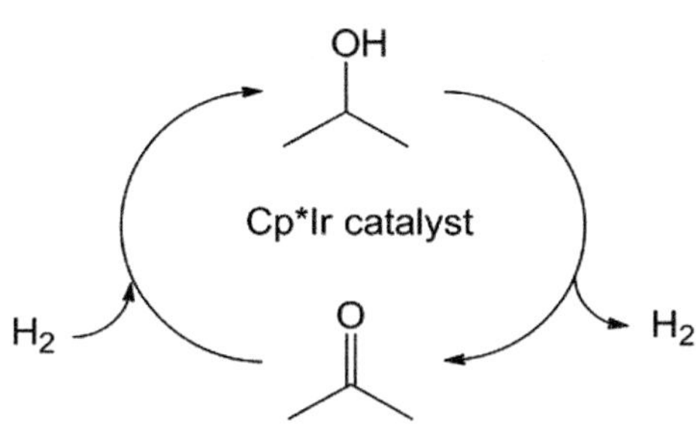

Figure 2.26: Inter-conversion between acetone and 2-propanol.

[Source: https://pubs.acs.org/doi/10.1021/om100901b]

The comprehensive discussion of the process of pH-reliant on alcohol dehydrogenation in the aqueous solution catalyzed by associated [C, C] or [C, N] Cp*Ir complexes was reported recently, normally assisting the mechanistic recommendations and demonstrating the vital dependency of the examined reaction systems on pH.

Structurally associated PCP-pincer Ir complexes (Figure. 2.27) were produced by the straightforward [4+2] cyclo-addition and engaged as the catalysts in acceptor-less dehydrogenation of the alcohols. Such type of complexes can be modified easily with the functional sidearm which is capable of interrelating with the catalytic position, consequently making them appropriate candidates for the catalytic studies including ligand–metal cooperation. The H_2 creation includes intra-molecular cooperation amongst the metal center and structurally isolated functionality.

The theoretical investigation on the process of acceptor-less alcohol dehydrogenation intermediated by the iridium (Ir) catalyst [Cp*Ir(bpyO)] (bpyO¼α,α⁰bipyridonate) recommends that the metal-ligand supportive work includes dearomatization/aromatization of bpyO ligand (Zeng et al., 2014; Zhang et al., 2014). Based on those outcomes, the fresh ruthenium (Ru) catalyst [(HMB)Ru(bpyO)(H_2O)] (HMB¼hexamethylbenzene) was prepared (Figure. 2.28). Another ruthenium (Ru) catalyst made from the Ruthenium-hydride precursor [$RuH_2(PPh_3)_3$CO] and the pincer PNP-ligands efficiently encourages dehydrogenative oxidation of the alcohols with Turn-over Frequencies as high as about 1.410^4 h^1 after twenty minutes, at the moderate temperatures and without any activation additive.

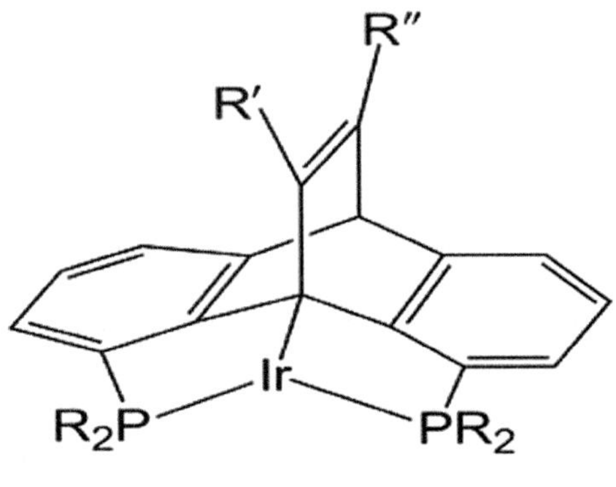

Figure 2.27: The complex of PCP-pincer Ir.

[Source: https://pubs.acs.org/doi/10.1021/om100901b]

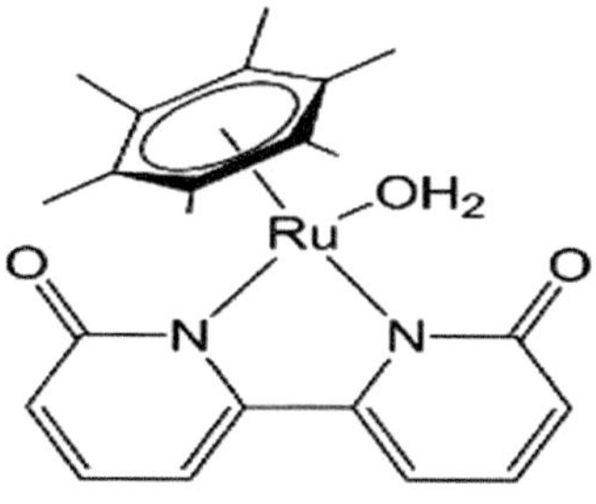

Figure 2.28: Ruthenium [(HMB)Ru(bpyO)(H2O)] catalyst.

[Source: https://pubs.acs.org/doi/10.1021/cs401101m]

The reaction is the instance of effective hydrogen production from the renewable source of alcohol. The recommended procedure might also include intra-molecular concentrated hydrogen loss from the dihydride Ruthenium species tracked by the outer-sphere dehydrogenative stage (Figure. 2.29).

Ruthenium (Ru) compounds can be utilized for transfer hydrogenations. Therefore, the series of water-soluble Ru(II) half sandwiched complexes with the 2,2^0-bipyridine ligands act as the catalysts in transfer hydrogenation of diverse ketones to the conforming alcohols, utilizing the mixture of formic acid/sodium formate as the source of hydrogen. Interestingly, some attained alcohols can be deuterated selectively at the benzylic C (carbon), if the catalytic transmission hydrogenation is executed in D_2O (Figure. 2.30). This point can be described by the rapidly adjustable deuteration of the hydride intermediate

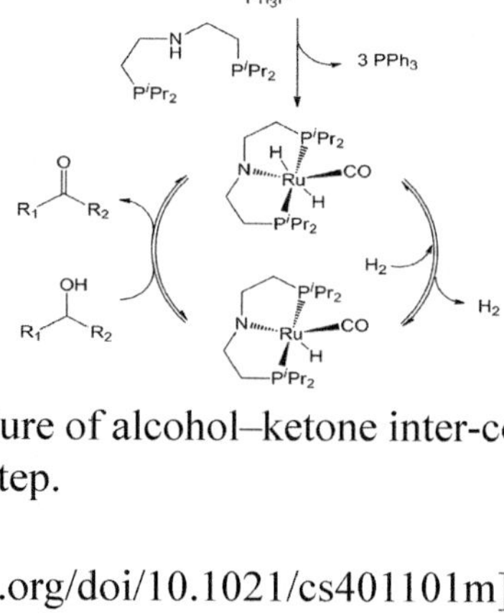

Figure 2.29: The procedure of alcohol–ketone inter-conversion joined with the hydrogen development step.

[Source: https://pubs.acs.org/doi/10.1021/cs401101m]

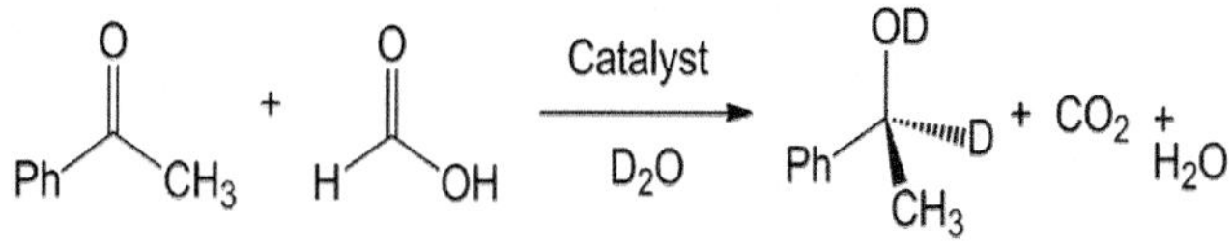

Figure 2.30: Catalytic relocation hydrogenation and deuteration of the acetophenone.

[Source: https://www.researchgate.net/publication/283763126_Catalytic_Oxidation_of_Alcohols_Recent_Advances]

Ru(X_2) (X¼H, D) in the existence of formic acid and D_2O. The deuteration is quicker than the hydrogenation procedure; as an outcome, selective deuteration of the phenyl ethanol at benzylic carbon takes place. Using this process, selective deuteration of diverse organic substrates can be performed (Shimizu et al., 2009; Aliende et al., 2012).The above-defined iridium- or ruthenium-PNP kind complexes catalyze the one-pot dehydrogenation /hydrogenation sequence/aldol condensation (Guerbet reaction), consequently permitting to make β-alkylated ketones (Yang et al., 2012; Musa et al., 2013). The dehydrogenative CdC bond creation takes place in this and associated cases; the offered catalytic sequence includes hydrogen abstraction by collaboration of the acidic sidearm with metal-hydride position, ligand interchange step with alkoxide creation, catalyst regeneration by the hydride eradication and creation of the carbonyl products; metal-free aldol condensation; and lastly, transfer of the H_2 to aldol condensation products to produce β-alkylated ketone (Figure. 2.31).

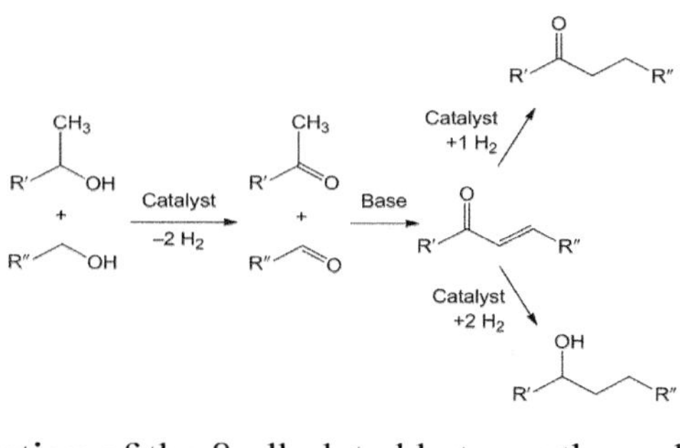

Figure 2.31: Preparation of the β-alkylated ketones through dehydrogenation / hydrogenation sequence/ aldol condensation (Guerbet reaction).

[Source: https://www.researchgate.net/publication/283763126_Catalytic_Oxidation_of_Alcohols_Recent_Advances]

Replacement of the noble metals in the catalyst for inexpensive and profuse materials is of apparent significance, and several works concern the study and preparation of the final systems. Therefore, Co(II) alkyl complexes of the aliphatic PNP pincer ligands (Figure. 2.32) are vigorous pre-catalysts for hydrogenation of the ketones and acceptor-less dehydrogenation of the alcohols under mild circumstances; in this situation the alcohol dehydrogenation possibly proceeds through the Co(I)/(III) redox cycle (Zhang & Hanson, 2013; Zhang et al., 2013).

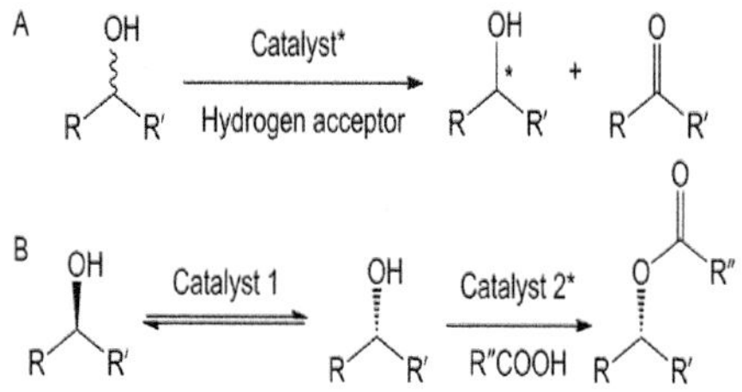

Figure 2.32: Co (II) complex with the PNP pincer ligand as the catalyst precursor for the acceptor-less dehydrogenation of the alcohols.

[Source: https://www.researchgate.net/publication/283763126_Catalytic_Oxidation_of_Alcohols_Recent_Advances]

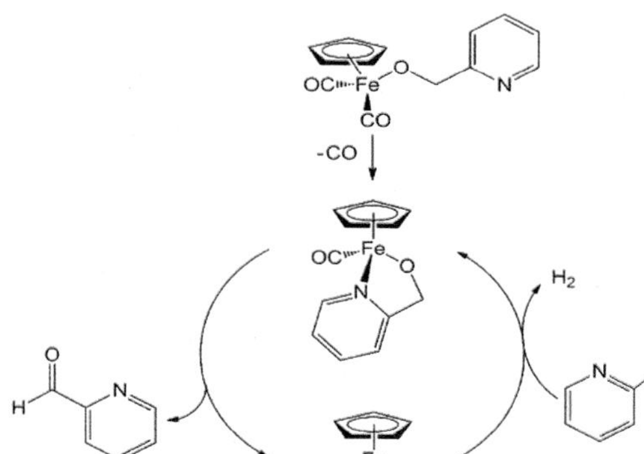

Figure 2.33: Dehydrogenation of the 2-pyridyl methanol utilizing [CpFeCl(CO)2] as the pre-catalyst.

[Source: https://www.researchgate.net/publication/283763126_Catalytic_Oxidation_of_Alcohols_Recent_Advances]

2-Pyridylmethanol byproducts can be efficiently dehydrogenated in the acceptor-less way utilizing a Fe catalyst, e.g., resultant of the [CpFeCl(CO)$_2$] (Kamitani et al., 2014). The co-catalyst NaH is utilized; its part is perhaps to offer the conforming sodium alkoxide, which reacts with the iron complex to produce the energetic Fe-alkoxide complex. The involved pyridine ligand then dislocates one of the carbon monoxide ligands, creating the Fe-N, O-metallacycle, from which 2-pyridine carboxy aldehyde and iron hydride complex are produced by the β-hydride elimination. Oxidative addition of the 2-pyridinyl methanol attached with H_2 reductive eradication finishes the catalytic cycle. It must be revealed that this type of work is somehow linked to the conclusion that in the Iron-catalyzed oxidation of the alkanes, pyrazine carboxylic acid, and the analogs have an important promoting effect (Kirillov & Shul'pin, 2013).

Generally, the iron catalysts appear to be very auspicious and fascinating works on Iron-catalyzed acceptor-less dehydrogenation of the alcohols are done every year. Therefore, an economical and operationally simple process can be revealed, engaging, as a catalyst, the mixture of available Iron(III) acetylacetonate, K_2CO_3, and 1,10-phenanthroline. Despite usually being active, the systems above experience the problems common for the homogeneous catalysts, and therefore the heterogeneous catalysts have been introduced (Mitsudome et al., 2008; Song et al., 2014). For example, the non-oxidative dehydrogenation of the benzyl alcohol was attained utilizing the hydrotalcite-assisted gold catalyst. Nano-crystalline rhenium particles worry another instance of the reusable heterogeneous catalyst to transform benzylic and secondary alcohols to aldehydes and ketones, via catalytic acceptor-less dehydrogenation through the novel γ-CH activation mechanism. Curiously, aldehydes and primary alcohols act as the inhibitors of dehydrogenation reaction (Chen et al., 2014; Yi et al., 2014).

The heterogeneous Rhodium-on-C systems catalyze the dehydrogenation of secondary and primary alcohols to the conforming ketones and carboxylic acids in water under the basic conditions (Sawama et al., 2014). Upon the formation of carboxylic acids, H_2O played a vital role in the source of oxygen. Pt nano-clusters, sustained on the oxides of metal, was also tried, and it was discovered that the assistance, e.g., amphoteric alumina, has an active part in the catalytic procedure, reacting with alcohols and producing a proton and an alkoxide (Kon et al., 2013). B-Hydrogen eradication from the alkoxide to the low coordinated Pt^0 position affords ketone and Pt-H; then protolysis of the Pt-H by adjacent proton redevelops the Pt^0 position with the release of the H_2 gas.

Nickel (Ni) is a well-known promoter of the hydrogen transferal reactions and therefore it is anticipated that nickel can be utilized in the reactions under examination. In concurrence, acceptor-free dehydrogenation of the secondary alcohols by Nickel NPs sustained on the alumina has been informed. Low-coordinated Ni^0 positions and metal/support boundaries are thought to play an important role in the catalytic process (Shimizu et al., 2013a; 2013b).

Nano-Cu(0) on the alumina [$Cu(0)/Al_2O_3$] NPs were made by a simple process from Cu aluminum hydrotalcite and displayed to be an effective catalyst for the oxidant- and acceptor-free dehydrogenation of amines and alcohols.

TiO_2-assisted Cobalt NPs are also active catalytically in the acceptor-less dehydrogenation of several aliphatic secondary alcohols to conforming ketones (Damodara et al., 2014).

The acceptor-less dehydrogenation can't only be used for simple conversion of the alcohols to conforming aldehydes or ketones or for hydrogen production and storage, but also as a vital part of more complex multistep processes or tandem reactions. Furthermore, apart from the typical model alcohols, other unusual alcohols are also introduced and some of these alcohols are mentioned below. Therefore, de-hydrogenative [Ir]-catalyzed combination of alcohols and aryl hydrazines can be utilized for the choosy synthesis of aryl hydrazones (Figure. 2.34). This way is simpler and hypothetically can strive with the customary synthetic methods to aryl hydrazones centered on the condensation of aryldiazonium or arylhydrazines salts with the carbonyl compounds (Li et al., 2014).

Dehydrogenative lactonization of the diols is an effective way for several lactones (Figure. 2.35). The lactone creation is discovered to be catalyzed by the recoverable steady dicationic Ir complex with the 6,6^0-dihydroxy-2,2^{0}bipyridine ligands and engages a variety of aliphatic and benzylic diols in the aqueous media (Fujita et al., 2014).

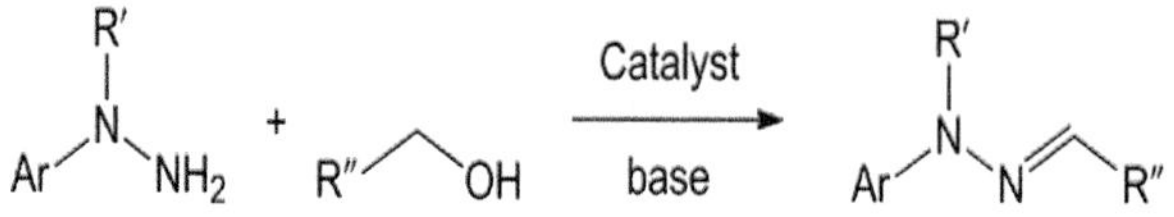

Figure 2.34: Direct synthesis of the aryl hydrazones via dehydrogenative [Ir] catalyzed the combination of alcohols and arylhydrazines.

[Source: https://pubs.acs.org/doi/abs/10.1021/jo501161u]

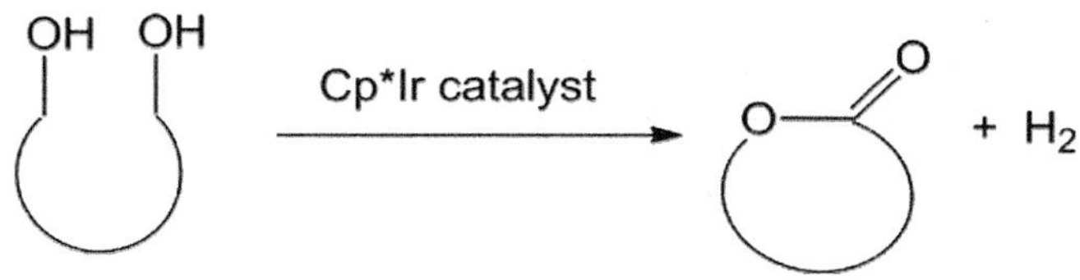

Figure 2.35: Dehydrogenative lactonization of the diols to lactones.

[Source: https://onlinelibrary.wiley.com/doi/abs/10.1002/cctc.201300717]

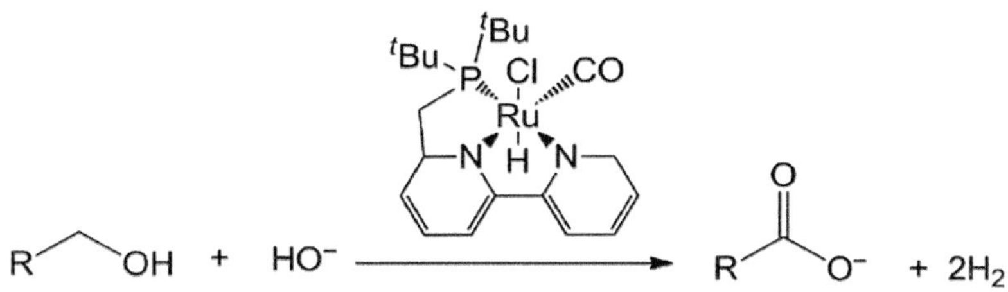

Figure 2.36: Exchange of alcohols to the carboxylic acids with H2O as the source of oxygen, catalyzed by the Ru complex.

[Source: https://www.ncbi.nlm.nih.gov/pubmed/23344447]

The conversion of alcohols to the carboxylic acids with no hydrogen acceptor or oxidant utilizes water as the source of an oxygen atom with a related discharge of di-hydrogen gas (Balaraman et al., 2013). The reaction is usually catalyzed by the ruthenium complex at the low loading in simple aqueous solution (Figure. 2.36). The same or associated complexes are lively in numerous tandem reactions which include dehydrogenative oxidation; the offered process of the catalysis includes the metal-ligand collaboration and both H_2 and O_2 generation at the single metal center (Gnanaprakasam & Milstein, 2011; Li & Hall, 2013).

2.4. OXIDATIVE DESYMMETRIZATION

Oxidation of the alcohols is widely used to familiarize asymmetry to preliminary compounds. For this, region-, chemo- or stereo-specific oxidations are engaged, e.g., in the Oppenauer reaction, where the substrate is generally oxidized by the transmission of hydrogen atoms to the sacrificial ketone, like acetone. Alcohols can be the source of H_2 and therefore the reducing agent in the asymmetric hydrogenations. For example, asymmetric transferal hydrogenation of the ketones is an effective and comparatively simple technique to familiarize asymmetry in the molecule (Bartoszewicz et al., 2013).

Other examples comprise of oxidative desymmetrization of meso-diols, OKR of the racemic secondary alcohols, etc. Generally, the kinetic resolution is defined as the procedure where two enantiomers of the racemic mixture are converted to products at dissimilar rates. Therefore, one enantiomer of the racemate is transformed selectively to the product, while the second is left behind. This approach permits to reach a maximum of 50% produce of the enantiopure residual sec-alcohol. To overcome this drawback, an adjustment of the approach, namely DKR (dynamic kinetic resolution), was introduced. In this situation, the kinetic resolution technique is joined with the racemization procedure, where the enantiomers are exchanged whereas one of them is used up. Thus, a 100% theoretical produce of one enantiomer out of two can be reached because of the continuous equilibrium shift. In the majority of the anticipated dynamic kinetic resolution processes, many catalytic systems, e.g., transition-metal catalysts and enzymes, work together. Both reactions can be encouraged by the catalyst. The process can contain the temporary oxidation of the substrate with hydrogen transferal to the transition-metal complex.

A typical example of the catalytic desymmetrization is the region-selective oxidation of the polyols, specifically 1,2-diols to give the conforming

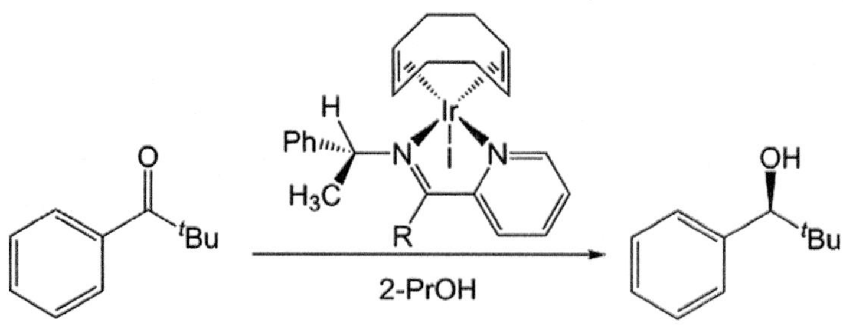

Figure 2.37: Example of the asymmetric transferal hydrogenation of ketones.

[Source: https://www.ncbi.nlm.nih.gov/pubmed/23606577]

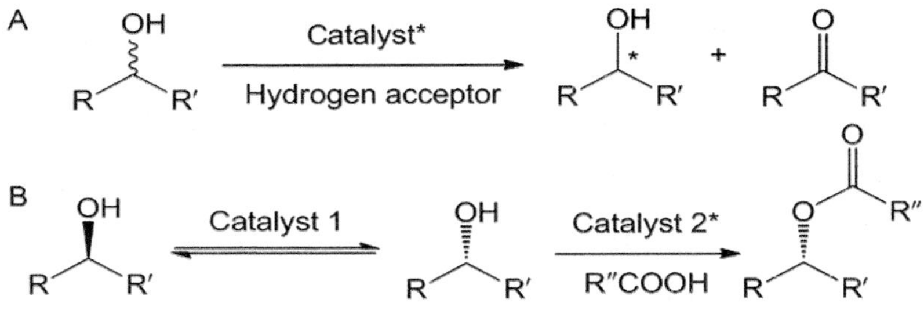

Figure 2.38. (A) Oxidative kinetic resolution (B) and dynamic kinetic resolution of the racemic secondary alcohols.

[Source: https://onlinelibrary.wiley.com/doi/abs/10.1002/ange.200705875]

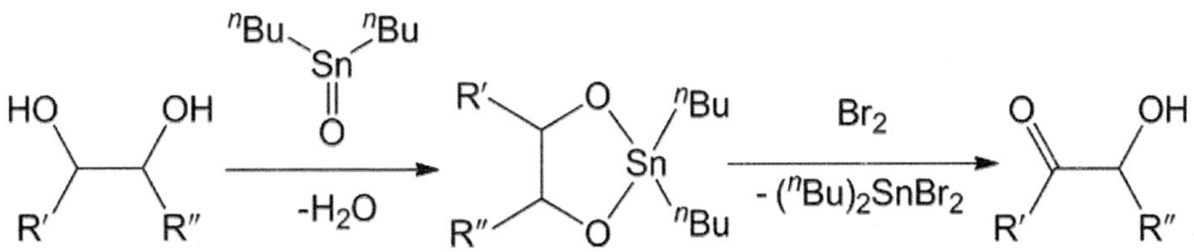

Figure 2.39: Regio-selective oxidation of the 1,2-diols utilizing organotin catalysts.

[Source: https://pubs.rsc.org/en/content/articlelanding/2013/ra/c3ra42754d]

α-hydroxy ketones, adaptable manmade intermediates, and the significant class of biologically vigorous products (Arita et al., 2008; William et al., 2013). Organotin compounds are well-known catalysts in this conversion, which continues through the creation of stannylene acetal as the intermediates (Figure. 2.40). Other metal catalysts, e.g., Pd complexes with C5H5N ligands are also utilized for the chemo-selective aerobic transformation of exposed diols into the conforming hydroxy ketones. Palladium–(NHC) complexes are active catalytically as well; propene development was noticed in this situation, demonstrating that the reaction likely continues through the reductive eradication from the Pd-allyl-hydride intermediate (Bettucci et al., 2010; 2011).

The wide variety of vicinal polyols and diols, including triols, tetraols, and 1,2-diols, were oxidized at the secondary alcohols selectively to give α-hydroxy ketones, utilizing chiral Pd complexes with the pyridinyl oxazoline consequential ligands as the catalysts and the benzoquinone as terminal oxidant (De Crisci et al., 2013). The ligand environment and geometry have an important influence on the chemo-selectivity and activity of the catalytic systems. Neocuproine can be used as the ligand to form efficient Palladium catalysts for the chemoselective oxidation of exposed vicinal polyols under trivial reaction conditions.

It is significant that normally the careful oxidation of exposed polyols and especially adjacent diols at the secondary alcohol is the difficult task because of the fierce CdC bond cleavage, poor chemoselectivity, overoxidation, etc. Concerning the procedure, the facile creation of the chelating protonated diolates from adjacent diols leads to higher rates and chemo-selectivity for the oxidation of secondary alcohols moiety because of the easy β-H eradication from the secondary Palladium alkoxide. On the other hand, for mixtures of secondary and primary alkoxides, the creation of the principal Palladium alkoxide is desirable and consequently, the choosiness for the oxidation of the primary alcohols to produce aldehydes is witnessed in this case. The available diamine -sparteine was applied as the chiral ligand for Palladium to form an efficient system for the asymmetric catalysis, especially kinetic resolution of cyclopropyl, benzylic, and allylic secondary alcohols and desymmetrizations of meso-diols. The associated oxidative desymmetrization of the meso-diols with the chiral Ir(Cp*) catalyst can be attained with enantiocontrol and good efficiency (Suzuki et al., 2003; Ebner et al., 2009). In some situations, an oxidative lactonization takes place if the primary meso-diols are utilized as starting materials (Figure. 2.42).

Chiral N-sulfonyl diamine ligands are utilized to create efficient chiral bi-functional amido iridium catalysts for asymmetric aerobic oxidations of prochiral and meso- diols to give around >99% of the hydroxyl ketones and 50% of the lactones (Moritani et al., 2013).

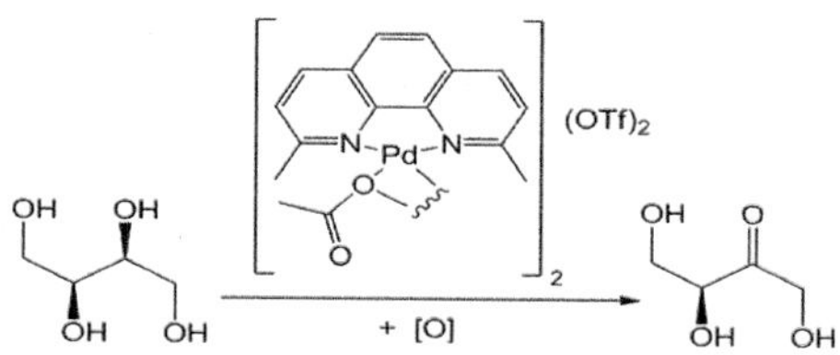

Figure 2.40: Stereospecific synthesis of the (S)-erythrulose.

[Source: https://www.sciencedirect.com/science/article/pii/S0065305515000052]

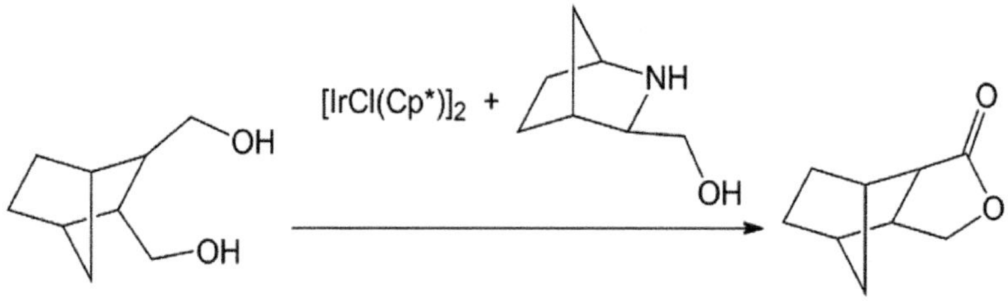

Figure 2.41: Enantiospecific making of lactones by the oxidative lactonization of the primary meso-diols.

[Source: https://www.sciencedirect.com/science/article/pii/S0065305515000052]

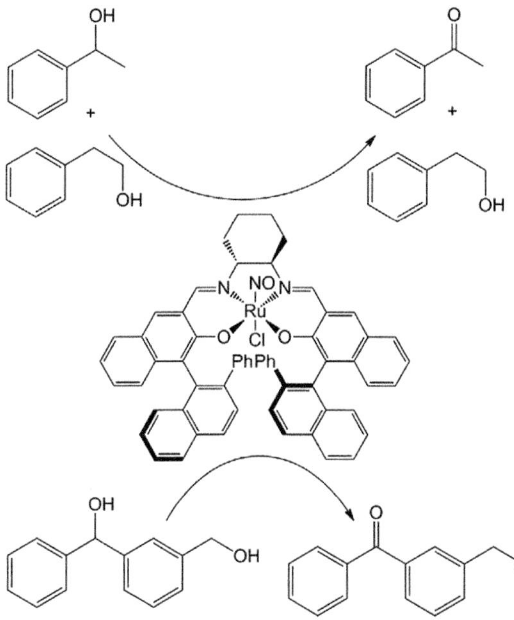

Figure 2.42: Chemo-selective oxidation of the secondary alcohol with Ru catalysts.

[Source: https://www.sciencedirect.com/science/article/pii/S0065305515000052]

Chemo-selective oxidation of the secondary alcohols moiety can be executed with ruthenium (Ru) catalysts having phenylindenyl ligand

(Manzini et al., 2013). The careful oxidation of 1-phenyl ethanol to the acetophenone from the blend of phenyl ethanol isomers, lacking the oxidation of other isomers, can then be accomplished (Figure. 2.42). Generally, only the secondary alcohols portions are oxidized and the catalysts can be utilized for the chemical parting of isomers or particular oxidation of functionalized molecules.

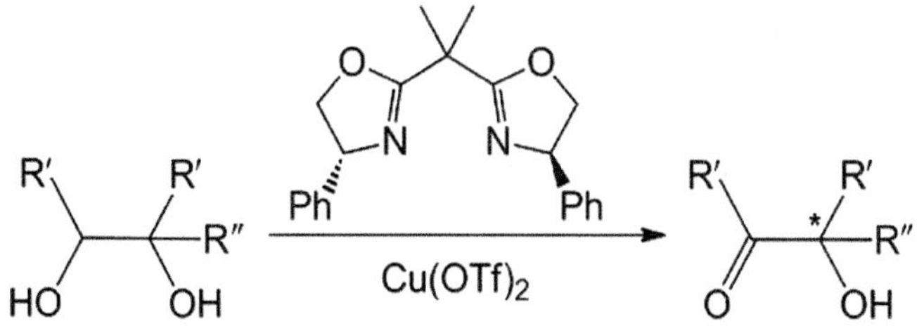

Figure 2.43: Conversion of the 2-diols to equivalent chiral α-keto alcohols.

[Source: https://www.sciencedirect.com/science/article/pii/S0065305515000052]

The use of inexpensive, more accessible and reusable catalysts is the common goal in the catalytic investigation. Under this viewpoint, several recent journals on the oxidative asymmetric resolution of the alcohols include 3d and other profuse metals. Therefore, in situ amalgamations of Cu (II) triflate and (R, R)-Ph-BOX [BOX¼bis(oxazoline)] with N-bromosuccinimide as an oxidant permitted to attain the desymmetrizations of 2-diols to manage the conforming chiral α-keto alcohols (Figure. 2.43).

An air-stable, well-known (NHC)–Nickel^0 complex is one more efficient 3d metal catalyst for trivial anaerobic catalytic oxidation of the secondary alcohols with functionalities like alkene, ether, and tertiary amine, utilizing non-anhydrous, degassed 2,4-dichlorotoluene as solvent and oxidant (Onomura et al., 2007; Berini et al., 2010). The use of stable, cheap, and easy to deal with chlorinated solvent as the oxidant makes the catalytic system striking for scaling-up and multistep synthesis; furthermore, potentially dangerous and difficult to eliminate species are less probable to be formed. Additionally, primary alkylic and benzylic alcohols are unreactive with the catalytic system and therefore can be recovered (Willemsen et al., 2013).

Jacobsen's chiral Manganese (III)-Salen complexes establish another example of inexpensive and existing catalysts for enantio-selective oxidation of the racemic secondary alcohols. In this situation, sodium hypochlorite was utilized as oxidant. Linked macrocyclic chiral Manganese (III) Salen complexes were made practical for OKR of the secondary alcohols with NBS co-oxidants and di acetoxy iodo benzene [PhI $(OAc)_2$] and, in the biphasic dichloromethane water solvent blend; catalyst can be recycled easily up to 7 times without reducing its performance (Bera et al., 2013; Zhang et al., 2014).

Transition metals can usually be eradicated from the catalytic system. Therefore, the quinine-derived urea (CH_4N_2O) organocatalyst is efficient in enantioselective oxidation of the wide variety of diaryl-replaced meso-1,2-diols utilizing bromination reagents as the oxidants (Figure. 2.44). The technique is simple, works at ambient temperature and uses accessible reagents to produce α-hydroxy ketones in decent produces up to 94% and enantioselectivities up to 95%.

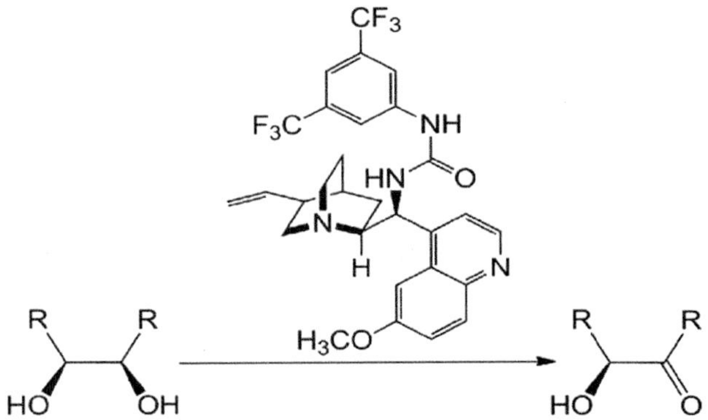

Figure 2.44: Enantioselective oxidation of the diaryl-replaced meso-1,2-diols.

[Source: https://www.sciencedirect.com/science/article/pii/S0065305515000052]

Furthermore, both enantio- and chemo-selectivities can synergistically add to the particular oxidation of the racemic diols having diverse substituents, managing hydroxy ketones as the only product.

REFERENCES

1. Al-Hunaiti, A., Niemi, T., Sibaouih, A., Pihko, P., Leskelä, M., & Repo, T. (2010). Solvent free oxidation of primary alcohols and diols using thymine iron (III) catalyst. *Chemical Communications, 46*(48), 9250–9252.
2. Al-Hunaiti, A., Räisänen, M., Pihko, P., Leskelä, M., & Repo, T. (2014). Organocatalytic Oxidation of Secondary Alcohols Using 1, 2-Di (1-naphthyl)-1, 2-ethanediamine (NEDA). *European Journal of Organic Chemistry,* 28, 6141–6144.
3. Aliende, C., Pérez-Manrique, M., Jalón, F. A., Manzano, B. R., Rodrıguez, A. M., & Espino, G. (2012). Arene ruthenium complexes as versatile catalysts in water in both transfer hydrogenation of ketones and oxidation of alcohols. Selective deuterium labeling of rac-1-phenylethanol. *Organometallics*, *31*(17), 6106–6123.
4. Allen, S. E., Walvoord, R. R., Padilla-Salinas, R., & Kozlowski, M. C. (2013). Aerobic copper-catalyzed organic reactions. *Chemical Reviews, 113*(8), 6234–6458.
5. Amberg, M., Doenges, M., Stapf, G., & Hartung, J. (2014). Formation of 3-acyloxy-γ-butyrolactones from 4-pentenols in vanadium-catalyzed oxidations. Tetrahedron, *70*(34), 5321–5331.
6. Arends, I. W., & Sheldon, R. A. (2010). Modern oxidation of alcohols using environmentally benign oxidants. *Modern Oxidation Methods, 1*(1), 147–185.
7. Arita, S., Koike, T., Kayaki, Y., & Ikariya, T. (2008). Aerobic oxidative kinetic resolution of racemic secondary alcohols with chiral bifunctional amido complexes. *Angewandte Chemie International Edition, 47*(13), 2447–2449.
8. Bäckvall, J. E. (Ed.). (2011). *Modern Oxidation Methods* (Vol. 1, pp. 1–25). John Wiley & Sons.
9. Balaraman, E., Khaskin, E., Leitus, G., & Milstein, D. (2013). Catalytic transformation of alcohols to carboxylic acid salts and H 2 using water as the oxygen atom source. *Nature Chemistry, 5*(2), 122.
10. Baratta, W., Bossi, G., Putignano, E., & Rigo, P. (2011). Pincer and diamine Ru and Os diphosphane complexes as efficient catalysts for the dehydrogenation of alcohols to ketones. *Chemistry–A European Journal, 17*(12), 3474–3481.

11. Bartlett, S. L., & Beaudry, C. M. (2011). High-yielding oxidation of β-hydroxyketones to β-diketones using o-iodoxybenzoic acid. *The Journal of Organic Chemistry, 76*(23), 9852–9855.
12. Bartoszewicz, A., Ahlsten, N., & Martín-Matute, B. (2013). Enantioselective Synthesis of Alcohols and Amines by Iridium-Catalyzed Hydrogenation, Transfer Hydrogenation, and Related Processes. *Chemistry–A European Journal, 19*(23), 7274–7302.
13. Beejapur, H. A., Campisciano, V., Franchi, P., Lucarini, M., Giacalone, F., & Gruttadauria, M. (2014). Fullerene as a platform for recyclable TEMPO organocatalysts for the oxidation of alcohols. *Chem Cat Chem, 6*(8), 2419–2424.
14. Ben-Daniel, R., Alsters, P., & Neumann, R. (2001). Selective aerobic oxidation of alcohols with a combination of a polyoxometalate and nitroxyl radical as catalysts. *The Journal of Organic Chemistry, 66*(25), 8650–8653.
15. Ben-David, H., Iron, M. A., & Neumann, R. (2013). Platinum complexes of cationic ligands for the aerobic oxidation of "inert" perfluoro-substituted alcohols. *Chemical Communications, 49*(17), 1720–1722.
16. Bera, P. K., Maity, N. C., Abdi, S. H., Noor-ul, H. K., Kureshy, R. I., & Bajaj, H. C. (2013). Macrocyclic Mn (III) salen complexes as recyclable catalyst for oxidative kinetic resolution of secondary alcohols. *Applied Catalysis A: General, 467*, 542–551.
17. Berini, C., Brayton, D. F., Mocka, C., & Navarro, O. (2009). (NHC)-Pd or-Ni-Catalyzed Oxidation of Secondary Alcohols by Chlorobenzene. *Synfacts, 12*, 1393–1393.
18. Berini, C., Winkelmann, O. H., Otten, J., Vicic, D. A., & Navarro, O. (2010). Rapid and Selective Catalytic Oxidation of Secondary Alcohols at Room Temperature by Using (N-Heterocyclic Carbene)–Ni0 Systems. *Chemistry – A European Journal, 16*(23), 6857–6860.
19. Besson, M., & Gallezot, P. (2000). Selective oxidation of alcohols and aldehydes on metal catalysts. *Catalysis Today, 57*(1–2), 127–141.
20. Bettucci, L., Bianchini, C., Filippi, J., Lavacchi, A., & Oberhauser, W. (2011). Chemoselective Aerobic Diol Oxidation by Palladium (II)–Pyridine Catalysis. *European Journal of Inorganic Chemistry, 11*, 1797–1805.
21. Bettucci, L., Bianchini, C., Oberhauser, W., Hsiao, T. H., & Lee, H.

M. (2010). Chemoselective aerobic oxidation of unprotected diols catalyzed by Pd–(NHC)(NHC= N-heterocyclic carbene) complexes. *Journal of Molecular Catalysis A: Chemical, 322*(1–2), 63–72.

22. Bianchini, C., & Shen, P. K. (2009). Palladium-based electrocatalysts for alcohol oxidation in half cells and in direct alcohol fuel cells. *Chemical Reviews, 109*(9), 4183–4206.
23. Borthakur, R., Asthana, M., Kumar, A., Koch, A., & Lal, R. A. (2013). Solvent free selective oxidation of alcohols catalyzed by a trinuclear complex with a dicopper (II)–monozinc (II) centre using hydrogen peroxide as an oxidant. *RSC Advances, 3*(45), 22957–22962.
24. Borthakur, R., Asthana, M., Saha, M., Kumar, A., & Pal, A. K. (2014). An efficient oxidation of alcohols using a new trinuclear copper complex as a reusable catalyst under solvent free conditions. *RSC Advances, 4*(41), 21638–21643.
25. Bowman, R. K., Brown, A. D., Cobb, J. H., Eaddy, J. F., Hatcher, M. A., Leivers, M. R., ... & Xie, S. (2013). Synthesis of HCV Replicase Inhibitors: Base-Catalyzed Synthesis of Protected α-Hydrazino Esters and Selective Aerobic Oxidation with Catalytic Pt/Bi/C for Synthesis of Imidazole-4, 5-dicarbaldehyde. *The Journal of Organic Chemistry, 78*(23), 11680–11690.
26. Cao, Q., Dornan, L. M., Rogan, L., Hughes, N. L., & Muldoon, M. J. (2014). Aerobic oxidation catalysis with stable radicals. *Chemical Communications, 50*(35), 4524–4543.
27. Cecchetto, A., Fontana, F., Minisci, F., & Recupero, F. (2001). Efficient Mn–Cu and Mn–Co–TEMPO-catalysed oxidation of alcohols into aldehydes and ketones by oxygen under mild conditions. *Tetrahedron Letters, 42*(38), 6651–6653.
28. Chen, J., Fang, W., Zhang, Q., Deng, W., & Wang, Y. (2014). A Comparative Study of Size Effects in the Au-Catalyzed Oxidative and Non-Oxidative Dehydrogenation of Benzyl Alcohol. *Chemistry–An Asian Journal, 9*(8), 2187–2196.
29. Chung, K., Banik, S. M., De Crisci, A. G., Pearson, D. M., Blake, T. R., Olsson, J. V., ... & Waymouth, R. M. (2013). Chemoselective Pd-catalyzed oxidation of polyols: synthetic scope and mechanistic studies. *Journal of the American Chemical Society, 135*(20), 7593–7602.
30. Ciriminna, R., & Pagliaro, M. (2009). Industrial oxidations with organocatalyst TEMPO and its derivatives. *Organic Process Research*

& Development, 14(1), 245–251.

31. Da Silva, J. A., da Silva, J. J. F., & Pombeiro, A. J. (2011). Oxovanadium complexes in catalytic oxidations. *Coordination Chemistry Reviews, 255*(19–20), 2232–2248.
32. Damodara, D., Arundhathi, R., & Likhar, P. R. (2014). Copper Nanoparticles from Copper Aluminum Hydrotalcite: An Efficient Catalyst for Acceptor-and Oxidant-Free Dehydrogenation of Amines and Alcohols. *Advanced Synthesis & Catalysis, 356*(1), 189–198.
33. De Crisci, A. G., Chung, K., Oliver, A. G., Solis-Ibarra, D., & Waymouth, R. M. (2013). Chemoselective oxidation of polyols with chiral palladium catalysts. *Organometallics, 32*(7), 2257–2266.
34. Dobereiner, G. E., & Crabtree, R. H. (2009). Dehydrogenation as a substrate-activating strategy in homogeneous transition-metal catalysis. *Chemical Reviews, 110*(2), 681–703.
35. Dönges, M., Amberg, M., Stapf, G., Kelm, H., Bergsträßer, U., & Hartung, J. (2014). cis-2, 6-Bis-(methanolate)-piperidine oxovanadium (V) complexes as catalysts for oxidative alkenol cyclization by tert-butyl hydroperoxide. *Inorganica Chimica Acta, 420*, 120–134.
36. Du, Z., Ma, J., Ma, H., Wang, M., Huang, Y., & Xu, J. (2010). Vanadyl sulfate: a simple catalyst for oxidation of alcohols with molecular oxygen in combination with 2, 2, 6, 6-tetramethyl-piperidyl-1-oxyl. *Catalysis Communications, 11*(8), 732–735.
37. Duschek, A., & Kirsch, S. F. (2011). 2-Iodoxybenzoic acid—a simple oxidant with a dazzling array of potential applications. *Angewandte Chemie International Edition, 50*(7), 1524–1552.
38. Ebner, D. C., Bagdanoff, J. T., Ferreira, E. M., McFadden, R. M., Caspi, D. D., Trend, R. M., & Stoltz, B. M. (2009). The palladium-catalyzed aerobic kinetic resolution of secondary alcohols: reaction development, scope, and applications. *Chemistry–A European Journal, 15*(47), 12978–12992.
39. Fei, H., Shin, J., Meng, Y. S., Adelhardt, M., Sutter, J., Meyer, K., & Cohen, S. M. (2014). Reusable oxidation catalysis using metal-monocatecholato species in a robust metal–organic framework. *Journal of the American Chemical Society, 136*(13), 4965–4973.
40. Fernandes, R. A., & Bethi, V. (2014). An expedient osmium (VI)/K 3 Fe (CN) 6-mediated selective oxidation of benzylic, allylic and propargylic alcohols. *RSC Advances, 4*(76), 40561–40568.

41. Fernandes, R. R., Lasri, J., da Silva, M. F. C. G., da Silva, J. A., da Silva, J. J. F., & Pombeiro, A. J. (2011a). Bis-and tris-pyridyl amino and imino thioether Cu and Fe complexes. Thermal and microwave-assisted peroxidative oxidations of 1-phenylethanol and cyclohexane in the presence of various N-based additives. *Journal of Molecular Catalysis A: Chemical, 351*, 100–111.
42. Fernandes, R. R., Lasri, J., da Silva, M. F. C. G., Da Silva, J. A., da Silva, J. J. F., & Pombeiro, A. J. (2011b). Mild alkane C–H and O–H oxidations catalysed by mixed-N, S copper, iron and vanadium systems. *Applied Catalysis A: General, 402*(1–2), 110–120.
43. Figiel, P. J., Kirillov, A. M., da Silva, M. F. C. G., Lasri, J., & Pombeiro, A. J. (2010). Self-assembled dicopper (II) diethanolaminate cores for mild aerobic and peroxidative oxidation of alcohols. *Dalton Transactions, 39*(41), 9879–9888.
44. Figiel, P. J., Kirillov, A. M., Karabach, Y. Y., Kopylovich, M. N., & Pombeiro, A. J. (2009). Mild aerobic oxidation of benzyl alcohols to benzaldehydes in water catalyzed by aqua-soluble multicopper (II) triethanolaminate compounds. *Journal of Molecular Catalysis A: Chemical, 305*(1–2), 178–182.
45. Fujita, K. I., Ito, W., & Yamaguchi, R. (2014). Dehydrogenative Lactonization of Diols in Aqueous Media Catalyzed by a Water-Soluble Iridium Complex Bearing a Functional Bipyridine Ligand. *Chem Cat Chem, 6*(1), 109–112.
46. Gartshore, C. J., & Lupton, D. W. (2010). Readily Accessible oxazolidine nitroxyl radicals: bifunctional cocatalysts for simplified copper based aerobic oxidation. *Advanced Synthesis & Catalysis, 352*(18), 3321–3328.
47. Giachi, G., Frediani, M., Oberhauser, W., & Passaglia, E. (2012). Aerobic alcohol oxidation catalyzed by polyester-based Pd (II) macrocomplexes. *Journal of Polymer Science Part A: Polymer Chemistry, 50*(13), 2725–2731.
48. Gnanaprakasam, B., & Milstein, D. (2011). Synthesis of amides from esters and amines with liberation of H2 under neutral conditions. *Journal of the American Chemical Society, 133*(6), 1682–1685.
49. Gowrisankar, S., Neumann, H., Goerdes, D., Thurow, K., Jiao, H., & Beller, M. (2013). A Convenient and Selective Palladium-Catalyzed Aerobic Oxidation of Alcohols. *Chemistry–A European Journal, 19*(47), 15979–15984.

50. Griffith, W. P. (2011). Catalysis by Metal Complexes. *Ruthenium Oxidation Complexes, 1*, 1–15.

51. Grßinne, M. A. (2010). Anionic N, O-ligated Pd (II) complexes: highly active catalysts for alcohol oxidation. *Chemical Communications, 46*(38), 7238–7240.

52. Guo, S., Yu, J. T., Dai, Q., Yang, H., & Cheng, J. (2014). The Bu 4 NI-catalyzed alfa-acyloxylation of ketones with benzylic alcohols. *Chemical Communications, 50*(47), 6240–6242.

53. Hage, R., & Feringa, B. L. (2010). Wesley R. Browne, Johannes W. de Boer, Dirk Pijper, Jelle Brinksma. *Modern Oxidation Methods, 1*(1), 371.

54. Hamasaki, A., Kuwada, H., & Tokunaga, M. (2012). tert-Butylnitrite as a convenient and easy-removable oxidant for the conversion of benzylic alcohols to ketones and aldehydes. *Tetrahedron Letters, 53*(7), 811–814.

55. Han, L., Xing, P., & Jiang, B. (2014). Selective aerobic oxidation of alcohols to aldehydes, carboxylic acids, and imines catalyzed by a Ag-NHC complex. *Organic Letters, 16*(13), 3428–3431.

56. Hanson, S. K., Wu, R., & Silks, L. P. (2011). Mild and selective vanadium-catalyzed oxidation of benzylic, allylic, and propargylic alcohols using air. *Organic Letters, 13*(8), 1908–1911. Ohde, C., & Limberg, C. (2010). From Surface-Inspired Oxovanadium Silsesquioxane Models to Active Catalysts for the Oxidation of Alcohols with O_2—The Cinnamic Acid/Metavanadate System. *Chemistry – A European Journal, 16*(23), 6892–6899.

57. Hayashi, M., Sasano, Y., Nagasawa, S., Shibuya, M., & Iwabuchi, Y. (2011). 9-Azanoradamantane N-oxyl (Nor-AZADO): a highly active organocatalyst for alcohol oxidation. *Chemical and Pharmaceutical Bulletin, 59*(12), 1570–1573.

58. Hayashi, M., Shibuya, M., & Iwabuchi, Y. (2012). Oxidation of alcohols to carbonyl compounds with diisopropyl azodicarboxylate catalyzed by nitroxyl radicals. *The Journal of Organic Chemistry, 77*(6), 3005–3009.

59. Holan, M., & Jahn, U. (2013). Anaerobic nitroxide-catalyzed oxidation of alcohols using the NO+/NO· redox pair. *Organic letters, 16*(1), 58–61.

60. Hoover, J. M., Ryland, B. L., & Stahl, S. S. (2013). Copper/TEMPO-

catalyzed aerobic alcohol oxidation: mechanistic assessment of different catalyst systems. *ACS Catalysis, 3*(11), 2599–2605.

61. Hu, Z., & Kerton, F. M. (2012). Room temperature aerobic oxidation of alcohols using $CuBr_2$ with TEMPO and a tetradentate polymer based pyridyl-imine ligand. *Applied Catalysis A: General, 413*, 332–339.
62. Ingram, A. J., Solis-Ibarra, D., Zare, R. N., & Waymouth, R. M. (2014). Trinuclear Pd_3O_2 intermediate in aerobic oxidation catalysis. *Angewandte Chemie International Edition, 53*(22), 5648–5652.
63. Ishida, T., Watanabe, H., Takei, T., Hamasaki, A., Tokunaga, M., & Haruta, M. (2012). Metal oxide-catalyzed ammoxidation of alcohols to nitriles and promotion effect of gold nanoparticles for one-pot amide synthesis. *Applied Catalysis A: General, 425*, 85–90.
64. Ishii, Y., Sakaguchi, S., & Obora, Y. (2004). Aerobic oxidations and related reactions catalyzed by n-hydroxyphthalimide. *Modern Oxidation Methods, 1*(1), 187–240.
65. John, L. C., Gunay, A., Wood, A. J., & Emmert, M. H. (2013). Catalysts for convenient aerobic alcohol oxidations in air: systematic ligand studies in Pd/pyridine systems. *Tetrahedron, 69*(27–28), 5758–5764.
66. Join, B., Möller, K., Ziebart, C., Schröder, K., Gördes, D., Thurow, K., ... & Beller, M. (2011). Selective Iron-Catalyzed Oxidation of Benzylic and Allylic Alcohols. *Advanced Synthesis & Catalysis, 353*(16), 3023–3030.
67. Kamitani, M., Ito, M., Itazaki, M., & Nakazawa, H. (2014). Effective dehydrogenation of 2-pyridylmethanol derivatives catalyzed by an iron complex. *Chemical Communications, 50*(59), 7941–7944.
68. Karimi, B., & Badreh, E. (2011). SBA-15-functionalized TEMPO confined ionic liquid: an efficient catalyst system for transition-metal-free aerobic oxidation of alcohols with improved selectivity. *Organic & Biomolecular Chemistry, 9*(11), 4194–4198.
69. Kirillov, A. M., & Shul'pin, G. B. (2013). Pyrazinecarboxylic acid and analogs: Highly efficient co-catalysts in the metal-complex-catalyzed oxidation of organic compounds. *Coordination Chemistry Reviews, 257*(3–4), 732–754.
70. Kirillov, A. M., Kirillova, M. V., Shul'pina, L. S., Figiel, P. J., Gruenwald, K. R., da Silva, M. F. C. G., ... & Shul'pin, G. B. (2011). Mild oxidative functionalization of alkanes and alcohols catalyzed by new mono-and dicopper (II) aminopolyalcoholates. *Journal of*

Molecular Catalysis A: Chemical, 350(1–2), 26–34.

71. Kon, K., Siddiki, S. H., & Shimizu, K. I. (2013). Size-and support-dependent Pt nanocluster catalysis for oxidant-free dehydrogenation of alcohols. *Journal of Catalysis, 304*, 63–71.
72. Kopylovich, M. N., Gajewska, M. J., Mahmudov, K. T., Kirillova, M. V., Figiel, P. J., da Silva, M. F. C. G., ... & Pombeiro, A. J. (2012). Copper (II) complexes with a new carboxylic-functionalized arylhydrazone of β-diketone as effective catalysts for acid-free oxidations. *New Journal of Chemistry, 36(*8), 1646–1654.
73. Kopylovich, M. N., Karabach, Y. Y., Mahmudov, K. T., Haukka, M., Kirillov, A. M., Figiel, P. J., & Pombeiro, A. J. (2011). Heterometallic copper (II)–potassium 3D coordination polymers driven by multifunctionalized azo derivatives of β-diketones. *Crystal Growth & Design, 11*(10), 4247–4252.
74. Kopylovich, M. N., Mahmudov, K. T., Guedes da Silva, M. F. C., Figiel, P. J., Karabach, Y. Y., Kuznetsov, M. L., ... & Pombeiro, A. J. (2011). Ortho-hydroxyphenylhydrazo-β-diketones: tautomery, coordination ability, and catalytic activity of their copper (II) complexes toward oxidation of cyclohexane and benzylic alcohols. *Inorganic Chemistry, 50*(3), 918–931.
75. Kopylovich, M. N., Mizar, A., Guedes da Silva, M. F. C., Mac Leod, T. C., Mahmudov, K. T., & Pombeiro, A. J. (2013). Template Syntheses of Copper (II) Complexes from Arylhydrazones of Malononitrile and their Catalytic Activity towards Alcohol Oxidations and the Nitroaldol Reaction: Hydrogen Bond-Assisted Ligand Liberation and E/Z Isomerization. *Chemistry–A European Journal, 19*(2), 588–600.
76. Kumpulainen, E. T., & Koskinen, A. M. (2009). Catalytic activity dependency on catalyst components in aerobic copper–TEMPO oxidation. *Chemistry–A European Journal, 15*(41), 10901–10911.
77. Lenze, M., & Bauer, E. B. (2013). Chemoselective, iron (ii)-catalyzed oxidation of a variety of secondary alcohols over primary alcohols utilizing H_2O_2 as the oxidant. *Chemical Communications, 49*(52), 5889–5891.
78. Li, F., Sun, C., & Wang, N. (2014). Catalytic acceptorless dehydrogenative coupling of arylhydrazines and alcohols for the synthesis of arylhydrazones. *The Journal of Organic Chemistry, 79*(17), 8031–8039.

79. Li, H., & Hall, M. B. (2013). Mechanism of the formation of carboxylate from alcohols and water catalyzed by a bipyridine-based ruthenium complex: a computational study. *Journal of the American Chemical Society, 136*(1), 383–395.

80. Li, X. H., Meng, X. G., Liu, Y., & Peng, X. (2013). Direct oxidation of secondary alcohol to ester by performic acid. *Green Chemistry, 15*(12), 3332–3336.

81. Liang, Z. X., & Zhao, T. S. (Eds.). (2012). *Catalysts for Alcohol-Fuelled Direct Oxidation Fuel Cells* (Vol. 6, pp. 1–20). Royal Society of Chemistry.

82. Liu, J., & Ma, S. (2013). Aerobic oxidation of indole carbinols using Fe (NO 3) 3· 9H 2 O/TEMPO/NaCl as catalysts. *Organic & Biomolecular Chemistry, 11*(25), 4186–4193.

83. Liu, J., & Ma, S. (2013a). Iron-Catalyzed Aerobic Oxidation of Allylic Alcohols: The Issue of C C Bond Isomerization. *Organic Letters, 15*(20), 5150–5153.

84. Liu, J., & Ma, S. (2013b). Room temperature Fe (NO_3) 3Ě9H_2O/ TEMPO/NaCl-catalyzed aerobic oxidation of homopropargylic alcohols. *Tetrahedron, 69*(47), 10161–10167.

85. Liu, L., Yu, M., Wayland, B. B., & Fu, X. (2010). Aerobic oxidation of alcohols catalyzed by rhodium (III) porphyrin complexes in water: reactivity and mechanistic studies. *Chemical Communications, 46*(34), 6353–6355.

86. Lucio Anelli, P., Biffi, C., Montanari, F., & Quici, S. (1987). Fast and selective oxidation of primary alcohols to aldehydes or to carboxylic acids and of secondary alcohols to ketones mediated by oxoammonium salts under two-phase conditions. *The Journal of Organic Chemistry, 52*(12), 2559–2562.

87. Ma, X., Li, Z., Liu, F., Cao, S., & Rao, H. (2014). Tetra-n-butylammonium Bromide: A Simple but Efficient Organocatalyst for Alcohol Oxidation under Mild Conditions. *Advanced Synthesis & Catalysis, 356*(8), 1741–1746.

88. Mahmudov, K. T., Kopylovich, M. N., da Silva, M. F. C. G., Figiel, P. J., Karabach, Y. Y., & Pombeiro, A. J. (2010). New copper (II) dimer with 3-(2-hydroxy-4-nitrophenylhydrazo) pentane-2, 4-dione and its catalytic activity in cyclohexane and benzyl alcohol oxidations. *Journal of Molecular Catalysis A: Chemical, 318*(1–2), 44–50.

89. Mahmudov, K. T., Kopylovich, M. N., Sabbatini, A., Drew, M. G., Martins, L. M., Pettinari, C., & Pombeiro, A. J. (2014). Cooperative metal–ligand assisted E/Z isomerization and cyano activation at CuII and CoII complexes of arylhydrazones of active methylene nitriles. *Inorganic Chemistry, 53*(18), 9946–9958.

90. Majee, A., Kundu, S. K., Santra, S., & Hajra, A. (2012). Combination of NH2OH· HCl and NaIO4: a new and mild oxidizing agent for selective oxidation of alcohols to carbonyl compounds. *Tetrahedron Letters, 53*(33), 4433–4435.

91. Mallat, T., & Baiker, A. (2004). Oxidation of alcohols with molecular oxygen on solid catalysts. *Chemical Reviews, 104*(6), 3037–3058.

92. Manzini, S., Urbina-Blanco, C. A., & Nolan, S. P. (2013). Chemoselective oxidation of secondary alcohols using a ruthenium phenylindenyl complex. *Organometallics, 32*(2), 660–664.

93. Mitsudome, T., Mikami, Y., Funai, H., Mizugaki, T., Jitsukawa, K., & Kaneda, K. (2008). Oxidant-Free Alcohol Dehydrogenation Using a Reusable Hydrotalcite-Supported Silver Nanoparticle Catalyst. *Angewandte Chemie International Edition, 47*(1), 138–141.

94. Miura, T., Nakashima, K., Tada, N., & Itoh, A. (2011). An effective and catalytic oxidation using recyclable fluorous IBX. *Chemical Communications, 47*(6), 1875–1877.

95. Mizoguchi, H., Uchida, T., & Katsuki, T. (2014). Ruthenium-Catalyzed Oxidative Kinetic Resolution of Unactivated and Activated Secondary Alcohols with Air as the Hydrogen Acceptor at Room Temperature. *Angewandte Chemie International Edition, 53*(12), 3178–3182.

96. Mizuno, N. (Ed.). (2009). *Modern Heterogeneous Oxidation Catalysis: Design, Reactions and Characterization* (vol. 1, pp. 1–10). John Wiley & Sons.

97. Moritani, J., Hasegawa, Y., Kayaki, Y., & Ikariya, T. (2014). Aerobic oxidative desymmetrization of meso-diols with bifunctional amidoiridium catalysts bearing chiral N-sulfonyldiamine ligands. *Tetrahedron Letters, 55*(6), 1188–1191.

98. Mugemana, C., Chen, B. T., Bukhryakov, K. V., & Rodionov, V. (2014). Star block-copolymers: enzyme-inspired catalysts for oxidation of alcohols in water. *Chemical Communications, 50*(58), 7862–7865.

99. Murahashi, S. I., & Komiya, N. (2010). Ruthenium-Catalyzed Oxidation for Organic Synthesis. *Modern Oxidation Methods, 1*(1),

241–275.

100. Musa, S., Ackermann, L., & Gelman, D. (2013). Dehydrogenative Cross-Coupling of Primary and Secondary Alcohols. *Advanced Synthesis & Catalysis, 355*(14–15), 3077–3080.

101. Nasani, R., Saha, M., Mobin, S. M., Martins, L. M., Pombeiro, A. J., Kirillov, A. M., & Mukhopadhyay, S. (2014). Copper–organic frameworks assembled from in situ generated 5-(4-pyridyl) tetrazole building blocks: Synthesis, structural features, topological analysis and catalytic oxidation of alcohols. *Dalton Transactions, 43*(26), 9944–9954.

102. Nielsen, M., Kammer, A., Cozzula, D., Junge, H., Gladiali, S., & Beller, M. (2011). Efficient hydrogen production from alcohols under mild reaction conditions. *Angewandte Chemie International Edition, 50*(41), 9593–9597.

103. Obermayer, D., Balu, A. M., Romero, A. A., Goessler, W., Luque, R., & Kappe, C. O. (2013). Nanocatalysis in continuous flow: supported iron oxide nanoparticles for the heterogeneous aerobic oxidation of benzyl alcohol. *Green Chemistry, 15*(6), 1530–1537.

104. Onomura, O., Arimoto, H., Matsumura, Y., & Demizu, Y. (2007). Asymmetric oxidation of 1, 2-diols using N-bromosuccinimide in the presence of chiral copper catalyst. *Tetrahedron Letters, 48*(49), 8668–8672.

105. Ouyang, X. H., Song, R. J., & Li, J. H. (2014). Iron-Catalyzed Oxidative 1, 2-Carboacylation of Activated Alkenes with Alcohols: A Tandem Route to 3-(2-Oxoethyl) indolin-2-ones. European *Journal of Organic Chemistry, 16*, 3395–3401.

106. Painter, R. M., Pearson, D. M., & Waymouth, R. M. (2010). Selective catalytic oxidation of glycerol to dihydroxyacetone. *Angewandte Chemie International Edition, 49*(49), 9456–9459.

107. Parmeggiani, C., & Cardona, F. (2012). Transition metal based catalysts in the aerobic oxidation of alcohols. *Green Chemistry, 14*(3), 547–564.

108. Prebil, R., Stavber, G., & Stavber, S. (2014). Aerobic Oxidation of Alcohols by Using a Completely Metal-Free Catalytic System. *European Journal of Organic Chemistry, 2*, 395–402.

109. Räisänen, M. T., Al-Hunaiti, A., Atosuo, E., Kemell, M., Leskelä, M., & Repo, T. (2014). Mn (II) acetate: an efficient and versatile oxidation catalyst for alcohols. *Catalysis Science & Technology, 4*(8), 2564–

2573.

110. Roberts, S. M., & Whittall, J. (Eds.). (2007). Catalysts for Fine Chemical Synthesis, *Regio-and Stereo-Controlled Oxidations and Reductions* (Vol. 5, pp. 1–10). Wiley-Interscience.

111. Rong, Z. Q., Pan, H. J., Yan, H. L., & Zhao, Y. (2013). Enantioselective oxidation of 1, 2-diols with quinine-derived urea organocatalyst. *Organic Letters, 16*(1), 208–211.

112. Royer, A. M., Rauchfuss, T. B., & Gray, D. L. (2010). Organoiridium pyridonates and their role in the dehydrogenation of alcohols. *Organometallics, 29*(24), 6763–6768.

113. Ryland, B. L., & Stahl, S. S. (2014). Practical aerobic oxidations of alcohols and amines with homogeneous copper/TEMPO and related catalyst systems. *Angewandte Chemie International Edition, 53*(34), 8824–8838.

114. Sabbatini, A., Martins, L. M., Mahmudov, K. T., Kopylovich, M. N., Drew, M. G., Pettinari, C., & Pombeiro, A. J. (2014). Microwave-assisted and solvent-free peroxidative oxidation of 1-phenylethanol to acetophenone with a CuII–TEMPO catalytic system. *Catalysis Communications, 48*, 69–72.

115. Sanmartín-Matalobos, J., García-Deibe, A. M., Briones-Miguéns, L., González-Bello, C., Portela-García, C., & Fondo, M. (2014). Dual role of 2-tosylaminomethylaniline as a ligand and a nucleophile in the copper-mediated oxidation of methanol. *Dalton Transactions, 43*(2), 722–728.

116. Sarathy, S. M., Thomson, M. J., Togbé, C., Dagaut, P., Halter, F., & Mounaim-Rousselle, C. (2009). An experimental and kinetic modeling study of n-butanol combustion. *Combustion and Flame, 156*(4), 852–864.

117. Sasano, Y., Nagasawa, S., Yamazaki, M., Shibuya, M., Park, J., & Iwabuchi, Y. (2014). Highly chemoselective aerobic oxidation of amino alcohols into amino carbonyl compounds. *Angewandte Chemie International Edition, 53*(12), 3236–3240.

118. Sawama, Y., Morita, K., Yamada, T., Nagata, S., Yabe, Y., Monguchi, Y., & Sajiki, H. (2014). Rhodium-on-carbon catalyzed hydrogen scavenger-and oxidant-free dehydrogenation of alcohols in aqueous media. *Green Chemistry, 16*(7), 3439–3443.

119. Schultz, M. J., & Sigman, M. S. (2006). Recent advances in

homogeneous transition metal-catalyzed aerobic alcohol oxidations. Tetrahedron, 35(62), 8227–8241. Zhan, B. Z., & Thompson, A. (2004). Recent developments in the aerobic oxidation of alcohols. *Tetrahedron, 13*(60), 2917–2935.

120. Sedai, B., Diaz-Urrutia, C., Baker, R. T., Wu, R., Silks, L. P., & Hanson, S. K. (2013). Aerobic oxidation of β-1 lignin model compounds with copper and oxovanadium catalysts. *ACS Catalysis, 3*(12), 3111–3122.
121. Sedai, B., Díaz-Urrutia, C., Baker, R. T., Wu, R., Silks, L. P., & Hanson, S. K. (2011). Comparison of copper and vanadium homogeneous catalysts for aerobic oxidation of lignin models. *ACS Catalysis, 1*(7), 794–804.
122. Sheldon, R. A., & Arends, I. W. (2004). Organocatalytic oxidations mediated by nitroxyl radicals. *Advanced Synthesis & Catalysis, 346*(9-10), 1051–1071.
123. Sheldon, R. A., & Arends, I. W. (2006). Catalytic oxidations mediated by metal ions and nitroxyl radicals. *Journal of Molecular Catalysis A: Chemical, 251*(1–2), 200–214.
124. Sheldon, R. A., Arends, I. W. C. E., & Dijksman, A. (2000). New developments in catalytic alcohol oxidations for fine chemicals synthesis. *Catalysis Today, 57*(1–2), 157–166.
125. Sheldon, R. A., Arends, I. W., ten Brink, G. J., & Dijksman, A. (2002). Green, catalytic oxidations of alcohols. *Accounts of Chemical Research, 35*(9), 774–781.
126. Shibuya, M., Osada, Y., Sasano, Y., Tomizawa, M., & Iwabuchi, Y. (2011). Highly efficient, organocatalytic aerobic alcohol oxidation. *Journal of the American Chemical Society, 133*(17), 6497–6500.
127. Shimizu, K. I., Kon, K., Seto, M., Shimura, K., Yamazaki, H., & Kondo, J. N. (2013b). Heterogeneous cobalt catalysts for the acceptorless dehydrogenation of alcohols. *Green Chemistry, 15*(2), 418–424.
128. Shimizu, K. I., Kon, K., Shimura, K., & Hakim, S. S. (2013a). Acceptor-free dehydrogenation of secondary alcohols by heterogeneous cooperative catalysis between Ni nanoparticles and acid–base sites of alumina supports. *Journal of Catalysis, 300*, 242–250.
129. Shimizu, K. I., Sato, R., & Satsuma, A. (2009). Direct C–C Cross-Coupling of Secondary and Primary Alcohols Catalyzed by a γ-Alumina-Supported Silver Subnanocluster. *Angewandte Chemie International Edition, 48*(22), 3982–3986.

130. Sigman, M. S., & Jensen, D. R. (2006). Ligand-modulated palladium-catalyzed aerobic alcohol oxidations. *Accounts of Chemical Research, 39*(3), 221–229.

131. Son, S., & Toste, F. D. (2010). Non-Oxidative Vanadium-Catalyzed C–O Bond Cleavage: Application to Degradation of Lignin Model Compounds. *Angewandte Chemie International Edition, 49*(22), 3791–3794.

132. Song, H., Kang, B., & Hong, S. H. (2014). Fe-catalyzed acceptorless dehydrogenation of secondary benzylic alcohols. *ACS Catalysis, 4*(9), 2889–2895.

133. Spatareanu, A., Bercea, M., Budtova, T., Harabagiu, V., Sacarescu, L., & Coseri, S. (2014). Synthesis, characterization and solution behaviour of oxidized pullulan. *Carbohydrate Polymers, 111*, 63–71.

134. Steves, J. E., & Stahl, S. S. (2013). Copper (I)/ABNO-catalyzed aerobic alcohol oxidation: alleviating steric and electronic constraints of Cu/TEMPO catalyst systems. *Journal of the American Chemical Society, 135*(42), 15742–15745.

135. Suzuki, T., Morita, K., Matsuo, Y., & Hiroi, K. (2003). Catalytic asymmetric oxidative lactonizations of meso-diols using a chiral iridium complex. *Tetrahedron Letters, 44*(10), 1–20.

136. Taketoshi, A., Beh, X. N., Kuwabara, J., Koizumi, T. A., & Kanbara, T. (2012). Aerobic oxidative dehydrogenation of benzyl alcohols to benzaldehydes by using a cyclometalated ruthenium catalyst. *Tetrahedron Letters, 53*(28), 3573–3576.

137. Thottumkara, A. P., Bowsher, M. S., & Vinod, T. K. (2005). In situ generation of o-iodoxybenzoic acid (IBX) and the catalytic use of it in oxidation reactions in the presence of oxone as a co-oxidant. *Organic Letters, 7*(14), 2933–2936.

138. Tojo, G., & Fernández, M. I. (2006). *Oxidation of Alcohols to Aldehydes and Ketones: A Guide to Current Common Practice* (Vol. 1, pp. 1–30). Springer Science & Business Media.

139. Tong, X., Sun, Y., Yan, Y., Luo, X., Liu, J., & Wu, Z. (2014). Transition metal-free catalytic oxidation of aromatic alcohols with molecular oxygen in the presence of a catalytic amount of N-bromosuccinimide. *Journal of Molecular Catalysis A: Chemical, 391*, 1–6.

140. Trincado, M., Banerjee, D., & Grützmacher, H. (2014). Molecular catalysts for hydrogen production from alcohols. *Energy &*

Environmental Science, 7(8), 2464–2503.

141. Wang, L., Li, J., Lv, Y., Zhao, G., & Gao, S. (2012). Selective aerobic oxidation of alcohols catalyzed by iron chloride hexahydrate/TEMPO in the presence of silica gel. *Applied Organometallic Chemistry, 26*(1), 37–43.
142. Wang, L., Li, J., Zhao, X., Lv, Y., Zhang, H., & Gao, S. (2013). An efficient and scalable room temperature aerobic alcohol oxidation catalyzed by iron chloride hexahydrate/mesoporous silica supported TEMPO. *Tetrahedron, 69*(30), 6041–6045.
143. Wang, N., Liu, R., Chen, J., & Liang, X. (2005). NaNO 2-activated, iron–TEMPO catalyst system for aerobic alcohol oxidation under mild conditions. *Chemical Communications, 42*, 5322–5324.
144. Wang, S. S., Zhang, J., Zhou, C. L., Vo-Thanh, G., & Liu, Y. (2012). An ionic compound containing Ru (III)-complex cation and phosphotungstate anion as the efficient and recyclable catalyst for clean aerobic oxidation of alcohols. *Catalysis Communications, 28*, 152–154.
145. Wigington, B. N., Drummond, M. L., Cundari, T. R., Thorn, D. L., Hanson, S. K., & Scott, S. L. (2012). A Biomimetic Pathway for Vanadium-Catalyzed Aerobic Oxidation of Alcohols: Evidence for a Base-Assisted Dehydrogenation Mechanism. *Chemistry–A European Journal, 18*(47), 14981–14988.
146. Willemsen, J. S., van Hest, J. C., & Rutjes, F. P. (2013). Potassium formate as a small molecule switch: controlling oxidation–reduction behaviour in a two-step sequence. *Chemical Communications, 49*(30), 3143–3145.
147. William, J. M., Kuriyama, M., & Onomura, O. (2013). An efficient method for selective oxidation of 1, 2-diols in water catalyzed by Me 2 SnCl 2. *RSC Advances, 3*(42), 19247–19250.
148. William, J. M., Kuriyama, M., & Onomura, O. (2013). Electrochemical oxidation of 1, 2-diols to α-hydroxyketones in water. *Electrochemistry, 81*(5), 374–376.
149. William, JM, Kuriyama, M., & Onomura, O. (2014). Boronic Acid-Catalyzed Selective Oxidation of 1, 2-Diols to α-Hydroxy Ketones in Water. *Advanced Synthesis & Catalysis, 356*(5), 934–940.
150. Xu, S., Itto, K., Satoh, M., & Arimoto, H. (2014). Unexpected dehomologation of primary alcohols to one-carbon shorter carboxylic

acids using o-iodoxybenzoic acid (IBX). *Chemical Communications, 50*(21), 2758–2761.

151. Yang, J., Liu, X., Meng, D. L., Chen, H. Y., Zong, Z. H., Feng, T. T., & Sun, K. (2012). Efficient Iron-Catalyzed Direct β-Alkylation of Secondary Alcohols with Primary Alcohols. *Advanced Synthesis & Catalysis, 354*(2-3), 328–334.
152. Yi, J., Miller, J. T., Zemlyanov, D. Y., Zhang, R., Dietrich, P. J., Ribeiro, F. H., ... & Abu-Omar, M. M. (2014). A Reusable Unsupported Rhenium Nanocrystalline Catalyst for Acceptorless Dehydrogenation of Alcohols through γ-C–H Activation. *Angewandte Chemie International Edition, 53*(3), 833–836.
153. Zakzeski, J., Bruijnincx, P. C., Jongerius, A. L., & Weckhuysen, B. M. (2010). The catalytic valorization of lignin for the production of renewable chemicals. *Chemical Reviews, 110*(6), 3552–3599.
154. Zeng, G., Sakaki, S., Fujita, K. I., Sano, H., & Yamaguchi, R. (2014). Efficient catalyst for acceptorless alcohol dehydrogenation: interplay of theoretical and experimental studies. *ACS Catalysis, 4*(3), 1010–1020.
155. Zhang, D. D., Chen, X. K., Liu, H. L., & Huang, X. R. (2014). A computational mechanistic study of pH-dependent alcohol dehydrogenation catalyzed by a novel [C, N] or [C, C] cyclometalated Cp* Ir complex in aqueous solution. *New Journal of Chemistry, 38*(8), 3862–3873.
156. Zhang, G., & Hanson, S. K. (2013). Cobalt-catalyzed acceptorless alcohol dehydrogenation: synthesis of imines from alcohols and amines. *Organic Letters, 15*(3), 650–653.
157. Zhang, G., Han, X., Luan, Y., Wang, Y., Wen, X., & Ding, C. (2013). L-Proline: an efficient N, O-bidentate ligand for copper-catalyzed aerobic oxidation of primary and secondary benzylic alcohols at room temperature. *Chemical Communications, 49*(72), 7908–7910.
158. Zhang, G., Proni, G., Zhao, S., Constable, E. C., Housecroft, C. E., Neuburger, M., & Zampese, J. A. (2014). Chiral tetranuclear and dinuclear copper (II) complexes for TEMPO-mediated aerobic oxidation of alcohols: are four metal centres better than two?. *Dalton Transactions, 43*(32), 12313–12320.
159. Zhang, G., Scott, B. L., Wu, R., Silks, L. P., & Hanson, S. K. (2012). Aerobic oxidation reactions catalyzed by vanadium complexes of bis

(phenolate) ligands. *Inorganic Chemistry, 51*(13), 7354–7361.

160. Zhang, G., Vasudevan, K. V., Scott, B. L., & Hanson, S. K. (2013). Understanding the mechanisms of cobalt-catalyzed hydrogenation and dehydrogenation reactions. *Journal of the American Chemical Society, 135*(23), 8668–8681.
161. Zhang, Y., Zhou, Q., Ma, W., & Zhao, J. (2014). Enantioselective oxidation of racemic secondary alcohols catalyzed by chiral Mn (III)-salen complex with sodium hypochlorite as oxidant. *Catalysis Communications, 45*, 114–117.
162. Zhu, J., Wang, P. C., & Lu, M. (2012). Synthesis of novel magnetic silica supported hybrid ionic liquid combining TEMPO and polyoxometalate and its application for selective oxidation of alcohols. *RSC Advances, 2*(22), 8265–8268.
163. Zhu, Y., Zhao, B., & Shi, Y. (2013). Highly efficient Cu (I)-catalyzed oxidation of alcohols to ketones and aldehydes with diaziridinone. *Organic Letters, 15*(5), 992–995.

3

Covalent Organic Frameworks

CONTENTS

3.1. INTRODUCTION

Because of the remarkable performance and wide range application of the nano-porous materials in the fields of superhydrophobic interfaces, energy conversion, gas storage, energy storage, gas separation, catalysis and optoelectronics of the nano-porous materials over the last decade, has increased the interest in this material enormously. There are extensive ways in which chemist have been able to prepare a number of porous materials, the topological building block was designed by chemistry for the construction of these porous materials.

Using reticular chemistry, the primary family of porous materials to be manufactured as metal-organic frameworks (MOFs), which by the use of using polytopic linkers, networks polyatomic inorganic metal-containing clusters (Long & Yaghi, 2009; O'Keeffe, 2009). To create immense possibilities, the coordination adaptability of the constituent metal ions syndicates with the functional variety of the organic linker molecules. In 2005, in the production of porous organic frameworks linked via covalent bonds, Yaghi and co-workers established the usefulness of the topological design principle, which was the primary fruitful samples of covalent organic frameworks (COFs) (Cote et al., 2005). COFs are covalent porous crystalline polymers that allow the intricate incorporation of organic building blocks into a well-ordered structure with atomic accuracy (Perry et al., 2009; Tranchemontagne, 2009).

The chemical manufacturing of COFs has developed suggestively, since this landmark paper, and they display great potential for functional examination. Because they are made up pf of light-weight elements connected by durable covalent bonds, COFs have high thermal stabilities, possess low mass densities and deliver perpetual porosity. COFs can be considered into either three- (3D) or two- dimensional(2D) depending on the building block dimensions. In 2D COFs, the covalently bound framework is limited to 2D sheets, which heap additional to form a layered covered structure that gives intermittently associated columns. This columnar assembling structure offers an exclusive resource to synthesize ordered p systems that are problematic to generate via conservative covalent and/ or noncovalent methods. The well-organized columns in 2D COFs could enable charge carrier transport in the piling direction, which suggests that 2D COFs have the potential for emerging new photo-functional and type p-electronic materials for photovoltaics and optoelectronics. On the other hand, 3D COFs, which spread this framework three-dimensionally through a building block comprising of an sp^3 silane or carbon atom, typically have high specific surface areas (in some situations greater than 4000 m^2 g^{-1}), low densities (as low as 0.17 g cm^3) and frequent open sites, and. These structures make 3D COFs perfect applicants for gas storage.

This tutorial appraisal is separated into three chief segments that emphasis on the synthesis and design, the functions and the structure, and properties of this developing material.

3.2. DESIGN AND SYNTHESIS

3.2.1. Dynamic Covalent Chemistry

The manufacturing of polymeric materials has normally been conquered by kinetically measured reactions, which irrevocably form covalent bonds. It is problematic to crystallize connected organic polymers into solids using irretrievable reactions. On the other hand, dynamic covalent chemistry (DCC) causes the reversible development of covalent bonds, which can be created, broken, and recreated (Rowan et al., 2002). Consequently, unlike traditional covalent bond creation, DCC is thermodynamically measured and thus provides reversible reaction systems with "proof-reading" and "error checking" features, causing the creation of the most thermodynamically steady structures. By using the DCC idea for the formation of COFs, the polymer skeleton creation occurs alongside the crystallization procedure, while the self-healing response decreases the occurrence of structural flaws and helps in the creation of an ordered structure. As a consequence, the final COF product has a well-ordered crystalline structure with high thermodynamic constancy.

The synthesis and design of COFs have two key problems that must be taken into account to attain thermodynamic control in reversible reactions: the primary is the structure of the building blocks and the second is the artificial technique, counting the reaction conditions and reaction media.

3.2.2. Structure of Building Blocks

To attain an ordered and crystalline COF, the structure of the building blocks must meet two necessities: (1) the creation reaction of the COF should be a reversible reaction, and (2) the geometry of the building blocks should be well conserved in the COF. Concerning (1), the building blocks should comprise of reactive groups that activate dynamic covalent bond development, i.e., (a) there are no irreversible side reactions, and (b) the reaction system comprises of only oligomers, monomers, and polymers that are substitutable under thermodynamic control.

Concerning the second necessity, the building blocks should be conformationally stiff, and the bond creation direction must be distinct.

Numerous reversible reactions have been discovered in the energy to meet the first condition for fruitful COF formation (Figure 3.1). A mainstream of the identified frameworks depends on boron chemistry because boronic acids can be either co-condensed and self-condensed with di-alcohols to attain five-membered boronate ester linkages and six-membered boroxine (Figures 3.4 and 3.5). Acetonide-protected catechol has been reacted with boronic acids instead of di-alcohols in the occurrence of the Lewis acid catalyst BF_3OEt_2 to form COFs (Figure 3.1) (Spitler & Dichtel, 2010; Spitler et al., 2011). This in situ deprotection protocol delivers an applied resource of evading both the solubility and oxidation difficulties with the building blocks, particularly large aromatic di-alcohols. Comparable to the situation of either boronate or boroxine esters, borosilicate linkages have been applied to form COFs through the condensation of a tetraboronic acid with tert-butylsilanetriol (Figure 3.1–3.3) (Hunt et al., 2008).

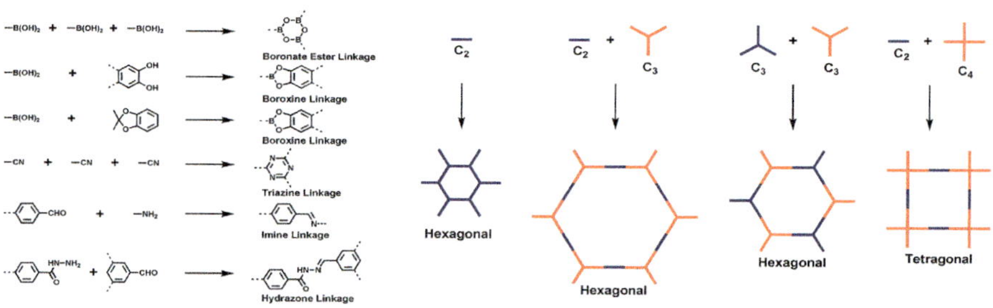

Figure 3.1: Schematic representation of the dynamic chemical reactions for the preparation of COFs.

[Source: https://pubs.rsc.org/en/content/articlelanding/2012/cs/c2cs35157a/]

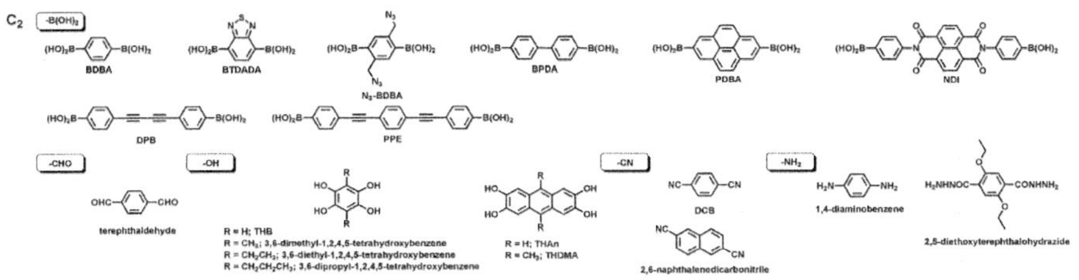

Figure 3.2: The geometry of the building blocks categorized into different reactive groups.

[Source: https://pubs.rsc.org/en/content/articlelanding/2012/cs/c2cs35157a/]

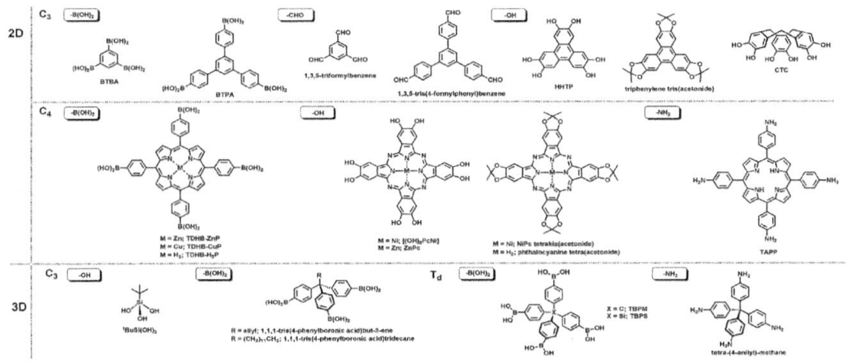

Figure 3.3: The combination of building blocks with different geometries to design 2D COFs.

[Source: https://pubs.rsc.org/en/content/articlelanding/2012/cs/c2cs35157a/]

Regardless of being thermally vigorous, boron-based COFs are liable to attack and are even hydrolyzed by water vapor in the air (Uribe-Romo et al., 2009; Lanni et al., 2011). Yaghi's group founded new linking chemistries for COFs, connecting dynamic pH-dependent, reversible condensation reactions that form both hydrazine and imine connections (Figure 3.4) (Wan et al., 2011; Uribe-Romo et al., 2011). Additional class of COFs, known as covalent triazine-based frameworks (CTFs), has been arranged via the cyclical trimerization of cyano reaction groups under isothermal circumstances (Kuhn et al., 2008; Bojdys et al., 2010). CTFs concurrently deliver high chemical, thermal, and mechanical constancies along with a high degree of conjugation; though, they characteristically have low crystallinity as a consequence of the deprived reversibility of the trimerization reaction.

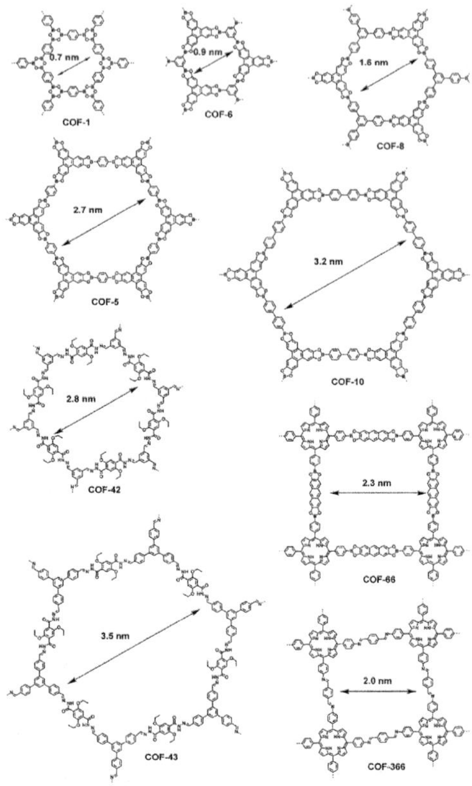

Figure 3.4: Schematic representation of COFs with boronate ester, boroxine, imine, and hydrazone linkages.

[Source: https://pubs.rsc.org/en/content/articlelanding/2012/cs/c2cs35157a/]

As stated above, the stiff conformation of the building blocks allows the topological strategy of the COFs. For clearness, these building blocks were categorized as either 3D or 2D based on basic regularity representation, which mentions the maneuvering balance of the reactive groups, as abridged in Figure 1b (2D-C_2, 2D-C_3, 2D-C_4, and 3D-T_d). As revealed in Figures 3.1–3.3, the geometry of the building blocks regulates the subsequent COF structure. Consequently, the combinations such as 3D-T_d + 3D-T_d, 3D-T_d + 2D-C_2, or 3D-T_d + 2D-C_3 can cause the formation of 3D COFs with dissimilar crystalline space groups. Contrary to this, designated combinations of 2D blocks (e.g., 2D-C_2 + 2D-C_3, 2D-C_3 + 2D-C_3 or 2D-C_2 + 2D-C_4) will afford 2D COFs with intended topology and pore structure (Figure 3.3). The inflexible nature and distinct bonding direction of arenas make aromatic p systems appropriate building blocks for COFs. The variety of aromatic systems permits many building block combinations, which donates the COFs with high litheness in their molecular design.

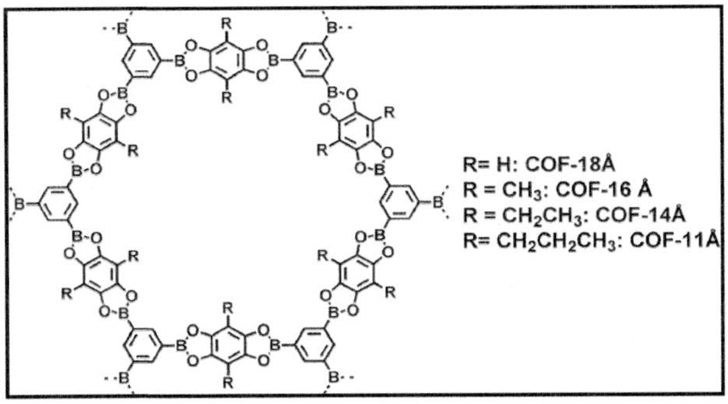

Figure 3.5: Schematic representation of boronate ester-linked COFs with alkyl chains in the walls.

[Source: https://pubs.rsc.org/en/content/articlelanding/2012/cs/c2cs35157a/]

3.2.3. Synthetic Methods

Variable the thermodynamic symmetry during covalent bond creation is important in making extremely well-ordered covalent networks. In specific, the reaction conditions (pressure, temperature, and absence or presence of template) and media should be considered for the creation of thermodynamically stable polymeric crystalline architectures.

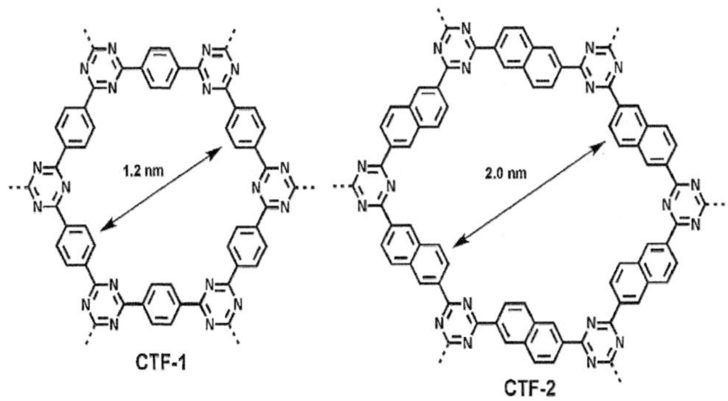

Figure 3.6: Schematic representation of COFs with triazine linkages.

[Source: https://pubs.rsc.org/en/content/articlelanding/2012/cs/c2cs35157a/]

Concerning the reaction media, molten metal salts and mixed solvent systems have been established to offer ionothermal and solvothermal circumstances, respectively, for the formation of COFs. Microwave reactions have been applied under solvothermal circumstances to rapidly form COFs. Contrary to these bulky approaches, COF monolayers or films have been discovered via reactions on substrates, such as graphene sheets and metal surfaces.

3.2.3.1. Solvothermal Synthesis

A characteristic solvothermal technique for forming COFs is given below. Mixed solvents and monomers are positioned in a Pyrex tube and degassed via numerous freeze–pump–thaw series. The tube is then closed and heated to a chosen temperature for some reaction time. The precipitate is composed, eroded with appropriate solvents, and desiccated under vacuum to harvest the COF as a solid powder. Matters such as the reaction rate, crystal growth rate, solubility, crystal nucleation, and 'self-healing' structure are significant points to ponder when choosing the reaction conditions and media.

- Solvent ratios and combinations are significant aspects in balancing between crystallization and framework formation when forming extremely crystalline COFs. For boroxine and boronate ester connected COFs, solvent combinations such as mesitylene -dioxane, THF–methanol, and DMAc–o-dichlorobenzene have been applied, while a dioxane–toluene solvent combination has been applied to produce borosilicate COFs (Tilford et al., 2006; Ding et al., 2011). Contrary to this, dioxane–aqueous acetic acid has been applied as the solvent for imine-connected COFs, while mesitylene–dioxane–aqueous acetic acid was active for hydrazone connected COFs. In all situations, an appropriate concentration of the building blocks in the solvent system is essential for the reaction to advance under thermodynamic control. Whether completely insoluble or completely solvable building blocks cannot cause the formation of COFs wants extra experimental researches (Feng et al., 2011).
- Similarly, an appropriate temperature is significant to confirm the reversibility of the reaction. Usually, COFs have been equipped at temperatures reaching between 85–120 1C reliant on the chemical reactivities of the building blocks. A closed reaction atmosphere is essential to permit the occurrence of water molecules that could trigger the reverse reaction in the system.

3.2.3.2. Microwave Synthesis

Cooper and co-workers established a high-quantity protocol for active covalent reactions in a warm reactor to quickly form boronate ester connected COFs (Campbell et al., 2009; Dogru et al., 2011). Microwave formation delivered numerous benefits over solvothermal approaches.

- Microwave formation yields COFs quickly. Thus, large-scale production is probable.
- A closed vessel is not mandatory for the microwave formation.
- The microwave solvent withdrawal procedure eliminates impurities and remains trapped in the frameworks more professionally, which endorses better porosity. The microwave protocol could deliver a powerful spare for the solvothermal technique.

3.2.3.3. Ionothermal Synthesis

Thomas and co-workers discovered an ionothermal formation technique to synthesize CTFs (Figure 3.6). The cyclotrimerization of aromatic nitrile building components (Figures 3.1–3.3, e.g., 1,4-dicyanobenzene) in melted $ZnCl_2$ at 400 1C afford crystalline conjugated CTFs with vigorous thermal and chemical solidities. $ZnCl_2$ acts as both the catalyst and the solvent for the trimerization reaction, which appears to be partly reversible. Though, such severe reaction circumstances boundary the building block accessibility. Most formed CTFs are amorphous materials that lack long-range molecular direction.

3.2.3.4. Synthesis of Monolayers on Metal Surfaces

Contrary to this bulky formation of COFs displayed above, monolayers of COF-5 and COF-1 are formed on a metal surface of by the condensation of the building blocks. The covalently bound SCOF-2 and SCOF-1 nano-architectures can be organized by transferring building blocks onto a pure Ag(111) surface from heated molybdenum crucible evaporators under ultrahigh vacuum.

In contrast to the bulky synthesis of COFs shown above, condensation of the building blocks onto a metal surface forms monolayers of COF-1 and COF-5 (Figure 3.4) (Zwaneveld et al., 2008). The covalently bound SCOF-1 and SCOF-2 nano-architectures can be organized by transferring the building blocks from heated molybdenum crucible evaporators onto a pure

Ag(111) surface under ultrahigh vacuum. Detecting the surface via scanning tunneling microscopy (STM) discloses the existence of hexagonal pores along with with slight numbers of uneven eight, seven and five-membered pores. The formation of defect-free monolayers on a metal surface may need fine-tuning of the reaction circumstances, refining the building block purity, and by means of an appropriate single-crystal metal, substrate border to direct the building block arrangement.

3.2.3.5. Synthesis of Monolayers on a HOPG Surface

In this experimentation, an extremely ordered pyrolytic graphite (HOPG) surface was used as an alternative of a metal surface to make monolayer COFs (Guan et al., 2012). 1,4-benzene boronic acid (BDBA), Biphenyldiboronic acid (BPDA and 9,9-dihexylfluorene-2,7-diboronic acid were placed onto a HOPG surface from their THF solutions, and the subsequent HOPG models were heated in a closed autoclave at 150 °C for 1 h to form COF monolayers. The sophistication of this technique is that the existence of $CuSO_45H_2O$ aids as a water "reservoir" to control the chemical symmetry, which is important in founding a high-quality monolayer. In the nonappearance of $CuSO_45H_2O$, the attention of the HOPG surface by the monolayer was only about 7%, which amplified intensely to 98% in the existence of $CuSO_45H_2O$. The water molecules unrestricted from $CuSO_45H_2O$ during the heating procedure can act as an asymmetry-working agent that thrusts the dehydration reaction backward and thus endorses the flaw remedy procedure, which forms the extremely ordered COF network. These water molecules can be reabsorbed by the $CuSO_4$ during the cooling procedure, which stops the disintegration of the SCOFs.

3.2.3.6. Synthesis of Oriented Thin Films on Graphene Surfaces

COFs equipped using these approaches are either monolayers or insoluble powders, which cannot be consistently incorporated into real devices or interfaced to electrodes. For this aim, the formation of COF thin films on a substrate is of comprehensive technical attention and important technological significance. Dichtel and co-workers have stated the characterizations and formation of layered oriented 2D COF sheets onto single-layer graphene (SLG on SiO_2) surfaces (Figure 3.7) (Cote et al., 2005; 2007; Colson et al., 2011; Spilter et al., 2012). By insertion of SLG/SiO_2 in the solvothermal reaction schemes, oriented COF thin films synthesized on the SLG. Numerous COF films, i.e., HHTP-DPB-COF, COF-5, NiPc-COF, TP-

COF, and ZnPc-PPE-COF have been effectively arranged on graphene, and their thicknesses can be adjusted by the use of the reaction time. Grazing incidence X-ray diffraction capacities disclose that the thin sheets have high crystallinity, and the layers in the thin sheets are vertically allied.

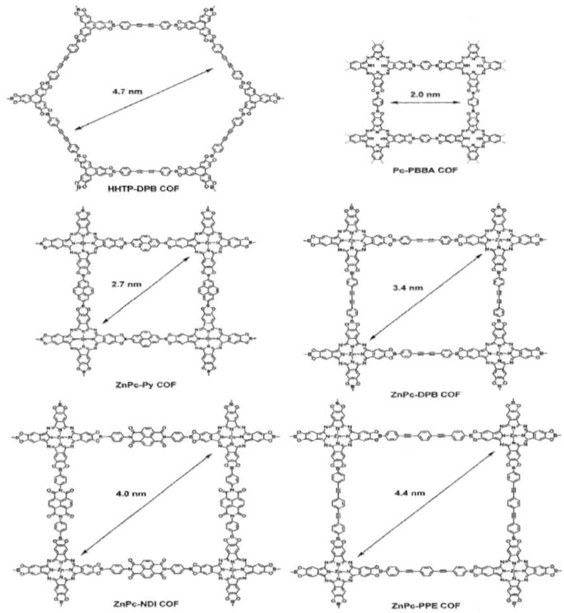

Figure 3.7: Schematic representation of the COFs with the largest pore size (HHTP-DPB COF) and tetragonal phthalocyanine COFs with different pore sizes prepared on graphene.

[Source: https://pubs.rsc.org/en/content/articlelanding/2012/cs/c2cs35157a/]

3.3. STRUCTURAL STUDIES OF COFS

COFs are extremely crystalline polymers with separate X-ray diffraction (XRD) designs. The crystalline structure of both 3D and 2D COFs can be inferred from XRD data. The structural imitation of COFs offers a significant tool in showing their piling structures, particularly for 2D COFs. In adding, elemental analysis, X-ray photoelectron spectroscopy, solid-state NMR spectroscopy, and infrared spectroscopy are all valuable for appraising the terminal groups, connections, and compositions of the COFs.

3.3.1. Structure of 3D COFs

In addition to the building block with 3D-T_d geometry, the geometry of the subordinate building blocks and their assembly designs govern the topology of 3D COFs. The self-condensation of tetrahedral (3D-T_d) nodes, or triangular (2D-C_3) building blocks or its co-condensation with linear (2D-C_2), forms the 3D COFs. The tetrahedral node can be either tetra(4dihydroxyborylphenyl) silane (TBPS), or tetra(4-dihydroxy boryl phenyl) methane (TBPM), while the linear block is terephthalaldehyde, and the triangular one is tert-butylsilane triol ($^tBuSi(OH)_3$) or 2,3,6,7,10,11hexahydroxy triphenylene (HHTP) (Figure 3.1). The self-condensation of the tetrahedral TBPS and TBPM forms COF-103 and COF-102, respectively, both of which assume a ctn topology (Figure 3.6). The ctn net was also practical for COF-202 and COF-105, which were manufactured via the co-condensation of the tertbutylsilane triol, respectively, with the tetrahedral TBPM (Figure 3.6) (Bunck & Dichtel, 2012). In contrast, the co-condensation of HHTP with TBPS and triangular HTTP produced COF-108 (Figure 3.8), which displays topology. The aim why one of these two structure kinds would be favored over another is still unclear. The distinguished changes between the two topologies are that bor is about 15% less thick than ctn and has larger pores than ctn. The imine-linked 3D COF has been organized by the use of a linear terephthalaldehyde and a tetrahedral tetra-(4anilyl) methane and as the building blocks (Figure 3.8). The resultant COF-300 has a dia topology containing an exclusive multiply interpenetrated construction. Cerius software was applied for structural imitation to draw the "blueprints" for the 3D COFs on the basis of the bor and ctn nets by fitting the molecular building blocks onto the tetrahedral nodes and fitting both the B_3O_3 linkages and the triangular unit onto the triangular nodes consistent to the particular cubic space group regularities for the P43m (bor) and I43d (ctn) nets. Energy minimizations byways of forcefield controls were applied to yield models with rational standards for all bond angles and lengths. The unit cell of each space group was then applied to feign the XRD patterns.

For 3D COFs, the major pore size is 3.1 nm in diameter for the boronate ester-linked COF-108 and the lowest pore size stated so far is 0.8 nm for imine-linked COF-300, and (Figure 3.8). The Brunauer–Emmett–Teller (BET) surface area, appraised by the use of the nitrogen sorption isotherm, ranges to 4210 m^2 g^{-1} (COF-103) from 1360 (COF-300). Dichtel and co-workers established a synthetic approach to functionalize the surface of the 3D COF pores by the use of truncated trigonal building blocks to attach organic groups such as allyl or alkyl chains (cote et al., 2007). The 3D COF materials stood high filling levels of the shortened building blocks while recalling the high porosity and the crystallinity of the parent 3D COFs. This technique is vital because it demonstrates the limits of the thermodynamic combination of COFs.

3.3.2. Structure of 2D COFs

3.3.2.1. Topology

The mixtures of 2D building blocks into 2D COFs can produce protracted networks with tetragonal and hexagonal topologies (Figure 3.1). Hexagonal 2D COFs have been arranged by joining the 2D-C_3 building blocks with either 2D-C_3 or 2D-C_2 blocks. The self-condensation of 2D-C_3 or 2D-C_2 building blocks also generates hexagonal COFs. The hexagonal honeycomb construction is an honeycomb assembler (HCA) network, which has two main types depending on the dissimilar piling mode of the 2D layers. One is the (binary neural network) bnn topology that brings into line the lattice straight with the layer set above the neighbor layer (also mentioned to as eclipsed stacking or AA stacking), and another is the Gray relational analysis (gra)network, which interprets layers comparative to each other by half a unit cell (also denoted to as AB stacking). In adding, tetragonal 2D COFs have been built by uniting 2D-C_4 blocks with either 2D-C_4 or 2D-C_2 blocks. The tetragonal topology of these 2D COFs is a single channel unit (scu) network, with the 2D-C_4 blocks positioned at the vertices and the 2D-C_2 units inhabiting the edges.

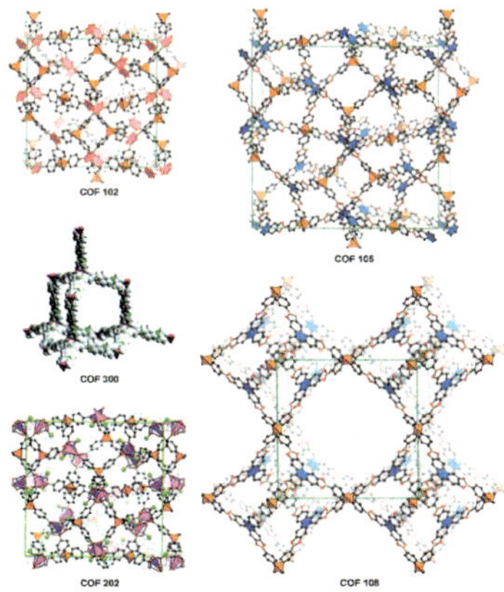

Figure 3.8: Schematic representation of 3D COFs.

[Source: https://pubs.rsc.org/en/content/articlelanding/2012/cs/c2cs35157a/]

To shape porous hexagonal architectures, the 2D-C_3 building blocks are tri-substituted offshoots such as1,3,5-triformylbenzene, hexahydroxytriphenylene tris(acetonide), 1,3,5-benzene boronic acid, 1,3,5-benzenetris(4-phenylboronic acid), 1,3,5-tris(4-formyl phenyl) benzene, and HHTP, whereas the 2D-C_2 blocks are di-substituted units such as BDBA, 1,2,4,5-tetra-hydroxybenzene (THB), 4,4^0-diphenyl butadiyne bis(boronic acid), 1,4-benzothiadiazole diboronic acid (BTBDBA), 1,4-dicyanobenzene terephthalaldehyde, and 1,4-diaminobenzene (Figure 3.2) (Ding et al., 2011a; 2011b). The size of the 2D-C_3 blocks and the length of the 2D-C_2 units control the pore size of the hexagonal 2D-COFs. Though, tetragonal 2D-COFs need a 2D-C_4 building block, which is chiefly given by phthalocyanine and porphyrin derivatives (Figure 3.2). In these situations, the pore size is contingent on the 2D-C_2 linkers, such as the larger the tetragonal pore when longer are the linkers. The variety of the metal types in the center of the phthalocyanine and porphyrin macrocycles agrees to the formation of alike frameworks by means of dissimilar transition metals, which therefore vagaries the optoelectronic or redo features of the frameworks (Wan et al., 2008; 2009; Feng et al., 2012).

3.3.2.2. Alignment of Layers

The 2D COFs comprise of planar films that pile to form layered constructions. High-resolution transmission electron microscopy (HRTEM) discloses that the 2D films are planar protracted and allied in parallel to generate the layered structures (Lukose et al., 2011; Yu et al., 2012). Though, this imagining does not designate the comparative locations of the adjacent films. Likewise, determining the arrangement of the layers in both tetragonal and hexagonal frameworks via XRD examination remains problematic. An influential tool for concerning the layer structure is ab initio procedure, which initially shapes up the monolayer structure and then concepts layered constructions via energy customization. The subsequent layered structures forecast the horizontal offset and distance between layers, which delivers visions into the exclusive arrangement of the 2D COF layers.

Lukose et al. (2011) designed the piling energy of numerous kinds of 2D COF (Figure 3.4, COF-8, COF-6, COF-5, and COF-1) lattices, counting serrated, AA, and inclined piling forms by the use of the density-functional founded tight-binding (DFT-B) technique. All of these lattice kinds forecast alike XRD profiles, which decide with the experimentally experiential designs; though, the piling energies were suggestively dissimilar on the basis of their offsets. For all of these COFs, the AB staggered structure has the lowermost creation energy, whereas the AA assembling mode has much greater generation energy. The alteration in the counterbalance values creates a curve between the offset distance the formation energy, which designates that a layered structure with a minor offset is the steadiest one in all situations. A counterbalance of 1.4 Å between the layers in both the armchair and zigzag directions is favored and consequences in more attractive Coulomb interlayer connections than eclipsed assembling. Therefore, it alleviates the materials. Dichtel and Clancy have used molecular mechanics to describe the potential energy surface (PES) allied with the offset between two HHTPDPB COF layers (Figure 3.7). The subsequent modification of this model by means of the DFT technique forecasts that the together layers in the HHTP-DPB COF have a counterbalance of 1.7–1.8 A° from the eclipsed loading structure.

For tetragonal 2D-COFs, we applied the DFT-B technique counting the Lennard-Jones (LJ) dispersion to compute the electronic and molecular structures of the layer 2D-NiPcBTDA-COF structure (Figure 3.7). The optimum single layer comprised of 332 atoms with a lattice constant of a = b = 46.30 A°. The four kinds of produced assembling forms comprise of flawless AA, slipped structures with minor shifts in the *a* and *b* directions, and AB. The crystal assembling energy for the AB assembling was only 178.31 kcal mol^{-1}, which amplified to 312.84 kcal mol^{-1} for the AA assembling mode. Associated to AA assembling, the slipped structures with counterbalances of 1.44 and 0.72 A° in the b and a direction displayed an amplified stacking energy of 340.55 and 340.60 kcal mol^{-1}, respectively.

Warped building blocks have been established for the building of 2D COFs. For instance, tetraphenylporphyrin offshoots comprising of four meso-substituted phenyl groups warped vertical to the porphyrin ring form the boronate ester connected tetragonal 2D COFs (Figure 3.9, H_2P-COF, ZnP-COF, CuP-COF).

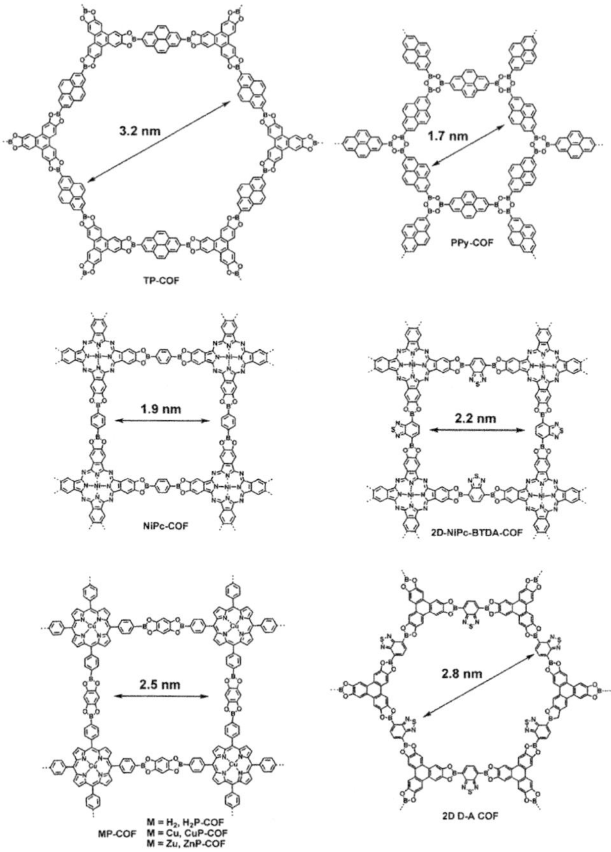

Figure 3.9: Schematic representation of p electronic and photoconductive COFs.

[Source: https://pubs.rsc.org/en/content/articlelanding/2012/cs/c2cs35157a/]

DFT-B imitations designate that the monolayer has a dihedral angle between the porphyrin and phenyl units as high as 901, which reduced to 451 in the layered structure. This consequence designates that the p–p assembling of the layers can overawe the steric repulsion between the ortho hydrogen atoms. The porphyrin film develops more planar in the assembled 2D-COF structure.

Lately, a bowl-shaped building block, cyclotricatechylene, has been applied in the formation of a 2D zigzag COF (Figure 3.10) (Tilford et al., 2008). It is astounding that this 2D zigzag COF still permits p–p assembling between layers with a layer-to-layer distance of 3.36 A°.

Structure resolution on the basis of the XRD patterns has typically been achieved on the basis of the above calculation consequences, which applied the unit cell of each space group to forecast the consistent XRD pattern. The competition between the experimentally observed and calculated XRD patterns has been applied to appraise the appropriateness of each space group. Typically, the space groups in the AB assembling mode finally gives an XRD pattern that diverges extensively from the detected curve. This result evidently rejects the likelihood of AB assembling in 2D COFs, which approves with the DFT-B calculations. Though the slightly slipped stacking and the concealed assembly in AA stacking modes provide alike XRD patterns to the investigational findings, and among the AA stacking modes, there were no substantial variations in the simulated XRD patterns, even though they have diverse assembling energies.

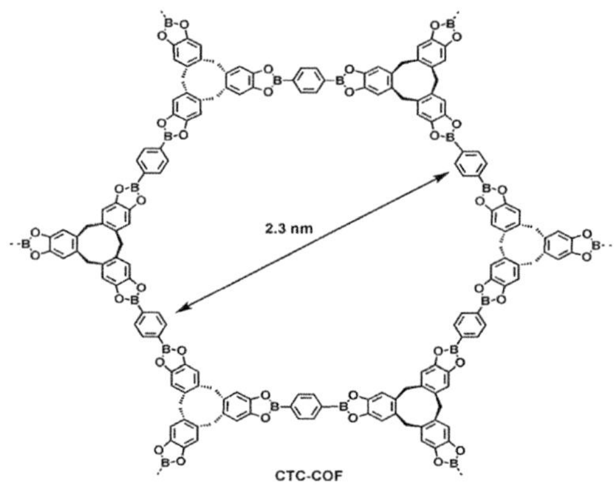

Figure 3.10: Schematic representation of a 2D zigzag COF.

[Source: https://pubs.rsc.org/en/content/articlelanding/2012/cs/c2cs35157a/]

3.3.2.3. Channel Structure

In the 2D COFs, the layered structure produces 1D channels. Consequently, the size of the channel is reliant on the length and size of the building blocks. In this background, 2D COFs are exclusive because they cause the construction of predesigned porous materials as they permit the detailed molecular design of 1D channels.

The minimum channel is 0.86 nm in size for the microporous COF-1, for hexagonal 2D COFs (Figure 3.4), which produced by the means of the self-condensation of BPDA. Contrary to this, for a mesoporous HHTP-

DPB COF, the maximum channel detected was 4.7 nm in diameter which was produced from 4,4^0-diphenylbutadiynebis (boronic acid) (DPB) and HHTP (Figure 3.7). 1D channels between 0.86 and 4.7 nm in size have been developed by adopting the length and size of the building blocks. The surface area of the hexagonal 2D COFs reached from 711 to 2160 $m^2 g^{-1}$. There is no straightforward relationship between the surface area and the channel size based on presently stated results, and the surface area may suggestively reliant on the superiority of the COF resources.

The smallest one was 2.0 nm for an imine-linked porphyrin COF-366 and the largest channel was 3.8 nm, for tetragonal 2D COFs for a metallophthalocyanine 2D COF with PPE linkages (Figure 3.4). By using linkers with different lengths, the channel size can be finely adjusted between 2.0 and 3.8 nm (Figure 3.7). The surface area of the tetragonal 2D COFs reached from 360 to 1894 $m^2 g^{-1}$.

The corresponding synthetic control of the microscopic porosity and macroscopic structure is probable for 2D COFs (Figure 3.9). The co-condensation of a zinc porphyrin derived with THB forms ZnP-COF with a pore size of 2.5 nm. Time reliant on monitoring of this reaction scheme discloses that during the early condensation period, the COFs had flaws that designate one or two kinds of pores with large pore size were existing. Upon spreading the reaction time, these flaws self-healed to produce a single kind of mesopore with a 2.5 nm width in COFs. Furthermore, spreading the reaction time caused the development of the 2D COFs in the x–y directions and amplified the surface areas of the resultant ZnP-COF resources.

Lavigne and co-workers applied alkyl chain replaced linkers in the formation of the boronate ester related COFs to include organic groups into the walls of these channels (Nagai et al., 2011). The existence of alkyl chains in the channels permits their size to be finely adjusted from 1.8 nm to 1.1 nm, as detected for the sequence of COF-18 A°, COF-16 A°, COF-14 A°, and COF-11 A° (Figure 3.5).

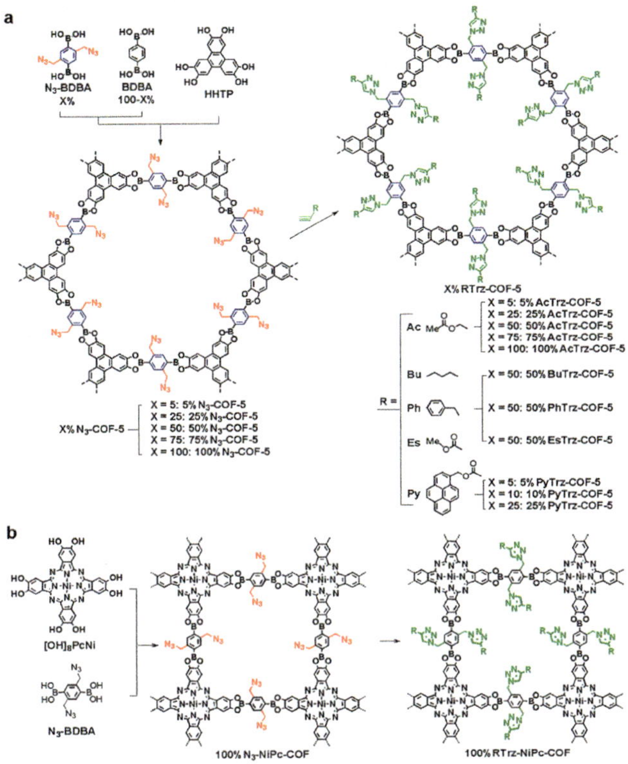

Figure 3.11: Schematic representation of the surface engineering of COFs of (a) hexagonal COF-5 and (b) tetragonal NiPc-COF.

[Source: https://pubs.rsc.org/en/content/articlelanding/2012/cs/c2cs35157a/]

The surface area reduced from 1263 to 105 $m^2 g^{-1}$ as the chain length amplified from consuming no alkyl replaced linker to having a propyl replaced linker. The existence of the alkyl chains generates a special microenvironment within the stations and therefore plays a noteworthy role in both gas adsorption and hydrolytic constancy.

Our group has established a simple and widespread approach for engineering the surface of 2D COF pores to permit the integration of a diversity of organic functionalities into the networks (Figure 3.11) (Han et al., 2008; Furukawa & Yaghi, 2009). This functionalization is made probable by the means of azide-appended building blocks for the formation of COFs with parapets to which a designable gratified of azide units is fastened. These azide units experience a measurable click reaction with alkynes to yield pore surfaces with anticipated groups and favored thicknesses. Numerous organic groups such as acetyl, alkyl chains, aromatic units, ester, and chromophoric moieties have been combined into the walls at nominated contents. This surface engineering approach is not only appropriate to hexagonal COFs but also to tetragonal ones (Figure 3.11). The pore size can be excellently adjusted from 1.2 nm to 3.0 nm for the hexagonal COF-5 family and 0.75 to 2.2 nm for the tetragonal NiPc-COF. The tailor-made wall surface establishes a stage towards the manufacturing of the COF pore surface to understand pre-designed arrangements, functionalities, and components.

3.4. FUNCTION EXPLORATION

The function of a COF arises from its porosity and molecular skeletons. As predesignate porous materials, COFs have emerged as a new media for gas adsorption and storage. These frameworks can be further functionalized to contain catalytic sites for use as heterogeneous catalysts. Because of their unique topology and stacking layer structures, 2D COFs produce columnar arrays throughout their building blocks, which are difficult to synthesize using traditional supramolecular or conventional porous material approaches. In this sense, 2D COFs offers a new platform for the molecular design of semiconducting materials with periodic and built-in molecular ordering (Klontzas et al., 2008, 2009).

3.4.1. Gas Adsorption and Storage

COFs have been touted as ideal materials for gas storage applications because they are composed of light elements linked by strong covalent bonds to form highly porous materials with high surface areas. The storage capabilities of COFs for gases, such as hydrogen, methane, and carbon dioxide, have been widely investigated. As a general rule, the gas adsorption capacity of a COF depends primarily on the components and topologies of its frameworks. Because of their larger surface area and pore volume, 3D COFs possess significantly higher uptake capacities than 2D COFs. Theoretical simulations predict that COFs doped with alkali metals could improve storage capacity, which demonstrates the usefulness of introducing functional sites for interactions with gases (Srepusharawoot et al., 2009; Assfour & Seifert, 2010).

3.4.1.1. Hydrogen Gas

Hydrogen gas storing certainly appeals public attention because it signifies a future energy reserve on the basis of its high energy density, high chemical profusion, and ecologically friendly features. The DOE roadmap mark for hydrogen storing is set to81 kg H_2 m^3 and 9 wt% at 253–323 K with a pressure of 100 atm by the year 2015. Of the resources that depend on the comparatively feeble physisorption at cryogenic (e.g., 100 K) temperatures, only permeable constituents with very high specific surface areas (43,000 m^2 g^{-1}) and enhanced pore sizes in the variety of 0.7 to 1.2 nm have the possibility to meet the DOE 2015 target.

Of the dissimilar COFs, the main hydrogen storing capacity is detected for 3D COF-102 (Figure 6; S_{BET}: 3620 m^2 g^{-1}, pore size: 1.2 nm) [38], which uptakes 72 mg per gram at 1 bar and 77 K. This size is analogous to those of MOF-5 (76 mg g^{-1}, S_{BET}: 3800 m^2 g^{-1}), MOF-177 (75 mg g^{-1}, S_{BET}: 4500 m^2 g^{-1}), and the permeable aromatic basis PAF-1 (75 mg g^{-1}, S_{BET}: 5600 m^2 g^{-1}). In adding, COF-10 (S_{BET}: 1760 m^2 g^{-1}, pore size: 3.2 nm) displayed the maximum hydrogen uptake, 39.2 mg g^{-1} at 1 bar and 77 K, of the 2D COF sequence. These sizes are important and validate the potential of COFs as hydrogen storing constituents.

Erudite multi-scale hypothetical protocols have been established to appraise the communication of hydrogen gas with the COF pores (Li et al., 2010). Primary, precise ab initio approaches, such as DFT, Couple Cluster (CC) and Møller–Plesset (MP) approaches, are applied to compute the binding energy of hydrogen to the building unit of the COF frames (Klontzas et al., 2010). Then, either DFT or mixed approaches connecting MM and QM (MM/QM) are applied to enhance the unit cell and to attain the material essential for fitting the ab initio founded interatomic potentials that define the contact between the COF pore and the hydrogen molecule. The consequences prophesied that the binding energies between the COF framework and the hydrogen molecules play nearly no part in the hydrogen uptake. In other ways, the hydrogen uptake is conquered by the free volume of the COFs and the surface area. Lastly, Grand Canonical Monte Carlo (GCMC) imitations that reason for thousands of atoms delivers a theoretic isotherm for hydrogen adsorption. Most of these theoretic predictions are in good contract with the measured hydrogen storing values (Garberoglio, 2007; Tylianakis et al., 2011).

Though COFs are valuable for hydrogen storing at low temperatures, their work at ambient temperature and pressure is still low and cannot encounter the DOE necessities. Adsorption imitations of the COFs display that doping with charged species can disturb their capacity. Lithium ions improve this storing capacity the most because of the amplified binding energy (Assfour & Seifert, 2010; Li & Yang, 2008). However no COF materials have really been arranged, these imitation studies forecast that pillared COFs, spillover COFs, and new 3D COF architectures are talented hydrogen storing applicants for the conference the DOE target (Furukawa et al., 2010; Kim et al., 2011).

3.4.1.2. Methane Gas

Methane gas is known as a potential vehicular fuel; though, the absence of an economic, effective, and safe on-board storing system is a main technical barricade that stops methane-driven vehicles from challenging with the conventional ones. The mark value for methane storing set by the DOE is 180 cm^3 (STP) at 35 bars. The uppermost COF storing capacity was detected for COF-102 (Figure 3.8), which had an amount of 187 mg /g. COF-103 also confirmed a high volume, which was 175 mg/g. These amounts are analogous to the maximum detected values for MOFs (MOF-210; 220 mg/g) (Mendoza-Cortés et al., 2010).

Hypothetical imitation researches have been made for both 3D and 2D COFs to assess their capacity for taking methane (Lan et al., 2009). Methane molecules interrelate powerfully with the faces of the aromatic or boroxine ring but softly with the edge units. COF-103 and COF-102 achieve well in both delivery amount (229 and 234 v(STP)/v at 100 bar) and total uptake (255 and 260 v(STP)/v at 100 bar) which proposes that they are auspicious applicants for methane storing. Astonishingly, doping the COFs with lithium ions folded their methane uptakes at comparatively low pressures (p 0–50 bar) (Keskin, 2012). The replicated sorption of CH_4–H_2 mixes by numerous COFs designates that the membrane selectivity of COF-6 and COF-5 are improved than the selectivity of zeolites and analogous to MOFs and zeolitic imidazolate frameworks (ZIFs) (Millward & Yaghi, 2005). This effect suggests new potential for COFs in chemical departures, particularly the mass transportation of adsorbed species.

3.4.1.3. Carbon Dioxide

A primary greenhouse gas that is accountable for global warming is carbon dioxide, the increasing acidity of the oceans and increasing sea levels. The likelihood of taking CO_2 from industrial release causes has involved extensive attention. Of the present methods for taking CO_2, adsorption by porous constituents is technically feasible and energetically efficient. Yaghi and co-workers have premeditated a family of COFs and stated that the CO_2 acceptance of COF-102 (Figure 3.8) spreads 27 mmol g^{-1} at 35 bar and 298 K, which is greater than the uptake of both zeolite (5–8 mmol/g) and MOF-5 (22 mmol/ g) (Cavenati et al., 2004; Babarao & Jiang, 2008).

In adding to these investigational researches, hypothetical studies on molecular imitations for CO_2 apprehension have been achieved. Contrary to methane and hydrogen uptake, the quantity of additional CO_2 uptake is straight connected to the entire pore volume. COF-103 and COF-102 display greater adsorption at low bars than COF-108 and COF-105 because of their dense atomic packing; though, larger pore volumes for COF-108 and COF-105 deliver greater saturation storing volumes at high pressures, (96 mmol/ g and 92 mmol/g for COF-108 and COF-105, respectively) that even exceed the investigational value stated for MOF-210 (54 mmol/ g). Among the numerous alkaline-earth, alkali and transition metals, the most effective at improving the carbon dioxide storage capacity are the lithium-ion doping [59, 60]. For instance, lithiumdoped COF-105 and COF-102 have a capacity of 344 and 409 mg/ g, respectively, which are four- and eight-fold higher, respectively, than the pristine COFs at 298 K and 1 bar. Precise calculations cause the suggestion of new COF designs for the adsorption of carbon dioxide (Lan et al., 2010; Choi et al., 2011).

3.4.1.4. Ammonia

For the adsorption of CH_4, H_2, a CO_2COFs have displayed a high capacity, though these gas molecules only feebly interrelate with the COFs. For adsorbing ammonia, COFs with boronate-ester linkages are valuable because the boron sites are Lewis acids and synchronize with Lewis base gases, such as ammonia (Doonan et al., 2010; Kalidindi et al., 2011). Among the sequence of boronate ester-linked COF-10, COFs (Figure 3.4) is one of the finest materials for ammonia adsorption. It has an uptake capacity of 15 mol kg^{-1} (298 K, 1 bar). Stimulatingly, the COF material can be pedaled numerous times without substantial worsening in performance. These consequences propose that molecularly calculated COFs comprising of functional groups on their walls that can interrelate with gas molecules are applied resources to advance new materials for the storing of a specific gas.

3.4.2. Heterogeneous Catalysis

In adding to the adsorption of gas molecules, COFs can load numerous molecules into their nanopores (Patwardhan et al., 2011; Feng et al., 2012). For instance, COF-102 (Figure 3.8) ensnares ferrocene molecules with load quantities as high as 42% of the entire free pore volume via the redirection of ferrocene into the COF networks under vacuum. Alternatively, imine-based COF-LZU1 can load Pd ions into its pores via the coordination reaction between Pd ions and nitrogen atoms in the COFs to afford Pd/COF-LZU1 (Figure 3.12). Fascinatingly, the Pd ions on the COF walls are catalytically vigorous and available to both the reactant and the substrate, creating a varied catalytic system. Pd/COF-LZU1 catalyzes the Suzuki–Miyaura link reaction and displays unresolved activity. The heterogeneous catalyst is actual for a comprehensive range of reactants, formed excellent products, and is vigorous for many series.

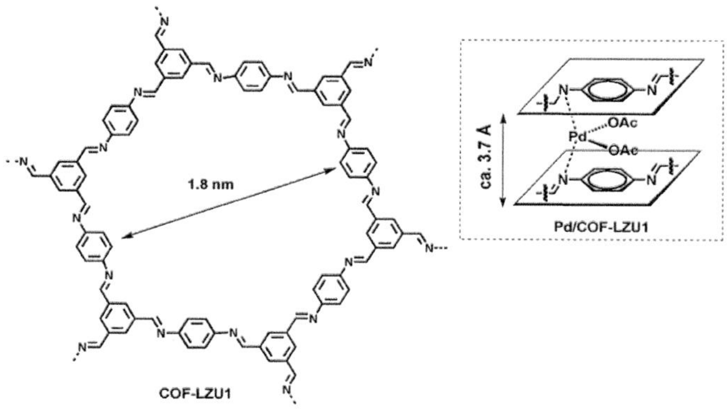

Figure 3.12: Schematic representation of imine-linked COF and its coordination with a Pd ion.

[Source: https://pubs.rsc.org/en/content/articlelanding/2012/cs/c2cs35157a/]

3.4.3. Semiconduction and Photoconduction

While the sturdy covalent bonds in 2D COFs form their backbone into tetragonal or hexagonal films, the out of plane p relations are the main driving force in the development of a layered structure. Such a structure encourages a large electronic link between the p-orbitals of the concealed assembling layers, which simplifies the transportation of charge carriers and photoexcited states through built-in and preorganized pathways (Figure

3.13). Particularly, this assembling structure is characteristic to 2D COFs, and the building blocks occasionally line up within the framework. In this logic, 2D COFs deliver a novel stage for the molecular strategy of photo conducting and semiconducting materials.

3.4.3.1. Semiconduction

The usage of COFs for the building of optoelectronic and electronic materials was established in 2008. We have planned a p-electronic 2D TP-COF containing linking hexagons made from pyrene-2,7-diboronic acid (PDBA) groups at the edges and HTTP molecules at the vertices (Figure 3.9).

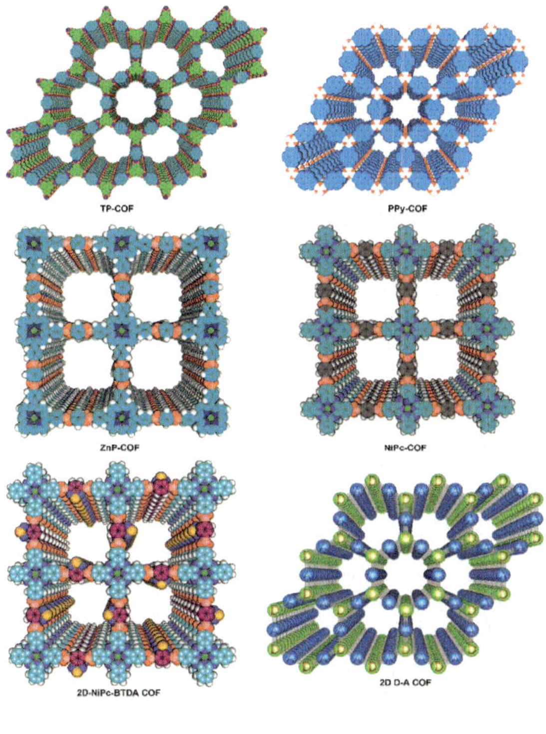

Figure 3.13: Schematic representation of the AA stacked structure of COFs with preorganized and built-in p columns.

[Source: https://pubs.rsc.org/en/content/articlelanding/2012/cs/c2cs35157a/]

TP-COF undertakes a belt-shaped structure with a thickness of 100 nm, a width of 300 nm, and length measured in microns. Structural measurements on the basis of the XRD pattern and imitations display that the planar films were layered in an obscured manner. Upon excitation of the pyrene units, TP-COF produced a robust blue luminescence. Fascinatingly, the excitation of triphenylene moieties also causes a blue release from just the pyrene units. This finding designates that the excitation energy of triphenylene was not restricted but could transmit to the pyrene units over the framework. As a consequence, TP-COF can yield an extensive variety of photons from the ultraviolet to the visible areas and change them to a vivid blue luminescence. Since of the covered assembling structure of the p-electronic apparatuses, TP-COF is semiconducting, which allows hole transportation and is proficient of on-off switching of the electric current as established by the I–V curve dimensions.

Substantial development towards the formation of p-electronic COFs has been accomplished by mixing metallophthalocyanine units into the skeletons since phthalocyanine results have plentiful p-electronic, redox, and photochemical features. The condensation of $[(OH)_8PcNi]$ with BDBA produces a 2D NiPc-COF with the phthalocyanine units inhabiting the apexes of the tetragonal framework (Figure 3.11). NiPc-COF assumes a flake-like morphology and contains covered 2D films. As an outcome, NiPc-COF is hole conducting and displays high carrier flexibility of 1.3 cm^2/ V.s. This carrier carriage feature is extremely subtle to the related units of the COFs. Contrary to the p-type features of NiPcCOF, a nickel phthalocyanine COF by means of the electron-lacking benzothiadiazole as the connection 2D-NiPc-BTDA-COF (Figure 3.13), displayed totally dissimilar charge carrier-carriage performances. The 2D-NiPc-BTDA-COF is an n-type semiconductor with electron movements as high as 0.6 $cm^2 V^1 s^1$. Variable the connection molecules cause a radical alteration in the electron deliveries of the metallophthalocyanine units, which accounts aimed at the reversal of the carrier-conveying mode. The obtainability of both n- and p-type conductors would make these materials appropriate for electronic applications.

Porphyrins are additional characteristic macrocycle analogous to phthalocyanines with protracted p conjugation and varied functionalities. The 2D polyporphyrin films are assembled in metal-over-metal and macrocycle-over macrocycle modes with separate partings to form pre-ordered porphyrin columns for charge carrier conveyance, in 2D porphyrin COFs (MP-COFs) (Figure 3.13). The kind of charge carriers that can change in 2D MP-COFs be contingent not only on the framework but also on the central metals in the macrocycles. Certainly, the freebase porphyrin H_2P-COF is hole conducting, though the copper (II) porphyrin CuP-COF is the differing and displays electron conveyance. Contrary to this, the zinc (II) porphyrin

ZnP-COF is ambipolar. Porphyrin COFs deliver two conducting networks in the porphyrin columns, metal-over-metal and macrocycle-over-macrocycle channels, which represent electron and hole transport, respectively. The focal metals act as a significant part both in founding channels for electron transmission and in controlling the carrier movement, which causes the development of quick hole, ambipolar and electron-conducting COFs.

Other two porphyrin COFs, i.e., COF-366 with an imine linkage and COF-66 with a boronate ester linkage (Figure 3.4), have been stated to be hole conducting with movements as high as 3.9 and 8.1 $cm^2/V·s$, respectively. Though the interlayer spacing in COF-366 (5.64 A°) is much greater than the spacing in COF-66 (3.81 A°), the carrier movement of COF-366 is double that of COF-66. These consequences propose that the charge transporters can change not only through the covered assembling layers but also over the intralayer planes in the p-conjugated interlinked COF-366.

All of the above investigational carrier movement values were gained by sources of the laser flash-photolysis time-determined microwave conductivity (FP-TRMC) method, which permits dimension of the charge carrier movement within numerous nanometers after laser pulse treatment under a quickly vacillating electric field. The FP-TRMC method specifies the characteristic carrier movement of materials dependable with those in the crystals because it can diminish the inspiration of influences such as filths, physical and chemical flaws, and electrode/organic edges, associated with the field-effect transistor (FET) and time-of-flight (TOF) approaches. Characteristically, the value of movement for COFs measured by the traditional TOF technique is more than two orders of magnitude lesser than that by TRMC determination, due to that the domain/grain restrictions interrupt the long-range (Bmm) translational movement of charge carriers in these materials. Therefore, the value of TRMC movement is only accountable in the devices formed within the nm scale.

Hypothetical imitations of the COF carrier movement specify that a band-like movement of delocalized charges instead of localized charge hopping accounts for the high charge carrier movement in 2D COFs. As demonstrated by triphenylene COF-5 (Figure 3.4), the comparative constancies of both the delocalized the localized changes in the p-stacked triphenylenes with variable intermolecular distance and warped angles and were evaluated. The consequences designate that the charges are local for the case of large twisted angles. Contrary to this, important charge delocalization happens in the covered assembling (twisted angle = 01) triphenylene units of the

COFs. The charge delocalization in triphenylene COFs was also delicate to the intermolecular distance. The bandlike movement of the delocalized charges designates movements on the order of 10 $cm^2 V^{-1} s^{-1}$ for the COFs. The movement of real COF materials was comparatively low because of boundary disorders and structural flaws.

3.4.3.2. Photoconduction

The examination of photo functional materials is significant for emerging light energy conversion, artificial photosynthesis, and optoelectronics. The molecular ordering of the p-electronic components acts as a significant role in measuring the working of these devices. Photoconductivity can be attained using high-quality single crystals of some p-conjugated arenas because of exciton movement over the lattice, as recognized by crystal engineering, followed by charge parting at the molecule–electrode line. The covered assembling structure and periodic arrangement of the p-columns donate COFs with a high likelihood of activating photoconductivity.

The primary illustration of a photoconductive COF is the pyrene based PPy-COF, which is arranged via the self-condensation of PDBA (Figure 3.13). PPy-COF undertakes a cubic form with its width, length, and thickness in the micrometer variety. These cubes are extremely blue luminescent because of the development of excimer in the assembled pyrenes. Fluorescence anisotropy dimensions disclose that the PPy-COF fluorescence is extremely depolarized, with an anisotropy of only 0.001, which is much lesser than the anisotropy of TP-COF (0.017). This consequence recommends that dissimilar TP-COF, which comprises of two consecutively connected apparatuses with dissimilar energy gaps, the excitons in the single constituent PPy-COF can movement not only over the film but also across the assembled layers. PPy-COF rapidly replies to treatment with visible light to generate a prominent photocurrent with an on-off ratio of $8{:}10^4$, which was only $2{:}10^4$ for TP-COF.

Contrary to the inadequate absorption of PPy-COF in the fine visible region, phthalocyanine and porphyrin COFs display broad absorbances straddling from the ultraviolet to the near-infrared area. These absorption features are significant for emerging organic photovoltaic and optoelectronics cells. Upon treatment with a xenon light basis (4400 nm), NiPc-COF (Figure 3.13) displays a substantial rise in current to 3 mA (photocurrent) from 20 nA (dark current) with an on-off ratio of 100. As of its wide absorbance, NiPc-COF is panchromatic and replies to photons with numerous wavelengths. Stimulatingly, NiPc-COF is enormously subtle and produces protuberant photocurrents upon radiation with a long wavelength near-infrared photons and visible light. Its amplified light-harvesting ability and improved carrier movement donate to the high sensitivity and photoconductivity. These consequences are the same for 2D-NiPc-BTDA-COF (Figure 3.13), which is panchromatically receptive to numerous photons [30]. Consequently, 2D-NiPc-BTDA-COF signifies a well-organized and photoconductive and thermally stable n-channel semiconductor.

Panchromatically, metalloporphyrin COFs are photoconductive; though, their photoconductivity is extremely reliant on the metal species of the porphyrin macrocycles. ZnP-COF displays an on-off proportion of $5:10^4$, which is 10000 and 150-fold greater than the on-off ratios of CuP-COF and H_2P-COF, respectively (Figure 3.11). An ambipolar and balanced charge carrier carriage plays a significant role in producing large photocurrents.

The crystallization of electron acceptors (A) and donors (D) into macroscopic heterojunctions with separated D and A field structures has been important to emerge organic electronics. Though, A and D molecules incline to pile on one another somewhat than separate into bi-continuous arrays, which provides disordered gatherings, where the charge transporters are stuck and willingly defeated via rapid blend. Lately, we have stated a COF with an integral D and A separated structure (Figures 3.9 and 3.13). This 2D D–A COF permits atomically exact D–A separation that is grading controlled at multiple levels from the atomic connection to the macroscopic crystalline structure. At the atomic level, the D and A molecules are covalently related rendering to their geometric incarceration. At the molecular level, the 2D polygon films are topologically linked polygons with all atoms in the same level. At the macroscopic level, the AA assembling mode forms the covalent 2D films into a background that brands up to the final D–A heterojunction structure with a comprehensive border and delivers pre-ordered, bi-continuous and periodic A-on-A and D-on-D columnar arrays. Such a vertically allied D–A heterojunction structure is ambipolar, displays a high-rate movement, and permits for a large on-off relation photoconductive retort.

These consequences validate that COFs gives an influential source of crystallizing periodic D–A arrays and separating D and A molecules. This approach can be applied to crystallize numerous D and A redox couples for the strategy of separated D–A systems functionally instead of just architecturally.

REFERENCES

1. Assfour, B., & Seifert, G. (2010). Adsorption of hydrogen in covalent organic frameworks: comparison of simulations and experiments. *Microporous and Mesoporous Materials*, *133*(1–3), 59–65.
2. Assfour, B., & Seifert, G. (2010). Hydrogen adsorption sites and energies in 2D and 3D covalent organic frameworks. *Chemical Physics Letters*, *489*(1–3), 86–91.
3. Babarao, R., & Jiang, J. (2008). Exceptionally high CO_2 storage in covalent-organic frameworks: Atomistic simulation study. *Energy & Environmental Science*, *1*(1), 139–143.
4. Bojdys, M. J., Jeromenok, J., Thomas, A., & Antonietti, M. (2010). Rational extension of the family of layered, covalent, triazine-based frameworks with regular porosity. *Advanced Materials, 22*(19), 2202–2205.
5. Bunck, D. N., & Dichtel, W. R. (2012). Internal Functionalization of Three-Dimensional Covalent Organic Frameworks. *Angewandte Chemie International Edition*, *51*(8), 1885–1889.
6. Campbell, N. L., Clowes, R., Ritchie, L. K., & Cooper, A. I. (2009). Rapid microwave synthesis and purification of porous covalent organic frameworks. *Chemistry of Materials*, *21*(2), 204–206.
7. Cavenati, S., Grande, C. A., & Rodrigues, A. E. (2004). Adsorption equilibrium of methane, carbon dioxide, and nitrogen on zeolite 13X at high pressures. *Journal of Chemical & Engineering Data*, *49*(4), 1095–1101.
8. Choi, Y. J., Choi, J. H., Choi, K. M., & Kang, J. K. (2011). Covalent organic frameworks for extremely high reversible CO 2 uptake capacity: a theoretical approach. *Journal of Materials Chemistry*, *21*(4), 1073–1078.
9. Colson, J. W., Woll, A. R., Mukherjee, A., Levendorf, M. P., Spitler, E. L., Shields, V. B., ... & Dichtel, W. R. (2011). Oriented 2D covalent organic framework thin films on single-layer graphene. *Science*, *332*(6026), 228–231.
10. Cote, A. P., Benin, A. I., Ockwig, N. W., O'Keeffe, M., Matzger, A. J., & Yaghi, O. M. (2005). Porous, crystalline, covalent organic frameworks. *Science*, *310*(5751), 1166–1170.
11. Cote, A. P., El-Kaderi, H. M., Furukawa, H., Hunt, J. R., & Yaghi,

O. M. (2007). Reticular synthesis of microporous and mesoporous 2D covalent organic frameworks. *Journal of the American Chemical Society, 129*(43), 12914–12915.

12. Ding, S. Y., Gao, J., Wang, Q., Zhang, Y., Song, W. G., Su, C. Y., & Wang, W. (2011a). Construction of covalent organic framework for catalysis: Pd/COF-LZU1 in Suzuki–Miyaura coupling reaction. *Journal of the American Chemical Society, 133*(49), 19816–19822.
13. Ding, X., Chen, L., Honsho, Y., Feng, X., Saengsawang, O., Guo, J., & Parasuk, V. (2011b). An n-channel two-dimensional covalent organic framework. *Journal of the American Chemical Society, 133*(37), 14510–14513.
14. Ding, X., Guo, J., Feng, X., Honsho, Y., Guo, J., Seki, S., & Jiang, D. (2011). Synthesis of metallophthalocyanine covalent organic frameworks that exhibit high carrier mobility and photoconductivity. *Angewandte Chemie International Edition, 50*(6), 1289–1293.
15. Dogru, M., Sonnauer, A., Gavryushin, A., Knochel, P., & Bein, T. (2011). A covalent organic framework with 4 nm open pores. *Chemical Communications, 47*(6), 1707–1709.
16. Doonan, C. J., Tranchemontagne, D. J., Glover, T. G., Hunt, J. R., & Yaghi, O. M. (2010). Exceptional ammonia uptake by a covalent organic framework. *Nature Chemistry, 2*(3), 235.
17. Feng, X., Chen, L., Dong, Y., & Jiang, D. (2011). Porphyrin-based two-dimensional covalent organic frameworks: synchronized synthetic control of macroscopic structures and pore parameters. *Chemical Communications, 47*(7), 1979–1981.
18. Feng, X., Chen, L., Honsho, Y., Saengsawang, O., Liu, L., Wang, L., ... & Jiang, D. (2012). An ambipolar conducting covalent organic framework with self-sorted and periodic electron donor-acceptor ordering. *Advanced Materials, 24*(22), 3026–3031.
19. Feng, X., Liu, L., Honsho, Y., Saeki, A., Seki, S., Irle, S., & Jiang, D. (2012). High-Rate Charge-Carrier Transport in Porphyrin Covalent Organic Frameworks: Switching from Hole to Electron to Ambipolar Conduction. *Angewandte Chemie International Edition, 51*(11), 2618–2622.
20. Furukawa, H., & Yaghi, O. M. (2009). Storage of hydrogen, methane, and carbon dioxide in highly porous covalent organic frameworks for clean energy applications. *Journal of the American Chemical Society,*

131(25), 8875–8883.

21. Furukawa, H., Ko, N., Go, Y. B., Aratani, N., Choi, S. B., Choi, E., ... & Yaghi, O. M. (2010). Ultrahigh porosity in metal-organic frameworks. *Science*, *329*(5990), 424–428.
22. Garberoglio, G. (2007). Computer simulation of the adsorption of light gases in covalent organic frameworks. *Langmuir*, *23*(24), 12154–12158.
23. Guan, C. Z., Wang, D., & Wan, L. J. (2012). Construction and repair of highly ordered 2D covalent networks by chemical equilibrium regulation. *Chemical Communications*, *48*(24), 2943–2945.
24. Han, S. S., Furukawa, H., Yaghi, O. M., & Goddard Iii, W. A. (2008). Covalent organic frameworks as exceptional hydrogen storage materials. *Journal of the American Chemical Society*, *130*(35), 11580–11581.
25. Hunt, J. R., Doonan, C. J., LeVangie, J. D., Côté, A. P., & Yaghi, O. M. (2008). Reticular synthesis of covalent organic borosilicate frameworks. *Journal of the American Chemical Society*, *130*(36), 11872–11873.
26. Kalidindi, S. B., Yusenko, K., & Fischer, R. A. (2011). Metallocenes@COF-102: organometallic host–guest chemistry of porous crystalline organic frameworks. *Chemical Communications, 47*(30), 8506–8508.
27. Keskin, S. (2012). Adsorption, diffusion, and separation of CH4/H2 mixtures in covalent organic frameworks: molecular simulations and theoretical predictions. *The Journal of Physical Chemistry C, 116*(2), 1772–1779.
28. Kim, D., Jung, D. H., Kim, K. H., Guk, H., Han, S. S., Choi, K., & Choi, S. H. (2011). Pillared covalent organic frameworks with balanced volumetric and gravimetric hydrogen uptake. *The Journal of Physical Chemistry C, 116*(1), 1479–1484.
29. Klontzas, E., Tylianakis, E., & Froudakis, G. E. (2008). Hydrogen storage in 3D covalent organic frameworks. A multiscale theoretical investigation. *The Journal of Physical Chemistry C, 112*(24), 9095–9098.
30. Klontzas, E., Tylianakis, E., & Froudakis, G. E. (2009). Hydrogen storage in lithium-functionalized 3-D covalent-organic framework materials. *The Journal of Physical Chemistry C, 113*(50), 21253–21257.

31. Klontzas, E., Tylianakis, E., & Froudakis, G. E. (2010). Designing 3D COFs with enhanced hydrogen storage capacity. *Nano letters, 10*(2), 452–454.
32. Kuhn, P., Antonietti, M., & Thomas, A. (2008). Porous, covalent triazine-based frameworks prepared by ionothermal synthesis. *Angewandte Chemie International Edition, 47*(18), 3450–3453.
33. Lan, J., Cao, D., & Wang, W. (2009). High uptakes of methane in Li-doped 3D covalent organic frameworks. *Langmuir, 26*(1), 220–226.
34. Lan, J., Cao, D., Wang, W., & Smit, B. (2010). Doping of alkali, alkaline-earth, and transition metals in covalent-organic frameworks for enhancing CO_2 capture by first-principles calculations and molecular simulations. *Acs Nano, 4*(7), 4225–4237.
35. Lanni, L. M., Tilford, R. W., Bharathy, M., & Lavigne, J. J. (2011). Enhanced hydrolytic stability of self-assembling alkylated two-dimensional covalent organic frameworks. *Journal of the American Chemical Society, 133*(35), 13975–13983.
36. Li, F., Zhao, J., Johansson, B., & Sun, L. (2010). Improving hydrogen storage properties of covalent organic frameworks by substitutional doping. *International Journal of Hydrogen Energy, 35*(1), 266–271.
37. Li, Y., & Yang, R. T. (2008). Hydrogen storage in metal-organic and covalent-organic frameworks by spillover. *AIChE Journal, 54*(1), 269–279.
38. Long, J. R., & Yaghi, O. M. (2009). The pervasive chemistry of metal–organic frameworks. *Chemical Society Reviews, 38*(5), 1213–1214.
39. Lukose, B., Kuc, A., & Heine, T. (2011). The structure of layered covalent-organic frameworks. *Chemistry–A European Journal, 17*(8), 2388–2392.
40. Mendoza-Cortés, J. L., Han, S. S., Furukawa, H., Yaghi, O. M., & Goddard III, W. A. (2010). Adsorption mechanism and uptake of methane in covalent organic frameworks: theory and experiment. *The Journal of Physical Chemistry A, 114*(40), 10824–10833.
41. Mendoza-Cortes, J. L., Pascal, T. A., & Goddard III, W. A. (2011). Design of covalent organic frameworks for methane storage. *The Journal of Physical Chemistry A, 115*(47), 13852–13857.
42. Millward, A. R., & Yaghi, O. M. (2005). Metal−organic frameworks with exceptionally high capacity for storage of carbon dioxide at room temperature. *Journal of the American Chemical Society, 127*(51),

17998–17999.

43. Nagai, A., Guo, Z., Feng, X., Jin, S., Chen, X., Ding, X., & Jiang, D. (2011). Pore surface engineering in covalent organic frameworks. *Nature communications*, *2*, 536.
44. O'Keeffe, M. (2009). Design of MOFs and intellectual content in reticular chemistry: a personal view. *Chemical Society Reviews*, *38*(5), 1215–1217.
45. Patwardhan, S., Kocherzhenko, A. A., Grozema, F. C., & Siebbeles, L. D. (2011). Delocalization and mobility of charge carriers in covalent organic frameworks. *The Journal of Physical Chemistry C*, *115*(23), 11768–11772.
46. Perry Iv, J. J., Perman, J. A., & Zaworotko, M. J. (2009). Design and synthesis of metal–organic frameworks using metal–organic polyhedra as supermolecular building blocks. *Chemical Society Reviews*, *38*(5), 1400–1417.
47. Rowan, S. J., Cantrill, S. J., Cousins, G. R., Sanders, J. K., & Stoddart, J. F. (2002). Dynamic covalent chemistry. *Angewandte Chemie International Edition*, *41*(6), 898–952.
48. Spitler, E. L., & Dichtel, W. R. (2010). Lewis acid-catalysed formation of two-dimensional phthalocyanine covalent organic frameworks. *Nature Chemistry*, *2*(8), 672.
49. Spitler, E. L., Colson, J. W., Uribe-Romo, F. J., Woll, A. R., Giovino, M. R., Saldivar, A., & Dichtel, W. R. (2012). Lattice expansion of highly oriented 2D phthalocyanine covalent organic framework films. *Angewandte Chemie International Edition*, *51*(11), 2623–2627.
50. Spitler, E. L., Giovino, M. R., White, S. L., & Dichtel, W. R. (2011). A mechanistic study of Lewis acid-catalyzed covalent organic framework formation. *Chemical Science*, *2*(8), 1588–1593.
51. Srepusharawoot, P., Scheicher, R. H., Moysés Araújo, C., Blomqvist, A., Pinsook, U., & Ahuja, R. (2009). Ab initio study of molecular hydrogen adsorption in covalent organic framework-1. *The Journal of Physical Chemistry C*, *113*(19), 8498–8504.
52. Tilford, R. W., Gemmill, W. R., zur Loye, H. C., & Lavigne, J. J. (2006). Facile synthesis of a highly crystalline, covalently linked porous boronate network. *Chemistry of Materials*, *18*(22), 5296–5301.
53. Tilford, R. W., Mugavero III, S. J., Pellechia, P. J., & Lavigne, J. J. (2008). Tailoring microporosity in covalent organic frameworks.

Advanced Materials, 20(14), 2741–2746.

54. Tranchemontagne, D. J. (2009). JL Mendoza-Corte s, M. O'Keeffe and OM Yaghi. *Chem. Soc. Rev*, *38*, 1257.
55. Tylianakis, E., Klontzas, E., & Froudakis, G. E. (2011). Multi-scale theoretical investigation of hydrogen storage in covalent organic frameworks. *Nanoscale*, *3*(3), 856–869.
56. Uribe-Romo, F. J., Doonan, C. J., Furukawa, H., Oisaki, K., & Yaghi, O. M. (2011). Crystalline covalent organic frameworks with hydrazone linkages. *Journal of the American Chemical Society, 133*(30), 11478–11481.
57. Uribe-Romo, F. J., Hunt, J. R., Furukawa, H., Klock, C., O'Keeffe, M., & Yaghi, O. M. (2009). A crystalline imine-linked 3-D porous covalent organic framework. *Journal of the American Chemical Society*, *131*(13), 4570–4571.
58. Wan, S., Gándara, F., Asano, A., Furukawa, H., Saeki, A., Dey, S. K.,& Seki, S. (2011). Covalent organic frameworks with high charge carrier mobility. *Chemistry of Materials*, *23*(18), 4094–4097.
59. Wan, S., Guo, J., Kim, J., Ihee, H., & Jiang, D. (2008). A belt-shaped, blue luminescent, and semiconducting covalent organic framework. *Angewandte Chemie International Edition, 47*(46), 8826–8830.
60. Wan, S., Guo, J., Kim, J., Ihee, H., & Jiang, D. (2009). A photoconductive covalent organic framework: self-condensed arene cubes composed of eclipsed 2D polypyrene sheets for photocurrent generation. *Angewandte Chemie International Edition, 48*(30), 5439–5442.
61. Yu, J. T., Chen, Z., Sun, J., Huang, Z. T., & Zheng, Q. Y. (2012). Cyclotricatechylene based porous crystalline material: Synthesis and applications in gas storage. *Journal of Materials Chemistry*, *22*(12), 5369–5373.
62. Zwaneveld, N. A., Pawlak, R., Abel, M., Catalin, D., Gigmes, D., Bertin, D., & Porte, L. (2008). Organized formation of 2D extended covalent organic frameworks at surfaces. *Journal of the American Chemical Society*, *130*(21), 6678–6679.

4

Recent Advances in the Synthesis of Carboxylic Acid Esters

CONTENTS

4.1. INTRODUCTION

In this chapter, new developments in the formation of carboxylic acid esters are summed up on the basis of consumption of carboxylic acids as electrophiles or nucleophiles in reactions. The carboxylic acids are linked up with the alcohols by condensation reagents and catalysts to afford the equivalent esters, together with initiation of 1 Equiv. of H_2O, in which the carboxylic acids can be considered as the electrophile. On the other hand, esters are produced by the reaction of the carboxylic acid with alkyl halides, carbocations, or their equivalents, which can be considered the carboxylate ions which derived from the carboxylic acids as the nucleophile. This chapter mostly familiarizes the new advancements in this field of the synthesis of esters, on the basis the arrangement of the role of carboxylic acids in reactions (Otera & Nishikido, 2009; Matsumoto et al., 2018).

In organic chemistry, drug molecules, the organic material compounds, and natural products often have ester unit as the functional group, so the production of the efficient synthesis of carboxylic acid esters by the use of carboxylic acids is still one of the central research topics (Mitsunobu & Yamada, 1967; Li & Corey, 2007). As for the viewpoint, the corresponding carboxylic acids are typically applied as the crucial initiating material and play a significant role in the synthesis of esters (Neises & Steglich, 1978).

So far, numerous types of approaches for the synthesis of esters from carboxylic acids are well known and applied, but a lot of researchers still have investigated the new approaches or aspects, from the point of view of green chemistry and industry the synthesis of esters also plays an important role. Novel developments in the formation of carboxylic acid esters are explained. The reactions are divided into two divisions, i.e., the reactions in the carboxylic acids are used as (1) nucleophiles and (2) electrophiles, in which many chemical reagents and catalysts as well as stimulating reaction media and methods are used for conducting the reactions. While many papers have come into being in this filed, we herein have presented key and particular examples (Mitsunobu & Yamada, 1967).

4.2. SYNTHESIS OF CARBOXYLIC ACID ESTERS USING CARBOXYLIC ACIDS AS ELECTROPHILES

The reaction of carboxylic acids with an additional amount of alcohols in the existence of a catalytic amount of H_2SO_4 by using Dean-Stark apparatus, in which H_2SO_4 catalyzes the adding of the alcohol to the carboxylic acid, and the H_2O thus produced is eliminated by Dean-Stark apparatus, is the characteristic and outdated method for the production of carboxylic acid esters (Figure 4.1 (a)). This reaction is known as Fischer esterification. Though, there are some downsides. The alcohols are used in surplus amount. The Dean-Stark apparatus is generally essential. Furthermore, the substrates bearing the functional group which reacts with the acid cannot remain exploited in this reaction. The use of DCC in the existence of a catalytic amount of DMAP is an unusual and steadfast method to be developed (Figure 4.1 (b)) DCC can assist as valuable condensation reagents. The amount of alcohols reduces the use of DCC as condensation reagent. Moreover, Mitsunobu reaction is also consistent technique (Zhao et al., 1997; But & Toy, 2006; Hirose et al., 2016).

The condensation reagents other than the use of DCC are also established. 2-Halo-pyridinium salts called Mukaiyama condensation reagent act as operative reagents. For the growth of beneficial condensation reagents, Mukaiyama et al. (1975) have widely underwritten this research area (Funasaka & Mukaiyama, 2008). Moreover, CDI (carbonyldiimidazole), BOP ((benzotriazole-1-yloxy)-tris(dimethylamino)phosphonium hexafluorophosphate)), methylmorpholinium chloride), and so on are well developed and used in the ester synthesis reactions (Paul & Anderson, 1960; Castro et al., 1975; Kim & Patel, 1994).

As mention above, many condensation reagents have been established so far. However, there are still rumors for this filed. Table 4.1 displays selected and recent reports of condensation reactions between alcohols and carboxylic acids using a stoichiometric amount of condensation reagents (Kunishima et al., 1999a; 1999b; 1999c; Engstrom et al., 2014).

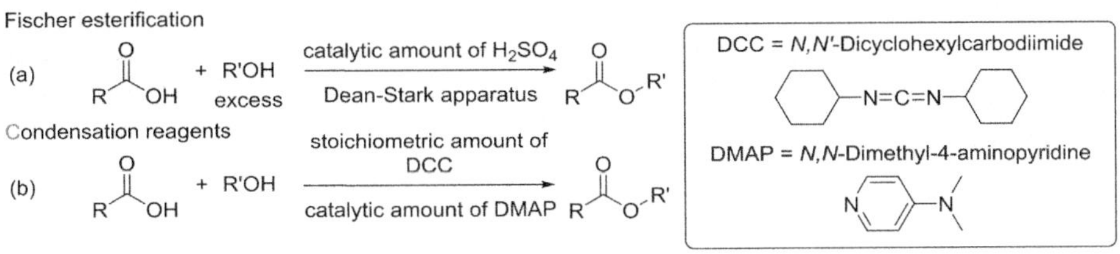

Figure 4.1: Typical and traditional procedures for the synthesis of carboxylic acid esters and their chemical structures of reagents.

[Source: https://cdn.intechopen.com/pdfs/60244.pdf]

Basically, the amount of alcohols is not excess (Table 4.1). For examples, polymerbounded Ph_3P (or Ph_3P)/I_2/microwave and $Ph_3P/I_2/Zn(OTf)_2$ condition have been reported, because the combination of PPh_3 and I_2 in the presence of the base has been well known so far (Choi et al., 2003; El-Faham & Albericio, 2011; Pathak et al., 2016). Moreover, the use of Ph_2PCl/I_2/imidazole, $POCl_3$/DMAP/Et_3N, Ph_3PBr_2, or PPh_3/BnN_3/microwave (Staudinger's phosphazene) systems was originated to be useful for the esterification (Chen et al., 2013; Dinesh et al., 2016). In coupling reactions, the hypervalent iodine reagents could be used. Trichloroisocyanuric acid and PPh_3 was also joint effectually (Taniguchi et al., 2011; März et al., 2017; Matsumoto et al., 2018). These reactions are built on the initiation of PPh_3 (Salomé & Kohn, 2009; Nowrouzi et al., 2010; Morcillo et al., 2011). The iron is used to achieve the catalytic activation of PPh_3. In the existence of PPh_3 with a catalytic quantity of flavin and azo compound under O_2, the Photo-irradiated process was established. XtalFluor-E and tropylium-based coupling reagents were beneficial for the esterification (Mamidi & Manna, 2013; Jaita et al., 2015; Pathak & Rokhum, 2015; Rodrigues et al., 2005; Tian et al., 2012; Boulogeorgou et al., 2015).

Table 4.1: Various Condensation Reagents and Conditions Used for the Synthesis of Carboxylic Acid Esters Under the Nearly Equimolar Carboxylic Acids and Alcohols (Recent and Selected Examples)

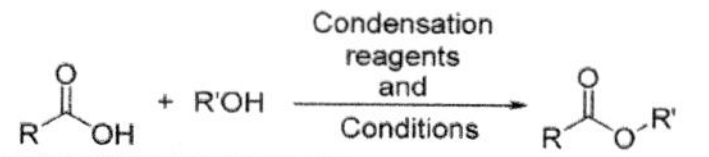

Condensation reagents and Conditions		
Ref. 20 Polymer-bounded Ph_3P (or Ph_3P) (1.2 eq), and I_2 (1.2 eq); CH_3CN, microwave	Ref. 30 Ph_3P (1 eq), and Ph⌒N_3 (1 eq); microwave	Ref. 34 PPh_3 (2 eq), cat. Fe(Pc) (5 mol%) and cat. MPO (10 mol%) under air*; CH_3CN, reflux
Ref. 24 Ph_3P (2 eq), I_2 (2 eq) and $Zn(OTf)_2$ (0.05 eq); CH_3CN, 60 °C	Ref. 31 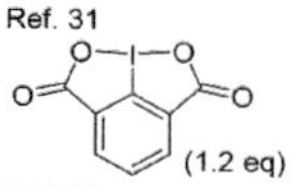 (1.2 eq) PPh_3 (1 eq), and DMAP (1.2 eq); $CHCl_3$, reflux	Ref. 35 PPh_3 (2 eq), flavin (10 mol%), azo comound (10 mol%) and MS4A under O_2 (1 atm) and hv; CH_3CN
Ref. 27 Ph_2PCl (0.5 eq), I_2 (0.5 eq) and imidazole (1.5 eq); CH_3CN, reflux	Ref. 32 PhIO (1.5 eq), PPh_3 (1.5 eq), and DMAP (0.1 eq); CH_2Cl_2, reflux PhIO reacts with carboxylic acids to form hypervalent iodines.	Ref. 36 Ref. 37 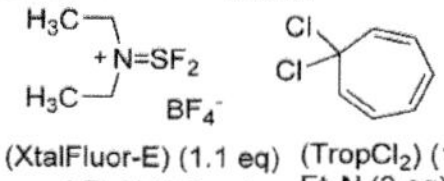 (XtalFluor-E) (1.1 eq) and Et_3N (1.5 eq); CH_2Cl_2 (TropCl$_2$) (1.05 eq) Et_3N (3 eq); CH_2Cl_2
Ref. 28 Ph_3PBr_2 (5 eq) and K_2CO_3 (11 eq) (The excess amount of alcohols were used.)	Ref. 33 PPh_3 (1 eq) and 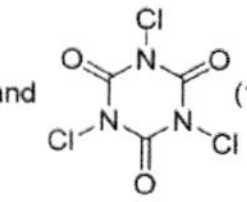 (1 eq); CH_2Cl_2	* Pc = phthalocyanine, MPO = 4-methoxypyridine
Ref. 29 $POCl_3$ (1 eq), DMAP (0.3 eq), and Et_3N (2.4 eq); CH_2Cl_2		

By the point of view of green chemistry, the application of a catalytic quantity of reagents is one of the eye-catching methods, which shows an almost equal ratio between alcohols and carboxylic acids. In 2000, it was reported by Ishihara et al. (2002) that at reflux condition the 0.1 to 1.0

mol% hafnium (IV) salts in toluene catalyzed the condensation reaction of an equimolar quantity of carboxylic acids and alcohols (Table 4.2, entry 1) (Ishihara et al., 2000; 2001; 2002; Nguyen & Lyons, 2015). Since then, for the effective esterification reactions, many kinds of catalysts have been observed. Designated cases are briefed in Table 4.2. $HClO_4$-SiO_2 (entry 4) (Nakayama et al., 2004; Mantri et al., 2005a; 2005b). Diphenylammonium triflate (entry 2), fluoroalkyldistannoxane (entry 3), Ti_4^+—mont (mont = montmorillonite) (entry 5), bulky diarylammonium arenesulfonates (entry 6), $Zn(ClO_4)_{2-}6H_2O$ (entry 7), pentafluorophenylammonium triflate (entry 8), TsOH or CSA (entry 9), *N,N*-diarylammonium pyrosulfates in H_2O (entry 11), phosohorofluoridic acid (entry 10), TfOH in Solkane365mfc (entry 12), zirconocene complex (entry 14), 2-oleamide-5-nitro-pyridinium *p*-toluenesulfonate (entry 13), and L-leucine as an organocatalyst (entry 15) have been described for the operative catalyst for the esterification by the use of equal or nearly equal quantity of alcohols or carboxylic acids (Chen & Munot, 2005; Sato et al., 2005; Nakamura et al., 2006).

Table 4.2: Representative Progresses for the Synthesis of Esters Using Equal or Nearly Equal Amount of Carboxylic Acids and Alcohols in the Presence of the Catalyst, Since 2000 (Selected Examples)

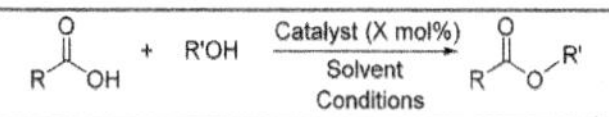

Entry	Authors	Published year	Catalyst (X mol%)	Solvent	Conditions	Ref.
1	H. Yamamoto et al.	2000	hafnium(IV) salts (0.1mol%-1.0 mol%)	toluene	reflux, 5-36 h	38
2	Y. Tanabe et al.	2000	diphenylammonium triflate (DPAT) (1 mol% - 10 mol%)	toluene	80 °C, 4-48 h	48
3	J. Otera et al.	2002	fluoroalkyldistannoxane (5 mol%)	FC-72	150 °C, 10-16 h	49
4	A. K. Chakraborti et al.	2003	$HClO_4$-SiO_2 (1 mol%)	solvent-free	80 °C, 3.5-20 h	50
5	K. Kaneda et al.	2003	Ti_4^+-mont (2 mol%)	solvent-free	110-150 °C, 3-6 h	53
6	K. Ishihara et al.	2005	bulky diarylammonium arenesulfonates (1 mol%-10 mol%)	heptane or solvent-free	r.t.-115 °C, 1-72 h	54
7	G. Bartoli et al.	2005	$Zn(ClO_4)_2$-$6H_2O$ (5 mol%)	solvent-free	80-100 °C, 6-48 h	59
8	Y. Tanabe et al.	2006	pentafluorophenylammonium triflate (PFPAT) (1 mol%)	toluene	80 °C, 2-48 h	60
9	K. Ishihara et al.	2008	*p*-toluenesulfonic acid (TsOH) or 10-camphorsulfonic acid (CSA) (5 mol%)	solvent-free	60-80 °C, 3-48 h	61
10	T. Murai et al.	2009	phosphorofluoridic acid (5 mol%)	solvent-free	100 °C, 6 h	62
11	K. Ishihara et al.	2012	*N,N*-diarylammonium pyrosulfates (5 mol%)	H_2O	60-80 °C, 6-47 h	63
12	N. Shibata et al.	2013	TfOH (0.2-1 mol%)	solkane365mfc	80 °C, 18-48 h	65
13	W. Wang et al.	2013	2-oleamido-5-nitro-pyridinium *p*-toluenesulfonate (1-10mol%)	isooctane	r.t.-reflux, 5-144 h	66
14	R. Qiu et al.	2017	zirconocene complex (1 mol%)	solvent-free	80-100 °C, 12 h	67
15	U. Sharma et al.	2018	*L*-Leucine (5 mol%)	solvent-free	80 °C, 12-24 h	68

The development by Kobayashi and coworkers for the esterification of carboxylic acids with alcohols (2 equiv.) was another unique method, in which *p*-dodecylbenzenesulfonic acid (DBSA) was applied as a surfactant-type catalyst in water (Figure 4.2). (Wakasugi et al., 2000; Xianget al., 2002; Chakraborti & Gulhane, 2003).

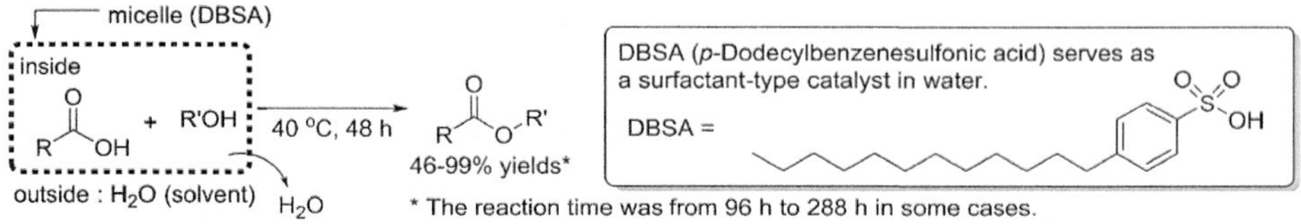

Figure 4.2: The use of DBSA in H_2O solvent (Manabe et al., 2001; 2002). The reactions were conducted in the ratio of RCOOH (1 equiv.) and R'OH (2 equiv.).

[Source: https://cdn.intechopen.com/pdfs/60244.pdf]

Because there is a production of micelles of DBSA, the esterification reactions between alcohol and carboxylic acid progressed in the micelles (Kawabata et al., 2003; Ishihara et al., 2005; Chakraborti et al., 2009).

After the reaction, the micelle issues H_2O. The alcohols and carboxylic acids with longer alkyl chains appear to be chosen because of the rise of the hydrophobicity. Therefore, the equilibrium lies on the right side between the product and starting materials (Sakakura et al., 2006; 2007a; 2007b; Ishihara, 2009).

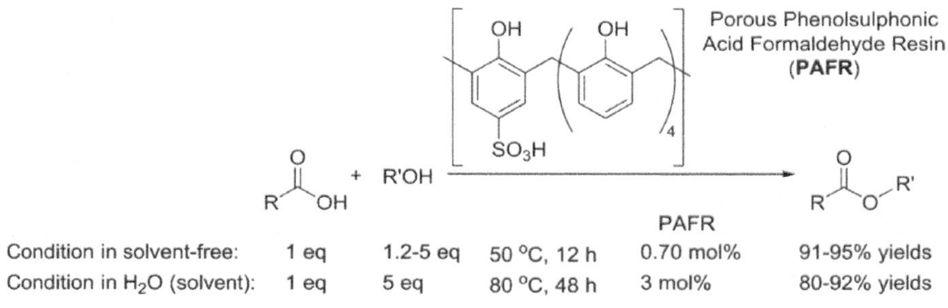

Figure 4.3: The use of PAFR as the condensation catalyst (Baek et al., 2016).

[Source: https://cdn.intechopen.com/pdfs/60244.pdf]

Baek et al. (2016) reported the utilization of resin who planned and produced the porous phenol sulfonic acid formaldehyde resin (PAFR) from and 4-hydroxybenzenesulfonic acid and formaldehyde (5 equiv.) in H_2O (Figure 4.3). The RCOOH and R'OH were allowed to react with solid resin to produce the corresponding esters in the good amount (Bartoli et al.,

2005; Funatomi et al., 2006; Murai et al., 2009). The recovery by the simple filtration and recycling without the significant loss of the catalytic activity is the benefits of resin. Other kinds of solid catalysts bearing SO_3H unit are also described. For instances, calix[*n*] arenes and *p*-sulfonic acid catalyst, SBA-15-functionalized propyl sulfonic acid catalyst, cyclodextrin-derived carbonaceous catalyst, β-polystyrene-supported sulfonic acid catalyst, and sulfonated hyperbranched poly(arylene oxindole) acid catalyst are established and applied for the esterification (Sakakura et al., 2008; 2011; Igarashi et al., 2012).

The use of the flow chemistry has emerged as a promising reaction tool for the esterification. For instance, Uozumi et al. used the PAFR to the flow technique. Fisher esterification by the flow system was verified, in which silica-bearing terminal -SO_3H group was applied (Xu et al., 2013; Wang et al., 2013; Tang et al., 2017; Manabe et al., 2001; 2002; Kumar et al., 2018).

4.3. NUCLEOPHILIC REACTIONS OF CARBOXYLATE ION INTERMEDIATES

4.3.1. The Use of Bases and Ionic Liquids

The production and addition of the carboxylate ion intermediates by the deprotonation of carboxylic acids is simple because the acidity of carboxylic acids is moderately high. One of the most popular methods is the S_N2 reaction of carboxylate ions with alkyl halides when carboxylate ions can be applied as the nucleophiles. It was observed by Miller and Clark (1977) that the user base toward carboxylic acids is CsF or KF. Later on, many reactions have been introduced in these fields. Current instances of this chemistry use the grouping of bases (such as Huning's base, Et_3N, and KF) and many ionic liquids (such as phosphonium salts and imidazolium salts), brief in Figure 4.4. The countercations of carboxylate ions are heavy cations such as imidazolium salts and phosphonium salts, which may raise the reactivity of the carboxylate ions toward electrophiles (Jing et al., 2008; Baek et al., 2016).

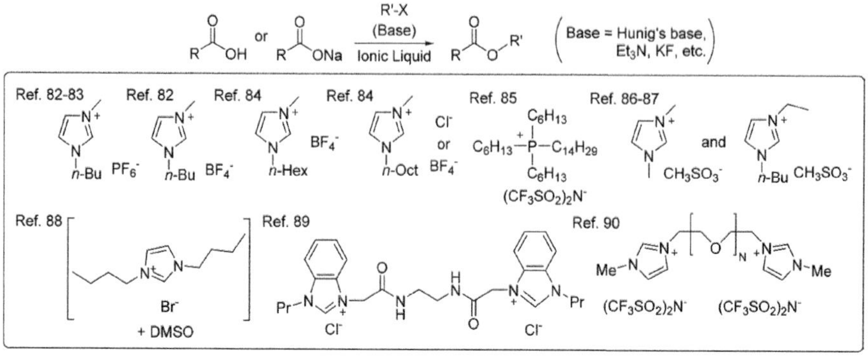

Figure 4.4: Recent and selected examples for nucleophilic reactions using carboxylate ion intermediates in various ionic liquids.

[Source: https://cdn.intechopen.com/pdfs/60244.pdf]

It has been well recognized that for carboxylic acids F^- sources such as CsF and KF can act as a good base, defined above. Due to this, for the deprotonation of carboxylic acid, Bu_4NF is also the eye-catching reagent. In 2001, Maruoka (2001) in situ formation of Bu_4NF was reported from the combination of a catalytic quantity of $KF\text{-}H_2O$ (5 equiv.) and Bu_4NHSO_4 (5 mol%), which takes the proton of carboxylic acids to produce reactive carboxylate ion intermediates, whose countercation is apparently bulky Bu4N+ (Figure 4.5 (a)). Moreover, it was also described by Matsumoto et al. that with a stoichiometric quantity of Bu_4NF the reactions of carboxylic acids efficiently produced and gathered reactive carboxylate ion intermediates, which reacted with many alkyl halides to produce the relative esters in modest to good amounts (Figure 4.5 (b)) (Minakawa et al., 2013).

In order to produce and gather extremely reactive carboxylate ion intermediates the use of good affinity of F^- and metals such as Si and Sn was also established. For instance, the reaction of Ph_3CF with silyl protected carboxylic acids produce the relative esters in the existence of an unimportant amount of SiF_4, in which SiF_4 might trigger both silyl-protected carboxylic acids (substrates) and $Ph_3C\text{-}F$ to create carboxylate ions and Ph_3C^+, respectively (Figure 4.6 (a)) Noyori et al. (1981). Maruoka et al also described the combination of silyl-protected carboxylic acids and Bu_4NF (Figure 4.6 (b)). The presence of CsF, alkyl halides reacted with intermediate bearing COO-Sn bond, as revealed in Figure 4.6 (c) (Maruoka, 2001).

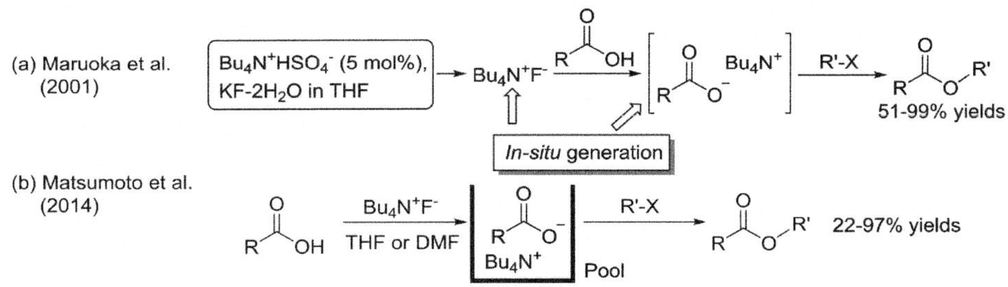

Figure 4.5: Nucleophilic reactions of carboxylate ion intermediates by using Bu4NF as the base. (a) In situ generation of a catalytic amount of Bu4NF (Maruoka, 2001). (b) The use of a stoichiometric amount of Bu4NF (Matsumoto et al., 2014).

[Source: https://cdn.intechopen.com/pdfs/60244.pdf]

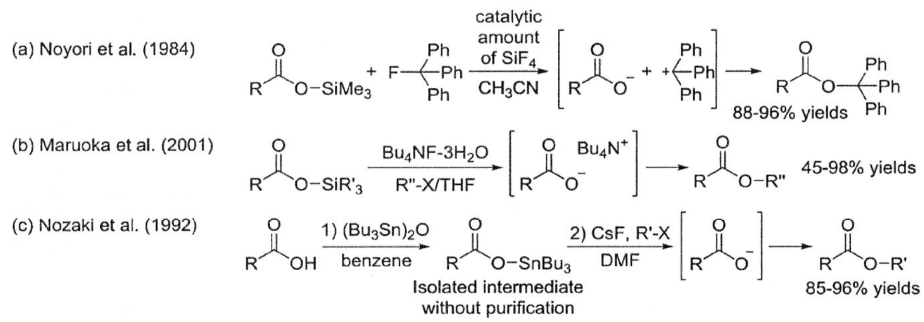

Figure 4.6: The use of affinity of F− and metals such as Si and Sn to generate carboxylate ion intermediates.

[Source: https://cdn.intechopen.com/pdfs/60244.pdf]

4.3.2. The Use of Electrochemical Reduction Methods

Electrochemistry is a spotless method, and principally in place of chemical reagents, the electron acts as the reagent (Manabe & Kobayashi, 2002). Thus, in organic formation, the waste resultant from reagents is not produced by electrochemistry and is familiar as one of the influential tools for green chemistry. The esterification of alkyl halides and carboxylic acids using electrochemistry, established by Nonaka et al. (1981) is original work (Figure 4.7). The solution holding carboxylic acids endured electrochemical reduction to create very reactive carboxylate ions, which reacted with alkyl halides to yield the corresponding esters. The complete reaction condition, the limitations, scope, and the mechanism of the electro reductive esterification reaction was investigated by Matsumoto et al. (2014) (Figure 4.7).

The production of reactive carboxylate ion intermediates by using electro-generated base (EGB) is also an effective method, developed by Shono et al. (Figure 4.8). 2-Pyrrolidone was condensed electrochemically and 2-pyrrolidone anion was produced and collected as the base in the solution phase as the base, which reacted with carboxylic acids to create carboxylate ions bearing the quaternary ammonium cation. The reaction was related to the development of macrolides (Fernandes et al., 2012; Karimi & Vafaeezadeh, 2012).

4.3.3. The Use of Electrophile Equivalents

When carboxylic acids aided as the nucleophile, some substrates were found to be active as the electrophile counterparts (Figure 4.9). For instance, use of 2-benzyloxy-1-methyl pyridinium triflate is one of the stimulating illustrations reported by Dudley (Figure 4.9 (a)). The benzyl cation was progressively produced, which was permitted to react with carboxylic acids. Cu-facilitated coupling reactions using aryl boronic acids or aryl trialkoxysilanes were established by Luo et al. (2010) (Figure 4.9 (b)).

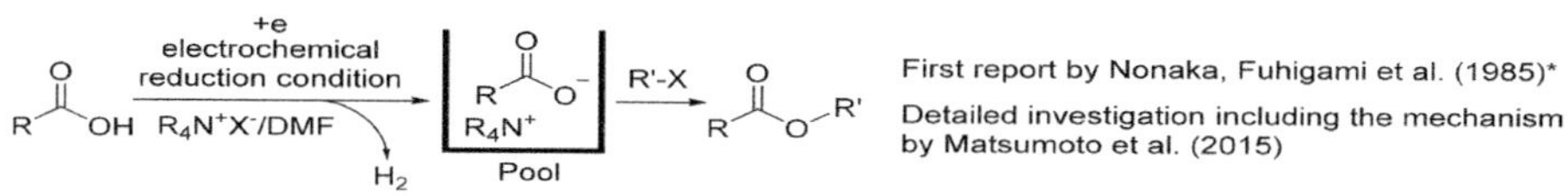

Figure 4.7: Electrochemical reduction condition to generate carboxylate ion intermediates followed by esterification (The electro-reductive esterification of carboxylic acids in the presence of alkyl halides was also examined).

[Source: https://cdn.intechopen.com/pdfs/60244.pdf]

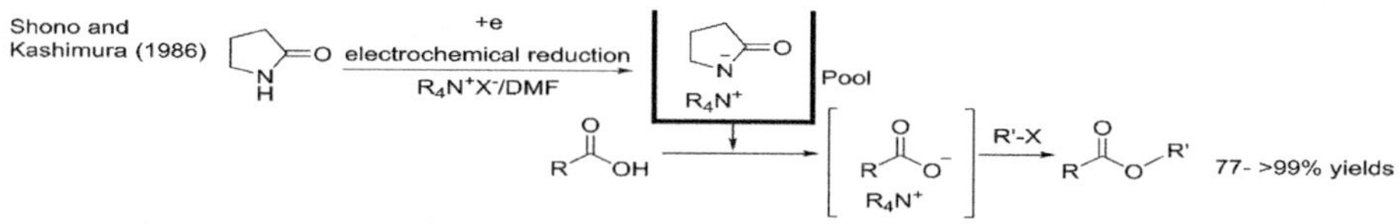

Figure 4.8: Electrochemical preparation of EGB (electro-generated base) and its utilization for the esterification (Shono and Kashimura).

[Source: https://cdn.intechopen.com/pdfs/60244.pdf]

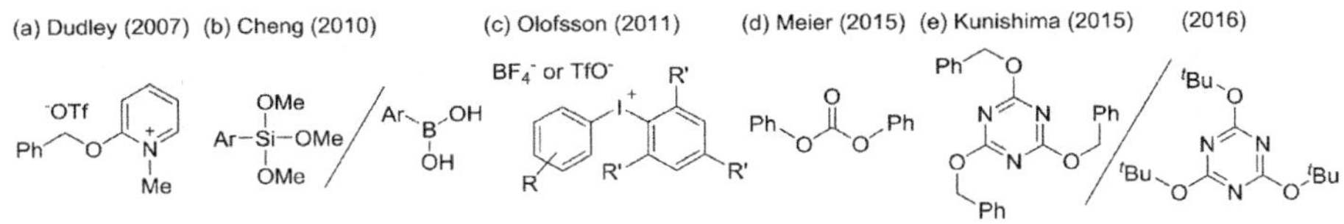

Figure 4.9: Selected examples for the use of electrophile equivalents.

[Source: https://cdn.intechopen.com/pdfs/60244.pdf]

For the esterification, diaryliodonium salts were seemed to be decent electrophiles (Figure 4.9 (c)). Kreye and Meier (2015) stated the application of diphenyl carbonate as the electrophile to yield phenyl esters (Figure 4.9 (d)). Kunishima et al. (1999a) observed that 2,4,6-tris(benzyloxy)-1,3,5-triazine (TriBOT) acts as the benzyl cation correspondent via the mechanism of S_N2 in the existence of a catalytic quantity of TfOH at ambient temperature (Figure 4.9 (e)). The reaction at high temperature also continued without TfOH. Therefore, *O*-benzylation of carboxylic acids happened easily (Thombal & Jadhav, 2015; Verheyen et al., 2017). This procedure was protracted to the application of 2,4,6-tris(tert-butoxy)-1,3,5-triazine (TriAT-*t*Bu), in which, via S_N1 mechanism, carboxylic acids can react with the *tert*-butyl cation to afford the conforming esters (Figure 4.9 (e)). Chisholm et al. also examined the reactions of the diphenyl methyl dichloroacetamide with carboxylic acids Shah et al. (2014). For the source of CH_3- unit dimethyl sulfoxide (DMSO) was utilized recently, in the reaction with carboxylic acids to offer the methyl esters revealed in Figure 4.10. The formation of methyl radical was shown (Furuta et al., 2017).

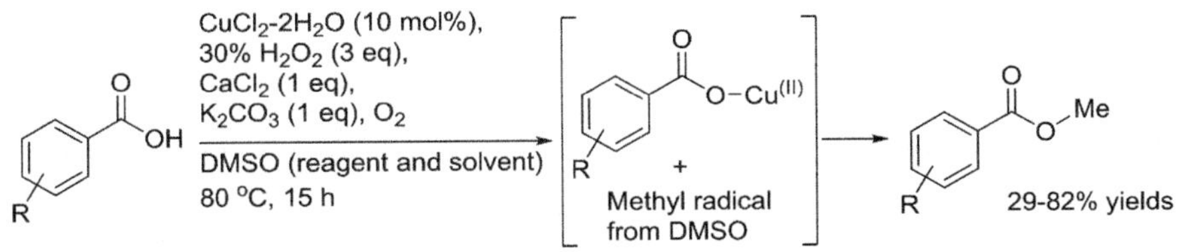

Figure 4.10: Methyl esterification of carboxylic acids using DMSO (Guo).

[Source: https://www.intechopen.com/books/carboxylic-acid-key-role-in-life-sciences/recent-advances-in-the-synthesis-of-carboxylic-acid-esters]

In situ production of benzyl bromide from toluene byproducts by means of $NaBrO_3/NaHSO_3$, subsequent by the nucleophilic reactions of the carboxylic acids could be attained by Khan et al. (Figure 4.11). Though aliphatic carboxylic acids were not appropriate, the aromatic carboxylic acids can be transformed into the conforming esters.

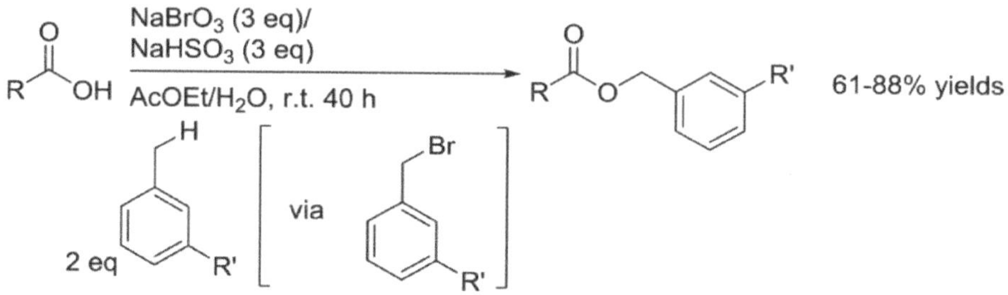

Figure 4.11: In situ formation of benzyl bromide as the electrophile equivalent (Khan).

[Source: https://www.intechopen.com/books/carboxylic-acid-key-role-in-life-sciences/recent-advances-in-the-synthesis-of-carboxylic-acid-esters]

4.4. THE USE OF THE FUNCTIONALIZATION OF C–H BONDS IN ESTERIFICATION

Lately, appropriate substrates via the functionalization of C-H bond and the esterification of carboxylic acids have been widely studied. For instance, Zhang et al. observed that the reaction of toluene and carboxylic acids in the existence CF_3SO_3H (10 mol%), $Pd(OAc)_2$ (10 mol%), and *N*, *N*-dimethylacetamide (1 equiv.) under O_2 (1 atm) afforded the conforming esters via benzylic C-H bond initiation (Figure 4.12). Other stimulating instances of this method have been widely studied.

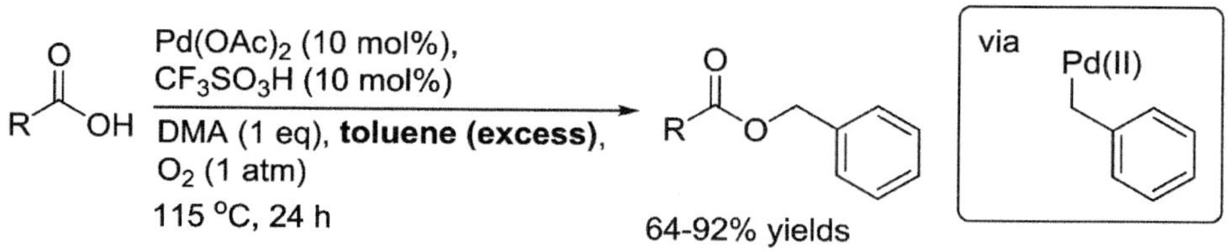

Figure 4.12: Esterification of carboxylic acids and toluene catalyzed by Pd(OAc)2 (Zhang).

[Source: https://www.intechopen.com/books/carboxylic-acid-key-role-in-life-sciences/recent-advances-in-the-synthesis-of-carboxylic-acid-esters]

4.5. THE USE OF METAL CATALYSTS IN ESTERIFICATION

The adding of carboxylic acids onto C-C multiple bonds begins with high atom efficacy to afford the equivalent alkyl or enol esters. Hg salts have been applied as the catalysts for these reactions for a long time; though, the application of these toxic salts should be evaded from the point of view of green chemistry. Ruthenium complexes have been rewarded much consideration for the substitute catalyst for the adding of carboxylic acids onto C-C multiple bonds. These ruthenium-catalyzed adding reactions of carboxylic acids to alkynes and numerous catalytic developments of alkyl esters by the accumulation of carboxylic acids to alkenes are summed up in this section.

4.5.1. The Addition of Carboxylic Acids onto Alkynes

Mitsuda and Dixneuf independently investigated Markovnikov's rule.

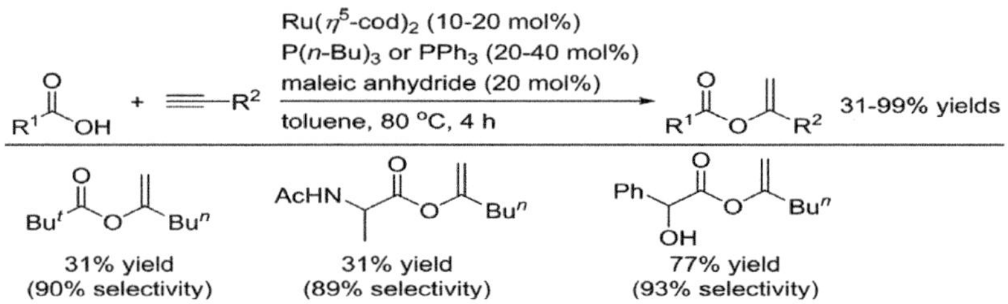

Figure 4.13: Markovnikov addition (Mitsudo et al., 1987).

[Source: https://www.intechopen.com/books/carboxylic-acid-key-role-in-life-sciences/recent-advances-in-the-synthesis-of-carboxylic-acid-esters]

In 1987, Mitsudo and coworkers (1987) stated that in the presence of Ru(η^5-cod)$_2$, maleic anhydride and phosphine ligands, the reaction of carboxylic acids counting *N*-protected α-amino acids with alkynes were conducted to afford the corresponding enol esters in 31–99% production (Figure 4.13). This reaction progresses with the great regioselectivity (>89% selectivity).

In contrast, for the Markovnikov addition, Dixneuf and coworkers established the simpler catalytic system. Therefore, (*p*-cymene) $RuCl_2(PPh_3)$ catalyzed the adding of carboxylic acids to the alkynes to provide the conforming adducts in high production (78–92%), as revealed in Figure 4.14. They also observed the extremely selective *anti*-Markovnikov adding by the use of Ru(η^3-methallyl)$_2$(diphosphine) to attain the *Z*-enol esters in high production with and *Z*- and regio-selectivities (Figure 4.15).

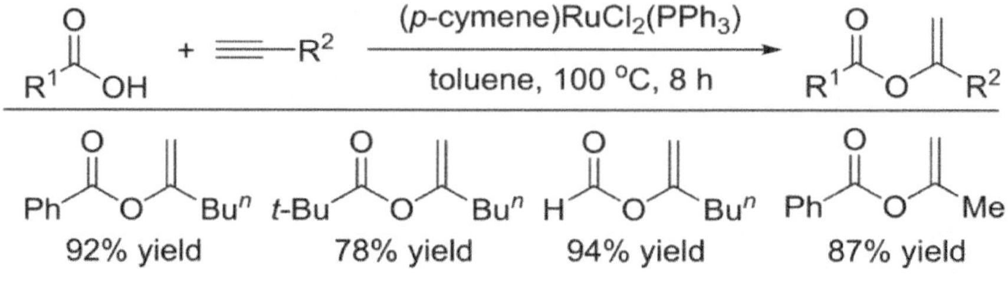

Figure 4.14: Markovnikov addition (Dixneuf).

[Source: https://www.intechopen.com/books/carboxylic-acid-key-role-in-life-sciences/recent-advances-in-the-synthesis-of-carboxylic-acid-esters]

Dixneuf et al. produced the reaction mechanisms for both regioselective additions (Figure 4.16), and for the *anti*-Markovnikov adding, the synthesis of vinylidene complex is critical.

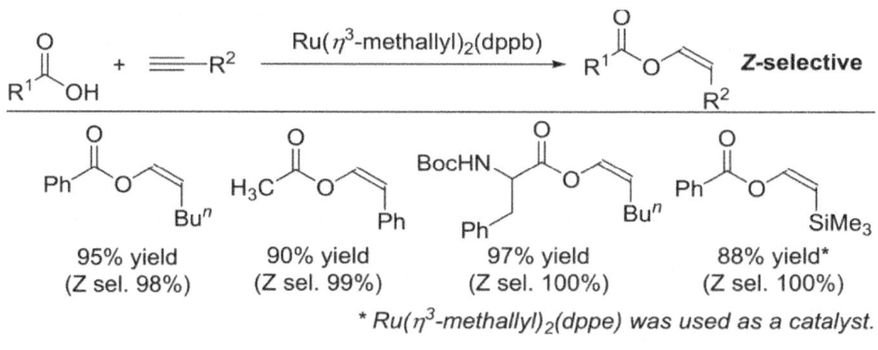

Figure 4.15: Anti- Markovnikov addition of carboxylic acids onto terminal alkynes (Dixneuf).

[Source: https://www.intechopen.com/books/carboxylic-acid-key-role-in-life-sciences/recent-advances-in-the-synthesis-of-carboxylic-acid-esters]

Therefore, through the activation of alkynes by the synthesis of π-complex with ruthenium catalyst, the Markovnikov adducts are produced, subsequent by the nucleophilic attack of the carboxylate ion onto the internal carbon atom of the protonolysis of Ru-C δ-bond and alkynes. In contrast, the *anti*-Markovnikov adducts are suitable for the vinylidene complex synthesis between the alkynes and the ruthenium catalyst, tracked by the nucleophilic outbreak of the carboxylate ion to the end carbon of the protonolysis and alkynes of the resultant intermediate.

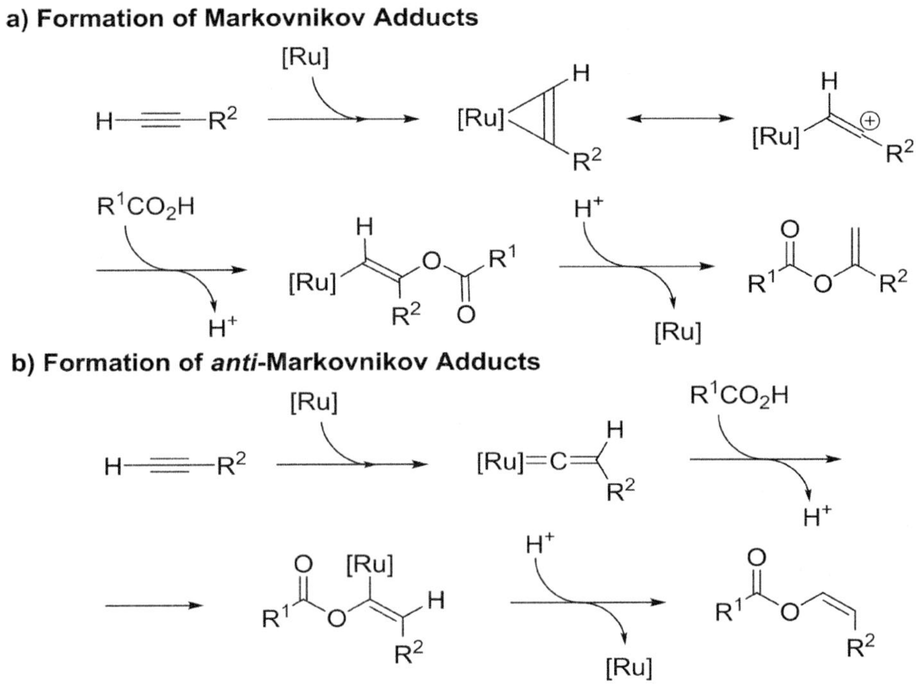

Figure 4.16: Reaction mechanisms of the present enol ester formations.

[Source: https://www.intechopen.com/books/carboxylic-acid-key-role-in-life-sciences/recent-advances-in-the-synthesis-of-carboxylic-acid-esters]

Two simple catalytic systems have been stated, in which the regioselectivity is simply controlled by the means of the similar or same catalyst. Goossen et al. (2003) observed that in the presence of K_2CO_3 the reaction of alkenes and carboxylic acids with $RuCl_2$(*p*-cymene)/phosphine catalyst produced the conforming Markovnikov adducts in good to outstanding production, while the use of DMAP as an alternative of K_2CO_3 produced the *anti*-Markovnikov (*Z*) esters in high production (Figure 4.17). The reactions with $RuClH(CO)(PCy)_3$ catalyst displayed the stimulating solvent-controlled regioselectivity (Figure 4.18). Therefore, the reaction of aryl acetylene with the benzoic acid in the existence of $RuClH(CO)(PCy)_3$ as a catalyst in CH_2Cl_2 produced the Markovnikov adducts in high production.

In contrast, the use of THF as a solvent as an alternative of CH_2Cl_2 caused the inverse of the selectivity to afford the conforming *anti*-Markovnikov (*Z*) esters in high production together with the great selectivity.

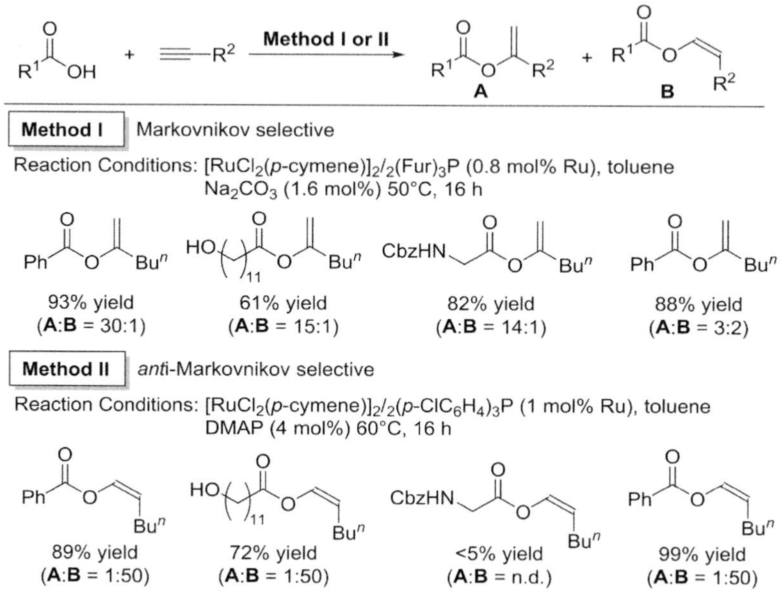

Figure 4.17: Regioselective addition (Goossen et al., 2003).

[Source: https://cdn.intechopen.com/pdfs/60244.pdf]

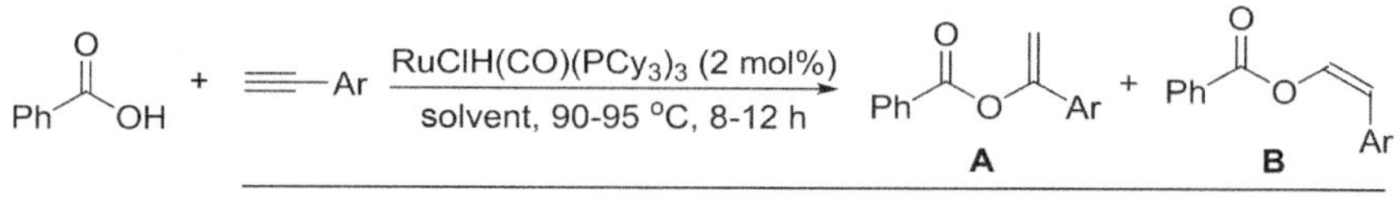

Ar	Solvent	Yield of **A**	Yield of **B**
Ph	CH_2Cl_2	98	0
	THF	13	85
p-MeOC$_6$H$_4$	CH_2Cl_2	91	0
	THF	0	95

Figure 4.18: Solvent-controlled selective additions (Yi).

[Source: https://cdn.intechopen.com/pdfs/60244.pdf]

It is thought that the use of THF enhanced the synthesis of vinylidene complex intermediates. Regrettably, this stimulating effect was not detected in the reaction of aliphatic alkynes.

The stereoselective synthesis of *anti*-Markovnikov (*E*) esters is so far restricted. It can be observed that (*E*)-selective synthesis of enol esters with their individual ruthenium catalyst (Figure 4.19), autonomously. Though, the *E*/*Z* ratios were reliant on the alkyne and/or substrates of carboxylic acids. The rhenium catalyst displayed the *E*-selectivity; however, the selectivity was also powerfully reliant on the substrates of alkynes. Alike selectivity was gained when $Re(CO)_5Br$ was applied as a catalyst, however, the *E*/*Z* ratios were moderate (Figure 4.20).

Two cases for the *E*- and regio-selective addition of carboxylic acids onto "internal" alkynes have been described (Figure 4.21). Lang et al. observed that the reaction of the carboxylic acid with symmetrical internal alkynes in the existence of $[Ru(CO)_2\{(p\text{-}CF_3C_6H_4)_3P\}(O_2CPh)_2]/B(C_6F_5)_3$ catalyst produced the corresponding *E*-enol esters in up to 99% production with the very high *E*-selectivity. This catalytic system also attained the stereoselective and regio-addition of carboxylic acids to asymmetrical internal alkynes, in which the *E*/*Z* ratio stretched up to 72:28. Kawatsura et al. (2011) described the reaction with alkynes having aryl and trifluoromethyl group. The reaction continued with great stereoselectivity and regioselectivity to give the conforming (*E*)-enol esters counting trifluoromethyl group in up to 92% production.

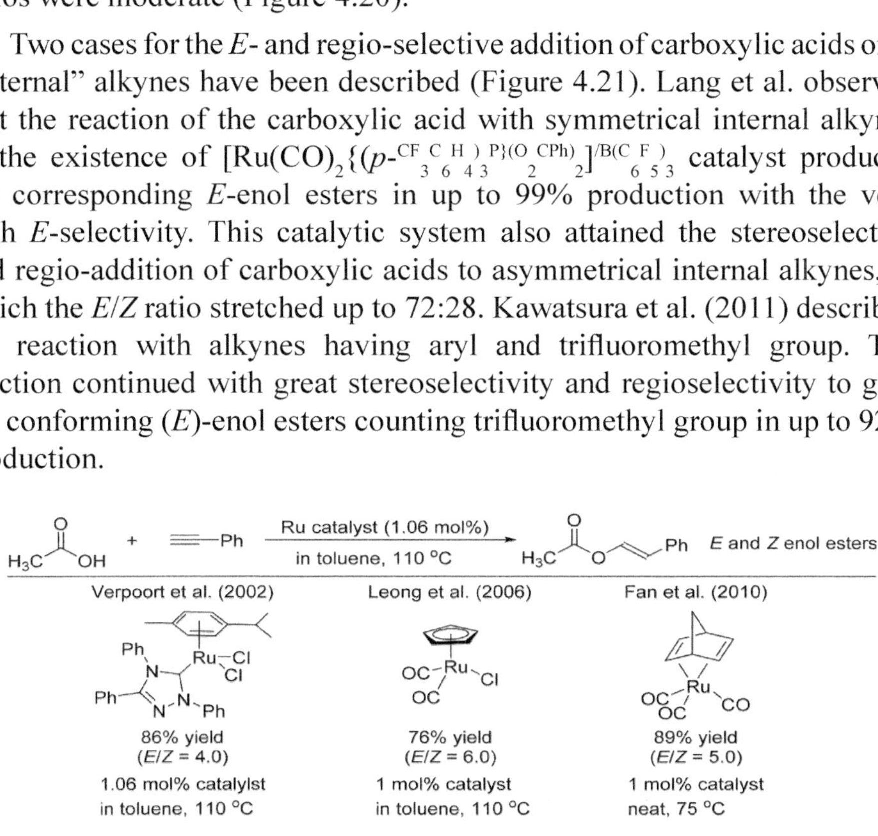

Figure 4.19: Ruthenium catalysts for E- selective additions.

[Source: https://cdn.intechopen.com/pdfs/60244.pdf]

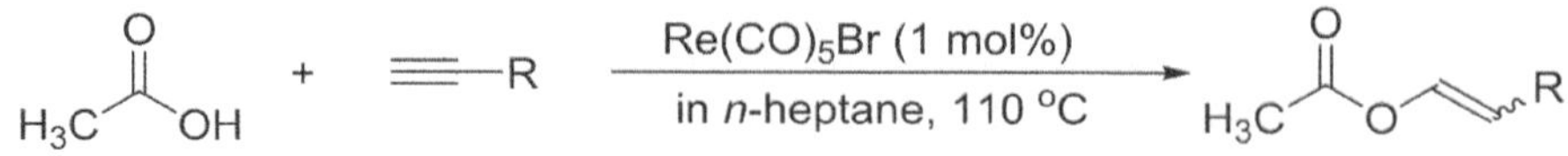

R	Yield	ratio (*E*/*Z*)
Ph	72	27/73
n-Bu	78	67/33

Figure 4.20: Rhenium catalysis for additions of carboxylic acids onto alkynes (Hua).

[Source: https://cdn.intechopen.com/pdfs/60244.pdf]

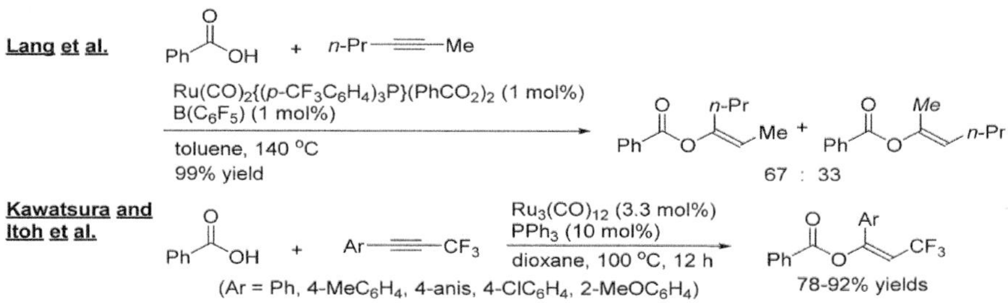

Figure 4.21: Regio- and stereoselective additions to unsymmetrical internal alkynes.

[Source: https://cdn.intechopen.com/pdfs/60244.pdf]

4.5.2. The Addition of Carboxylic Acids onto Alkenes

In 2004, the initial transition metal-catalyzed addition of carboxylic acids onto alkenes was discovered. Therefore, the reaction of norbornene with benzoic acids was conducted in the existence of $[Cp^{*}RuCl_2]_2$/AgOTf/Dppb catalyst at 85 °C in toluene to get the corresponding norbornyl benzoates in good to high production (Figure 4.22). Afterward, numerous metal catalysts for the addition of carboxylic acids to alkenes have been described and are potted in Table 4.3.

Norbornene, due to the strain of the C–C double bond, is normally a good substrate, so the reported catalyst produced the conforming esters in high production (Table 4.3). In 2005, the Au catalyst, where the four alkenes counting unstrained 1-octene were altered into the conforming esters in

75–95% production (entry 1). In these examples, under similar reaction conditions, the addition of phenols onto alkenes was also catalyzed. With only norbornene, $In(OTf)_3$ was also observed as a good catalyst under the solvent-free reaction circumstances (entry 3). Sakakura et al. (2008) reported omnipresent iron catalysis where the uneven ester such as acrylates can be formed under the solvent-free reaction situations (entry 4). Altered ruthenium(II) catalysis including Xanthos ligand enhanced the scope of substrates of alkenes equated to that with ruthenium(III) catalyst to produce the conforming esters in up to 99% production by Oe et al. (entry 5)

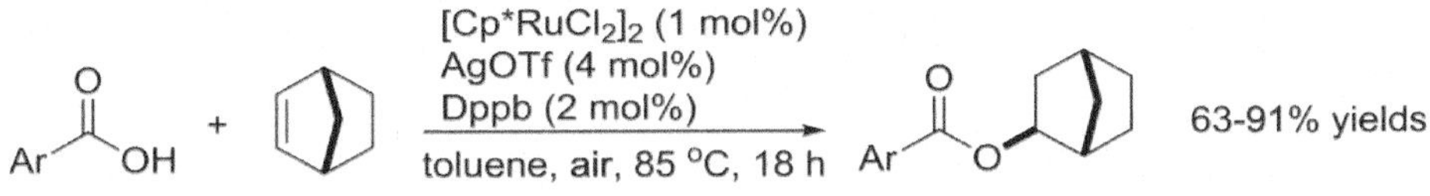

Figure 4.22: Ruthenium-catalyzed additions of carboxylic acids to norbornene (Oe).

[Source: https://cdn.intechopen.com/pdfs/60244.pdf]

It can be found that TfOH itself displayed the catalytic activity for the current addition reactions (Figure 4.23). Stimulatingly, the relatively large quantity of catalyst and/or greater reaction temperature reduced the chemical production of the product. It might be because of the polymerization of substrates of alkenes. Escorted with the significance of triflate in the metal-catalyzed reaction designated above, these metal catalysts might act as a TfOH basis.

Table 4.3. Recent Reported Catalysts and Generality of Alkene Substrates

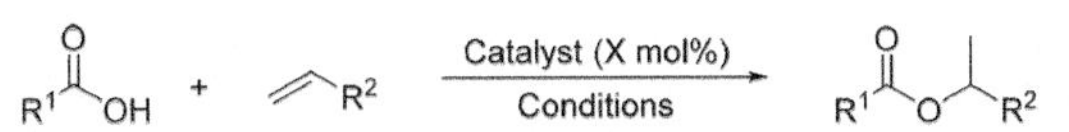

Entry	Authors	Published year	Catalyst (X mol%) & Conditions	Alkenes (yield, %)	Ref.
1	C. He et al.	2005	catalyst: $Ph_3PAuOTf$ (5 mol%) conditions: in toluene, 85 °C	norbornene (>95%) and 4 alkenes	142
2	K. K. Hii et al.	2005	catalyst: $Cu(OTf)_2$ (5 mol%) conditions: in dioxane, 80 °C	norbornene (80-98%)	143
3	W. Chen et al.	2007	catalyst: $In(OTf)_3$ (5 mol%) conditions: neat, 80 °C, 2 h	norbornene (80-92%)	144
4	T. Sakakura et al.	2008	catalyst: $Fe(OTf)_3$ (2 mol%) conditions: in Bu_2O, 80 °C, 18 h	norbornene (98-99%) cyclohexene (70-88%) 1-octene (78%)	145
5	Y. Oe et al.	2015	catalyst: $RuCl_2(xan)\{P(OPh)_3\}_3$ (5 mol%) AgOTf(10 mol%) conditions: in toluene, 80 °C	4-allylanisole (>99%) and 4 alkenes	146

Lately, Nicewicz et al. reported an organocatalytic addition under the LED light irradiation situations (Figure 4.24). The reaction of internal and/or cyclic alkenes and carboxylic acids progresses politely under mild circumstances to produce the conforming esters in up to 99% production regio-selectively.

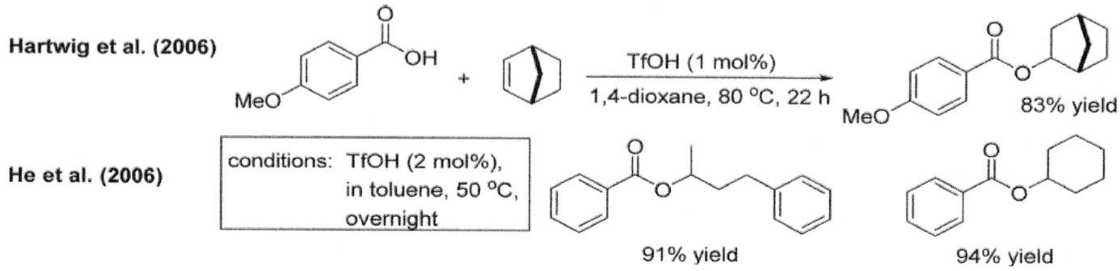

Figure 4.23: TfOH-catalyzed reactions.

[Source: https://cdn.intechopen.com/pdfs/60244.pdf]

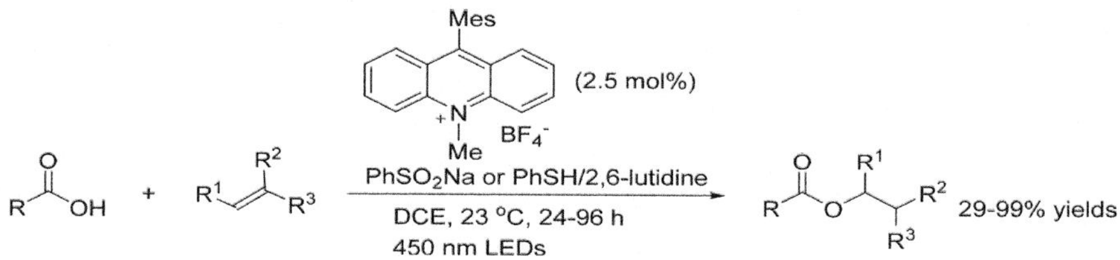

Figure 4.24: The additions of carboxylic acids onto alkenes in the presence of the organocatalyst under the LED irradiation (Nicewicz).

[Source: https://cdn.intechopen.com/pdfs/60244.pdf]

REFERENCES

1. Baek, H., Minakawa, M., Yamada, Y. M., Han, J. W., & Uozumi, Y. (2016). In-water and neat batch and continuous-flow direct esterification and transesterification by a porous polymeric acid catalyst. *Scientific Reports, 6*, 25925.
2. Bartoli, G., Boeglin, J., Bosco, M., Locatelli, M., Massaccesi, M., Melchiorre, P., & Sambri, L. (2005). Highly Efficient Solvent-Free Condensation of Carboxylic Acids with Alcohols Catalysed by Zinc Perchlorate Hexahydrate, Zn (ClO_4) 2.6 H_2O. *Advanced Synthesis & Catalysis, 347*(1), 33–38.
3. Boulogeorgou, M. A., Triantakonstanti, V. V., & Gallos, J. K. (2015). The system PhIO/Ph3P as an efficient reagent for mild and direct coupling of alcohols with carboxylic acids. *ARKIVOC, 3*, 244–251.
4. But, T. Y. S., & Toy, P. H. (2006). Organocatalytic mitsunobu reactions. *Journal of the American Chemical Society, 128*(30), 9636–9637.
5. Castro, B., Dormoy, JR, Evin, G., & Selve, C. (1975). Peptide coupling reagents I (1) -Benzotriazolyl hexafluorophosphate N-oxytrisdimethylamino phosphonium (BOP). *Tetrahedron Letters, 16* (14), 1219–1222.
6. Chakraborti, A. K., & Gulhane, R. (2003). Perchloric acid adsorbed on silica gel as a new, highly efficient, and versatile catalyst for acetylation of phenols, thiols, alcohols, and amines. *Chemical Communications, 15*, 1896–1897.
7. Chakraborti, A. K., Singh, B., Chankeshwara, S. V., & Patel, A. R. (2009). Protic acid immobilized on solid support as an extremely efficient recyclable catalyst system for a direct and atom economical esterification of carboxylic acids with alcohols. *The Journal of Organic Chemistry, 74*(16), 5967–5974.
8. Chen, C. T., & Munot, Y. S. (2005). Direct atom-efficient esterification between carboxylic acids and alcohols catalyzed by amphoteric, water-tolerant TiO (acac) 2. *The Journal of Organic Chemistry, 70*(21), 8625–8627.
9. Chen, H., Xu, X., Liu, L., Tang, G., & Zhao, Y. (2013). Phosphorus oxychloride as an efficient coupling reagent for the synthesis of esters, amides and peptides under mild conditions. *RSC Advances, 3*(37), 16247–16250.

10. Choi, M. K. W., He, H. S., & Toy, P. H. (2003). Direct radical polymerization of 4-styryldiphenylphosphine: Preparation of cross-linked and non-cross-linked triphenylphosphine-containing polystyrene polymers. *The Journal of Organic Chemistry, 68*(25), 9831–9834.
11. Clark, J. H., & Miller, J. M. (1977). Hydrogen bonding in organic synthesis V: potassium fluoride in carboxylic acids as an alternative to crown ether with acid salts in the preparation of phenacyl esters. *Tetrahedron Letters, 18*(7), 599–602.
12. Dinesh, M., Ranganathan, R., Archana, S., Sathishkumar, M., Roshan Banu, M. S., & Ponnuswamy, A. (2016). Staudinger's phosphazene as an efficient esterifying reagent. *Synthetic Communications, 46*(17), 1454–1460.
13. El-Faham, A., & Albericio, F. (2011). Peptide coupling reagents, more than a letter soup. *Chemical Reviews, 111*(11), 6557–6602.
14. Engstrom, K. M., Sheikh, A., Ho, R., & Miller, R. W. (2014). The stability of N, N-carbonyldiimidazole toward atmospheric moisture. *Organic Process Research & Development, 18*(4), 488–494.
15. Fernandes, S. A., Natalino, R., Gazolla, P. A. R., da Silva, M. J., & Jham, G. N. (2012). p-Sulfonic acid calix [n] arenes as homogeneous and recyclable organocatalysts for esterification reactions. *Tetrahedron Letters, 53*(13), 1630–1633.
16. Funasaka, S., & Mukaiyama, T. (2008). A versatile, practical, and inexpensive reagent, pyridine-3-carboxylic anhydride (3-PCA), for condensation reactions. *Bulletin of the Chemical Society of Japan, 81*(1), 148–159.
17. Funatomi, T., Wakasugi, K., Misaki, T., & Tanabe, Y. (2006). Pentafluorophenylammonium triflate (PFPAT): an efficient, practical, and cost-effective catalyst for esterification, thioesterification, transesterification, and macrolactone formation. *Green Chemistry, 8*(12), 1022–1027.
18. Furuta, A., Fukuyama, T., & Ryu, I. (2017). Efficient flow fischer esterification of carboxylic acids with alcohols using sulfonic acid-functionalized silica as supported catalyst. *Bulletin of the Chemical Society of Japan, 90*(5), 607–612.
19. Goossen, L. J., Paetzold, J., & Koley, D. (2003). Regiocontrolled Ru-catalyzed addition of carboxylic acids to alkynes: practical protocols for the synthesis of vinyl esters. *Chemical Communications, 6*, 706–

707.

20. Hirose, D., Gazvoda, M., Košmrlj, J., & Taniguchi, T. (2016). Advances and mechanistic insight on the catalytic Mitsunobu reaction using recyclable azo reagents. *Chemical Science, 7*(8), 5148–5159.
21. Igarashi, T., Yagyu, D., Naito, T., Okumura, Y., Nakajo, T., Mori, Y., & Kobayashi, S. (2012). Dehydrative esterification of carboxylic acids with alcohols catalyzed by diarylammonium p-dodecylbenzenesulfonates in water. *Applied Catalysis B: Environmental, 119*, 304–307.
22. Ishihara, K. (2009). Dehydrative condensation catalyses. *Tetrahedron, 6*(65), 1085–1109.
23. Ishihara, K., Nakagawa, S., & Sakakura, A. (2005). Bulky diarylammonium arenesulfonates as selective esterification catalysts. *Journal of the American Chemical Society, 127*(12), 4168–4169.
24. Ishihara, K., Nakayama, M., Ohara, S., & Yamamoto, H. (2001). A green method for the selective esterification of primary alcohols in the presence of secondary alcohols or aromatic alcohols. *Synlett, 7*, 1117–1120.
25. Ishihara, K., Nakayama, M., Ohara, S., & Yamamoto, H. (2002). Direct ester condensation from a 1: 1 mixture of carboxylic acids and alcohols catalyzed by hafnium (IV) or zirconium (IV) salts. *Tetrahedron, 58*(41), 8179–8188.
26. Ishihara, K., Ohara, S., & Yamamoto, H. (2000). Direct condensation of carboxylic acids with alcohols catalyzed by hafnium (IV) salts. *Science, 290*(5494), 1140–1142.
27. Jaita, S., Phakhodee, W., & Pattarawarapan, M. (2015). Ultrasound-assisted methyl esterification of carboxylic acids catalyzed by polymer-supported triphenylphosphine. *Synlett, 26*(14), 2006–2008.
28. Jing, L., Li, X. J., Han, Y. C., & Chu, Y. (2008). The esterification in cyclohexane/DBSA/water microemulsion system. *Colloids and Surfaces A: Physicochemical and Engineering Aspects, 326*(1–2), 37–41.
29. Karimi, B., & Vafaeezadeh, M. (2012). SBA-15-functionalized sulfonic acid confined acidic ionic liquid: a powerful and water-tolerant catalyst for solvent-free esterifications. *Chemical Communications, 48*(27), 3327–3329.
30. Kawabata, T., Mizugaki, T., Ebitani, K., & Kaneda, K. (2003). Highly efficient esterification of carboxylic acids with alcohols by

montmorillonite-enwrapped titanium as a heterogeneous acid catalyst. *Tetrahedron Letters, 44*(51), 9205–9208.

31. Kawatsura, M., Namioka, J., Kajita, K., Yamamoto, M., Tsuji, H., & Itoh, T. (2011). Ruthenium-catalyzed regio-and stereoselective addition of carboxylic acids to aryl and trifluoromethyl group substituted unsymmetrical internal alkynes. *Organic Letters, 13*(12), 3285–3287.
32. Kim, M. H., & Patel, D. V. (1994). "BOP" as a reagent for mild and efficient preparation of esters. *Tetrahedron Letters, 35*(31), 5603–5606.
33. Kreye, O., & Meier, M. A. (2015). Base catalyzed sustainable synthesis of phenyl esters from carboxylic acids using diphenyl carbonate. *RSC Advances, 5*(65), 53155–53160.
34. Kumar, M., Thakur, K., Sharma, S., Nayal, O. S., Kumar, N., Singh, B., & Sharma, U. (2018). Solvent-free, l-Leucine-Catalyzed Direct Dehydrative Esterification of Carboxylic Acids with Alcohols: Direct Synthesis of 3-Alkoxy 1 (3 H)-isobenzofuranone. *Asian Journal of Organic Chemistry, 7*(1), 227–231.
35. Kunishima, M., Kawachi, C., Iwasaki, F., Terao, K., & Tani, S. (1999a). Synthesis and characterization of 4-(4, 6-dimethoxy-1, 3, 5-triazin-2-yl)-4-methylmorpholinium chloride. *Tetrahedron Letters, 40*(29), 5327–5330.
36. Kunishima, M., Kawachi, C., Monta, J., Terao, K., Iwasaki, F., & Tani, S. (1999c). 4-(4, 6-dimethoxy-1, 3, 5-triazin-2-yl)-4-methyl-morpholinium chloride: an efficient condensing agent leading to the formation of amides and esters. Tetrahedron, 55(46), 13159–13170.
37. Kunishima, M., Morita, J., Kawachi, C., Iwasaki, F., Terao, K., & Tani, S. (1999b). Esterification of carboxylic acids with alcohols by 4-(4, 6-dimethoxy-1, 3, 5-triazin-2-yl)-4-methyl-morpholinium chloride (DMTMM). *Synlett, 8*, 1255–1256.
38. Li, J. J., & Corey, E. J. (2007). Name Reactions of Functional Group Transformations (Vol. 1, pp. 1–20). John Wiley & Sons.
39. Luo, F., Pan, C., Qian, P., & Cheng, J. (2010). Copper (II)-catalyzed esterification of arenecarboxylic acids with aryl-and vinyl-substituted trimethoxysilanes. *Synthesis*, *2010*(12), 2005–2010.
40. Mamidi, N., & Manna, D. (2013). Zn (OTf) 2-promoted chemoselective esterification of hydroxyl group bearing carboxylic acids. *The Journal of Organic Chemistry,* 78(6), 2386–2396.
41. Manabe, K., & Kobayashi, S. (2002). Dehydrative esterification of

carboxylic acids with alcohols catalyzed by polymer-supported sulfonic acids in water. *Advanced Synthesis & Catalysis, 344*(3–4), 270–273.

42. Manabe, K., Iimura, S., Sun, X. M., & Kobayashi, S. (2002). Dehydration reactions in water. Brønsted acid−surfactant-combined catalyst for ester, ether, thioether, and dithioacetal formation in water. *Journal of the American Chemical Society, 124*(40), 11971–11978.
43. Manabe, K., Sun, X. M., & Kobayashi, S. (2001). Dehydration reactions in water. Surfactant-type Brønsted acid-catalyzed direct esterification of carboxylic acids with alcohols in an emulsion system. *Journal of the American Chemical Society, 123*(41), 10101–10102.
44. Mantri, K., Komura, K., & Sugi, Y. (2005b). Efficient esterification of long chain aliphatic carboxylic acids with alcohols over ZrOCl2· 8H2O catalyst. *Synthesis, 2005*(12), 1939–1944.
45. Mantri, K., Nakamura, R., Komura, K., & Sugi, Y. (2005a). Esterification of long chain aliphatic acids with long chain alcohols catalyzed by multi-valent metal salts. *Chemistry Letters, 34*(11), 1502–1503.
46. Maruoka, K. (2001). Bidentate Lewis acids for organic synthesis. *Catalysis Today*, *66*(1), 33–45.
47. März, M., Chudoba, J., Kohout, M., & Cibulka, R. (2017). Photocatalytic esterification under Mitsunobu reaction conditions mediated by flavin and visible light. *Organic & Biomolecular Chemistry, 15*(9), 1970–1975.
48. Matsumoto, K., Shimazaki, H., Miyamoto, Y., Shimada, K., Haga, F., Yamada, Y., ... & Kashimura, S. (2014). Simple and convenient synthesis of esters from carboxylic acids and alkyl halides using tetrabutylammonium fluoride. *Journal of Oleo Science*, *63*(5), 539–544.
49. Matsumoto, K., Yanagi, R., & Oe, Y. (2018). Recent Advances in the Synthesis of Carboxylic Acid Esters. In Carboxylic Acid-Key Role in Life Sciences (Vol. 1, pp. 1–15). IntechOpen.
50. Minakawa, M., Baek, H., Yamada, Y. M., Han, J. W., & Uozumi, Y. (2013). Direct dehydrative esterification of alcohols and carboxylic acids with a macroporous polymeric acid catalyst. *Organic Letters, 15*(22), 5798–5801.
51. Mitsudo, T., Hori, Y., Yamakawa, Y., & Watanabe, Y. (1987). Ruthenium-catalyzed selective addition of carboxylic acids to alkynes. A novel synthesis of enol esters. *The Journal of Organic Chemistry,*

52(11), 2230–2239.

52. Mitsunobu, O., & Yamada, M. (1967). Preparation of esters of carboxylic and phosphoric acid via quaternary phosphonium salts. *Bulletin of the Chemical Society of Japan, 40*(10), 2380–2382.
53. Morcillo, S. P., Alvarez de Cienfuegos, L., Mota, A. J., Justicia, J., & Robles, R. (2011). Mild Method for the Selective Esterification of Carboxylic Acids Based on the Garegg–Samuelsson Reaction. *The Journal of Organic Chemistry, 76*(7), 2277–2281.
54. Mukaiyama, T., Usui, M., Shimada, E., & Saigo, K. (1975). A convenient method for the synthesis of carboxylic esters. *Chemistry Letters, 4*(10), 1045–1048.
55. Murai, T., Tanaka, N., Higashijima, S., & Miura, H. (2009). Phosphorofluoridic acid-catalyzed condensation reaction of carboxylic acids with 1-arylalkyl alcohols leading to esters under solvent-free conditions. *Chemistry Letters, 38*(7), 668–669.
56. Nakamura, Y., Maki, T., Wang, X., Ishihara, K., & Yamamoto, H. (2006). Iron (III)–Zirconium (IV) Combined Salt Immobilized on N-(Polystyrylbutyl) pyridinium Triflylimide as a Reusable Catalyst for a Dehydrative Esterification Reaction. *Advanced Synthesis & Catalysis, 348*(12-13), 1505–1510.
57. Nakayama, M., Sato, A., Ishihara, K., & Yamamoto, H. (2004). Water-Tolerant and Reusable Catalysts for Direct Ester Condensation between Equimolar Amounts of Carboxylic Acids and Alcohols. *Advanced Synthesis & Catalysis, 346*(11), 1275–1279.
58. Neises, B., & Steglich, W. (1978). Simple method for the esterification of carboxylic acids. *Angewandte Chemie International Edition in English, 17*(7), 522–524.
59. Nguyen, T. V., & Lyons, D. J. (2015). A novel aromatic carbocation-based coupling reagent for esterification and amidation reactions. *Chemical Communications, 51*(15), 3131–3134.
60. Nonaka, T., Kato, T., Fuchigami, T., & Sekine, T. (1981). Electroreduction of substituents on a pyridine ring in aqueous sulfuric acid. *Electrochimica Acta, 26*(7), 887–892.
61. Nowrouzi, N., Mehranpour, A. M., & Rad, J. A. (2010). A simple and convenient method for preparation of carboxylic acid alkyl esters, phenolic and thioesters using chlorodiphenylphosphine/I2 and imidazole reagent system. *Tetrahedron, 66*(50), 9596–9601.

62. Noyori, R., Murata, S., & Suzuki, M. (1981). Trimethysilyl triflate in organic synthesis. *Tetrahedron, 37*(23), 3899–3910.
63. Otera, J., & Nishikido, J. (2009). *Esterification: Methods, Reactions, and Applications* (vol. 1, pp. 1–15). John Wiley & Sons.
64. Pathak, G., & Rokhum, L. (2015). Selective monoesterification of symmetrical diols using resin-bound triphenylphosphine. *ACS Combinatorial Science, 17*(9), 483–487.
65. Pathak, G., Das, D., & Rokhum, L. (2016). A microwave-assisted highly practical chemoselective esterification and amidation of carboxylic acids. *RSC Advances, 6*(96), 93729–93740.
66. Paul, R., & Anderson, G. W. (1960). N,N'-Carbonyldiimidazole, a New Peptide Forming Reagent. *Journal of the American Chemical Society, 82*(17), 4596–4600.
67. Rodrigues, R. D. C., Barros, I. M., & Lima, E. L. (2005). Mild one-pot conversion of carboxylic acids to amides or esters with Ph3P/trichloroisocyanuric acid. *Tetrahedron Letters, 46*(35), 5945–5947.
68. Sakakura, A., Koshikari, Y., & Ishihara, K. (2008). Open-air and solvent-free ester condensation catalyzed by sulfonic acids. *Tetrahedron Letters, 49*(34), 5017–5020.
69. Sakakura, A., Koshikari, Y., Akakura, M., & Ishihara, K. (2011). Hydrophobic N, N-diarylammonium pyrosulfates as dehydrative condensation catalysts under aqueous conditions. *Organic Letters, 14*(1), 30–33.
70. Sakakura, A., Nakagawa, S., & Ishihara, K. (2006). Bulky diarylammonium arenesulfonates as mild and extremely active dehydrative ester condensation catalysts. *Tetrahedron, 62*(2–3), 422–433.
71. Sakakura, A., Nakagawa, S., & Ishihara, K. (2007). Direct ester condensation catalyzed by bulky diarylammonium pentafluorobenzenesulfonates. *Nature Protocols, 2*(7), 1746.
72. Sakakura, A., Watanabe, H., Nakagawa, S., & Ishihara, K. (2007). Unusual Rate Acceleration in Brønsted Acid Catalyzed Dehydration Reactions: Local Hydrophobic Environment in Aggregated N-(2, 6-diphenylphenyl)-N-mesitylammonium Pentafluorobenzenesulfonates. *Chemistry–An Asian Journal, 2*(4), 477–483.
73. Salomé, C., & Kohn, H. (2009). Triphenylphosphine dibromide: a simple one-pot esterification reagent. *Tetrahedron, 65*(2), 456–460.

74. Sato, A., Nakamura, Y., Maki, T., Ishihara, K., & Yamamoto, H. (2005). Zr (IV) Fe (III), Ga (III), and Sn (IV) Binary Metal Complexes as Synergistic and Reusable Esterification Catalysts. *Advanced Synthesis & Catalysis, 347*(10), 1337–1340.

75. Shah, J. P., Russo, C. M., Howard, K. T., & Chisholm, J. D. (2014). Spontaneous formation of PMB esters using 4-methoxybenzyl-2, 2, 2-trichloroacetimidate. *Tetrahedron Letters, 55*(10), 1740–1742.

76. Tang, Z., Jiang, Q., Peng, L., Xu, X., Li, J., Qiu, R., & Au, C. T. (2017). Zirconocene-catalyzed direct (trans) esterification of acyl acids (esters) and alcohols in a strict 1: 1 ratio under solvent-free conditions. *Green Chemistry, 19*(22), 5396–5402.

77. Taniguchi, T., Hirose, D., & Ishibashi, H. (2011). Esterification via iron-catalyzed activation of triphenylphosphine with air. *ACS Catalysis, 1*(11), 1469–1474.

78. Thombal, R. S., & Jadhav, V. H. (2015). Biomass derived β-cyclodextrin-SO3H carbonaceous solid acid catalyst for catalytic conversion of carbohydrates to 5-hydroxymethylfurfural. *Applied Catalysis A: General, 499,* 213–216.

79. Tian, J., Gao, W. C., Zhou, D. M., & Zhang, C. (2012). Recyclable hypervalent iodine (III) reagent iodosodilactone as an efficient coupling reagent for direct esterification, amidation, and peptide coupling. *Organic Letters, 14*(12), 3020–3023.

80. Verheyen, T., Smet, M., & De Borggraeve, W. M. (2017). Water Tolerant and Reusable Sulfonated Hyperbranched Poly (aryleneoxindole) Acid Catalyst for Solvent-Free Esterification. *Chemistry Select, 2*(30), 9822–9828.

81. Wakasugi, K., Misaki, T., Yamada, K., & Tanabe, Y. (2000). Diphenylammonium triflate (DPAT): efficient catalyst for esterification of carboxylic acids and for transesterification of carboxylic esters with nearly equimolar amounts of alcohols. *Tetrahedron Letters, 41*(27), 5249–5252.

82. Wang, W., Liu, H., Xu, S., & Gao, Y. (2013). Esterification Catalysis by Pyridinium p-Toluenesulfonate Revisited—Modification with a Lipid Chain for Improved Activities and Selectivities. *Synthetic Communications, 43*(21), 2906–2912.

83. Xiang, J., Orita, A., & Otera, J. (2002). Fluorous biphasic esterification directed towards ultimate atom efficiency. *Angewandte Chemie*

International Edition, 41(21), 4117–4119.

84. Xu, X. H., Azuma, A., Taniguchi, M., Tokunaga, E., & Shibata, N. (2013). Efficient direct ester condensation between equimolar amounts of carboxylic acids and alcohols catalyzed by trifluoromethanesulfonic acid (TfOH) in Solkane365mfc. *RSC Advances, 3*(12), 3848–3852.
85. Zhang, L., Zhang, G., Zhang, M., & Cheng, J. (2010). Cu (OTf) 2-mediated Chan-Lam reaction of carboxylic acids to access phenolic esters. *The Journal of Organic Chemistry, 75*(21), 7472–7474.
86. Zhao, Y., Bordwell, F. G., Cheng, J. P., & Wang, D. (1997). Equilibrium Acidities and Homolytic Bond Dissociation Energies (BDEs) of the Acidic H–N Bonds in Hydrazines and Hydrazides. *Journal of the American Chemical Society, 119*(39), 9125–9129.

5

Carboxylic Acid Reactions and Uses

CONTENTS

5.1. INTRODUCTION

A carboxylic acid is an organic complex having a (COOH) carboxyl functional group. They come about broadly in nature and are also unnaturally manufactured by humans. Upon deprotonation (removal of a proton), carboxylic acids produce a carboxylate anion with the general formula R-COO–, which can give a variety of beneficial salts like soaps (Sáenz-Galindo et al., 2018).

The most important functional group that present are the carboxylic acids C=O. This sort of organic compounds can be attained by different ways, some carboxylic acids, for example, fumaric acid, citric acid or lactic acid are formed from by fermentation most of these kinds of carboxylic acids are useful in the food industry (Oguz et al., 2017).

The general formula of the carboxylic acid is R-COOH, were as COOH indicates the carboxyl group, while R indicates the rest of the particle to which this group is connected. In this carboxyl group, there occurs a carbon which has a double bond with an oxygen atom and a single bond with a hydroxyl group (OH). The general arrangement of a carboxylic acid (COOH) is shown below.

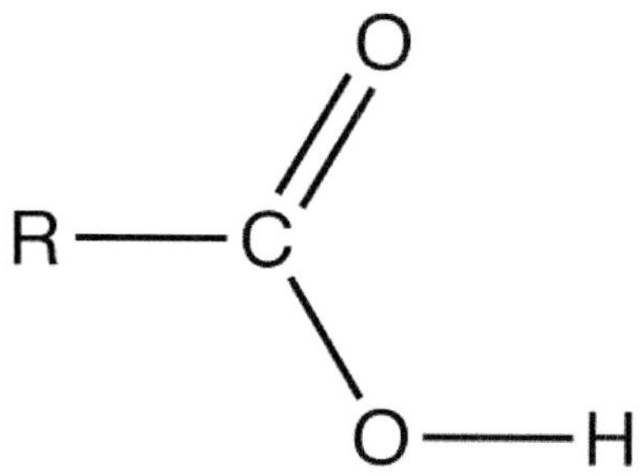

Figure 5.1: General structural arrangement of a carboxylic acid.

[Source: https://byjus.com/chemistry/carboxylic-acid-properties/]

From the illustration provided above, it can be perceived that a carboxylic acid has a hydroxyl group attached to a carbonyl carbon. Because of the electronegativity of the oxygen atom, this functional group can experience ionization and release a proton (Gedye et al., 1986; Yasa et al., 2017).

The carboxylate ion, formed from the elimination of a proton from the carboxyl group, is stabilized by the existence of two oxygen atoms (through which the (-) negative charge can transfer). Some common examples of carboxylic acids contain Formic acid and acetic acid (a component of vinegar).

5.2. NOMENCLATURE OF CARBOXYLIC ACIDS

Normally, these organic compounds are discussed by their insignificant names, which hold the suffix "-ic acid." An instance of an *insignificant name for a carboxylic acid is acetic acid (CH_3COOH).* In the IUPAC terminology of these compounds, the postfix "-oic acid" is given.

The recommendations that must be followed in the IUPAC terminology of carboxylic acids are listed beneath (Roughley & Jordan, 2011).

- The postfix "e" in the name of the analogous alkane is interchanged with "oic acid."
- When the aliphatic chain has only one carboxyl group, the carboxylic carbon is always referred to as one. For instance, CH_3COOH is termed as ethanoic acid.
- When the aliphatic chain holds more than one carboxyl group, the overall number of carbon atoms is counted and the amount of carboxyl groups is represented by Greek number prefixes like "di-," "tri-," etc.
- A carboxylic acid is termed by the addition of the suffixes and prefixes to the parental alkyl chain. Arabic numbers are used for specifying the positions of the carboxyl group.
- The name "carboxy" or "carboxylic acid" can also be allocated for a carboxyl substituent on a carbon chain. An instance of such terminology is the term 2-carboxyfuran for the compound 2-Furoic acid.

5.3. CARBOXYLIC ACID EXAMPLES

Some examples explaining the nomenclature of carboxylic acids according to IUPAC recommendations are providing below.

Table 5.1: IUPAC Nomenclature of Carboxylic Acids

S.No.	IUPAC Name of the Carboxylic Acid	Trivial Name and Formula
1	Methanoic acid	Formic acid, H-COOH
2	But-2-enoic acid	Crotonic acid, $CH_3CH=CH\text{-}COOH$
3	Carbonic acid	Carbonic acid, OH-COOH
4	Butanoic acid	Butyric acid, $CH_3(CH_2)_2COOH$

5.4. PROPERTIES OF CARBOXYLIC ACIDS

Maximum of the properties of carboxylic acids is due to the existence of the carboxyl group. Some of the chemical and physical properties of these mixtures are explained in this subsection.

5.4.1. Physical Properties of Carboxylic Acids

- Carboxylic acid particles are polar because of the presence of two electronegative oxygen atoms.
- They also contribute to hydrogen bonding because of the presence of the hydroxyl group and the carbonyl group (C=O).
- When positioned in nonpolar solvents, these compounds form ***dimers through hydrogen bonding*** between the carbonyl group of one carboxylic acid and the hydroxyl group of the other.

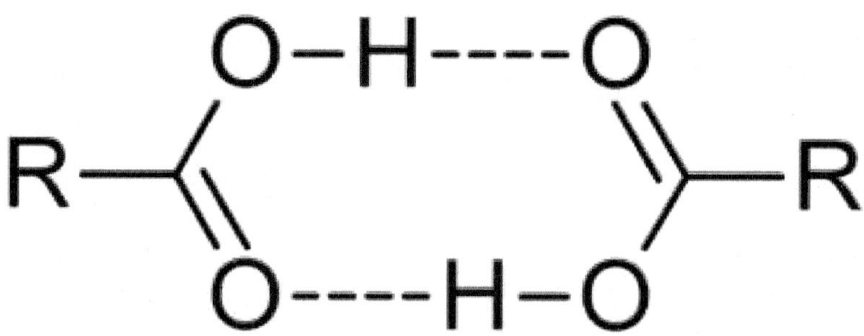

Figure 5.2: Hydrogen bonding in carboxylic acid.

[Source: https://byjus.com/chemistry/carboxylic-acid-properties/]

- The solubility of compounds having the carboxyl functional group in the water relies on the extent of the compound. The smaller the compound (the shorter the R group), the greater the solubility.
- The boiling point of a carboxylic acid is usually greater than that of water.
- These compounds have the capability to contribute protons and are therefore ***Bronsted-Lowry acids.***
- They normally have a strong sour fragrance. Though, their esters have sweet smells and are therefore used in perfumes.

5.4.2. Chemical Properties of Carboxylic Acids

- The ***α-carbon related to a carboxylic acid can simply be halogenated through the Hell-Volhard-Zelinsky reaction.***
- These compounds can be transformed into amines by using the Schmidt response.
- A carboxylic acid can be deduced to alcohol by reacting it with hydrogen to origin a hydrogenation reaction.
- By reaction with alcohols, these compounds produce esters.

5.5. PHYSICAL PROPERTIES OF CARBOXYLIC ACIDS

The table at the start of this page provided the boiling and melting points for a homologous group of carboxylic acids containing one to ten carbon atoms. The boiling points increased with the extent in a regular way, but the melting points did not increase like the boiling point. Unbranched acids made of an even number of carbon particles have melting points greater than the odd-numbered homologs containing one more or one less carbon. This indicates the variances in intermolecular attractive forces in the translucent state. In the table of fatty acids, we observe that the existence of a cis-double bond expressively drops the melting point of a compound. So, palmitoleic acid melts above 60° lesser than palmitic acid, and the same decreases happen for the C_{18} and C_{20} compounds. Again, changes in crystal filling and intermolecular forces are reliable (Carey et al., 2006; Constable et al., 2007).

The factors that affect the comparative boiling points and water solubilities of several kinds of compounds were discussed before. In common, dipolar attractive forces between molecules act to raise the boiling point of a particular compound, with hydrogen bonds being a great example (Ghose et al., 1999; Amarnath et al., 2012). Hydrogen bonding is also a main aspect in the water solubility of covalent compounds the next table lists a few instances of these properties for some like-sized polar compounds (the non-polar hydrocarbon hexane is given for comparison).

Table 5.2: Physical Properties of Some Organic Compounds

S.No.	IUPAC Name	Molecular Weight	Water Solubility	Formula	Boiling Point
1	Butanoic acid	88	Very soluble	$CH_3(CH_2)_2CO_2H$	164°C
2	1-pentanol	88	Slightly soluble	$CH_3(CH_2)_4OH$	138°C
3	Pentanal	86	Slightly soluble	$CH_3(CH_2)_3CHO$	103°C
4	Ethyl ethanoate	88	Moderately soluble	CH3CO2C2H5	77°C
5	Methyl propanoate	88	Slightly soluble	$CH_3CH_2CO_2CH_3$	80°C
6	Butanamide	87	Soluble	$CH_3(CH_2)_2CONH_2$	216°C
7	N,N-dimethylethanamide	87	Very soluble	$CH_3CON(CH_3)_2$	165°C
8	1-aminobutane	87	Very soluble	$CH_3(CH_2)_4NH_2$	103°C
9	Pentanenitrile	83	Slightly soluble	$CH_3(CH_2)_3CN$	140°C
10	Hexane	86	Insoluble	$CH_3 (CH_2)_4CH_3$	69°C

The first five records all have oxygen functional groups, and the comparatively great boiling points of the first two are obviously because of hydrogen bonding. Carboxylic acids have extremely high boiling points, due to a large part to dimeric relations containing two hydrogen bonds. The structural formula for the dimer of acetic acid is unknown here. When the mouse indicator passes above the drawing, an electron cloud figure will seem (Allen et al., 2012; Houlding et al., 2013; Lanigan et al., 2013). The great boiling points of the nitriles and amides are due to a large part to strong dipole attractions, enhanced in some circumstances by hydrogen bonding.

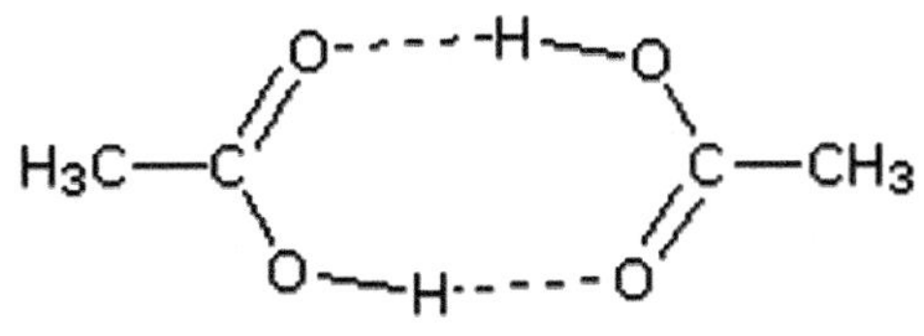

Figure 5.3: Strong dipole attractions in carboxylic acid.

[Source: https://chem.libretexts.org/Bookshelves/Organic_Chemistry/Map%3A_Organic_Chemistry_(Bruice)/17%3A_Carbonyl_Compounds_I%3A_Reactions_of_Carboxylic_Acids_and_Carboxylic_Derivatives/17.2____The_Structures_of_Carboxylic_Acids_and_Carboxylic_Derivatives]

5.6. THE ACIDITY OF CARBOXYLIC ACIDS

The pK_a's of some usual carboxylic acids are recorded in the next table. When we relate these values with those of related alcohols, like 2methyl-2-propanol (pK_a = 19) and ethanol (pK_a = 16), it is clear that carboxylic acids are sturdier acids by over 10 powers of 10! Moreover, electronegative substituents close to the carboxyl group act to raise the acidity (Varma & Naicker, 1999; Lukasik & Wagner-Wysiecka, 2014).

Table 5.3. pK_a of Different Organic Compounds

Compound	pK_a	Compound	pK_a
$CH_3CH_2CH_2CO_2H$	4.82	HCO_2H	3.75
$ClCH_2CH_2CH_2CO_2H$	4.53	CH_3CO_2H	4.74
$CH_3CHClCH_2CO_2H$	4.05	FCH_2CO_2H	2.65
$CH_3CH_2CHClCO_2H$	2.89	$ClCH_2CO_2H$	2.85
$C_6H_5CO_2H$	4.20	$BrCH_2CO_2H$	2.90
p-$O_2NC_6H_4CO_2H$	3.45	ICH_2CO_2H	3.10
p-$CH_3OC_6H_4CO_2H$	4.45	Cl_3CCO_2H	0.77

Why should the existence of a carbonyl group next to a hydroxyl group have such a deep effect on the acidity of the hydroxyl proton? To respond to this question, we must have to go back to the nature of acid-base symmetries and the description of pK_a, shown by the general equations given below. These relations were defined in a previous subdivision of this text.

$$H{-}A + H_2O \rightleftharpoons H_3O^{\oplus} + A:^{\ominus} \qquad K_{eq} = \frac{[H_3O^{\oplus}][A:^{\ominus}]}{[HA][H_2O]}$$

$$K_a = \frac{[H_3O^{\oplus}][A:^{\ominus}]}{[HA]} \qquad pK_a = -\log K_a = \log\left(\frac{1}{K_a}\right)$$

We know that an equilibrium supports the thermodynamically more constant side and that the extent of the equilibrium constant mirrors the energy difference between the constituents of each side. In an acid-base equilibrium, the stability always supports the weaker acid and base (these are the more constant components). Water is the standard base used for pK_a dimensions; subsequently, anything that balances the conjugate base ($A:^{(-)}$) of acid will essentially make that acid (H–A) stronger and move the equilibrium to the right. Both the carboxylate anion and the carboxyl group are equalized by resonance, but the steadiness of the anion is much better than that of the neutral function, as presented in the following figure. In the carboxylate anion, the two contributing buildings have equal weight in the mixture, and the C–O bonds are of equal size (between a single and a double bond). This stabilization guides to a significantly increased acidity, as shown by the energy diagram revealed by ticking the "Toggle Display" button (Perreux et al., 2002; Yaragorla et al., 2014; Ojeda-Porras et al., 2015).

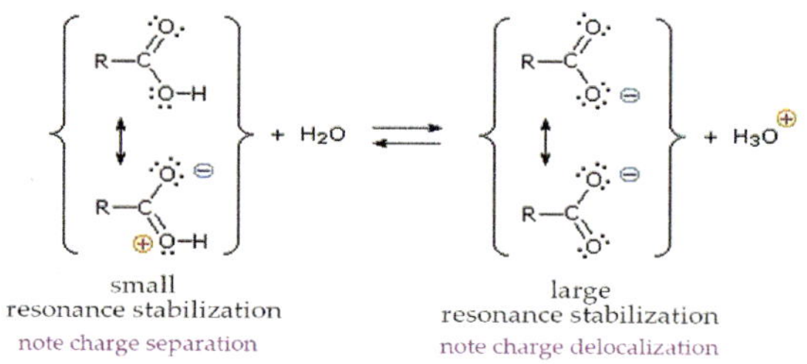

Figure 5.4: Effect of stabilization on acidity of carboxylic acids.

[Source:https://chem.libretexts.org/Bookshelves/Organic_Chemistry/Map%3A_Organic_Chemistry_(Bruice)/17%3A_Carbonyl_Compounds_I%3A_Reactions_of_Carboxylic_Acids_and_Carboxylic_Derivatives/17.2____The_Structures_of_Carboxylic_Acids_and_Carboxylic_Derivatives]

The resonance effect defined here is clearly the major contributor to the excellent acidity of carboxylic acids. Nevertheless, inductive effects also play a part. For instance, alcohols have pK_a's of 16 or greater but their acidity is increased by electron removing substituents on the alkyl group. The next diagram shows this factor for numerous simple inorganic and organic compounds (row #1) and displays how inductive electron removal may also surge the acidity of carboxylic acids (rows #2 & 3). The acidic hydrogen is colored red in all cases (Ugi, 1959; 1962).

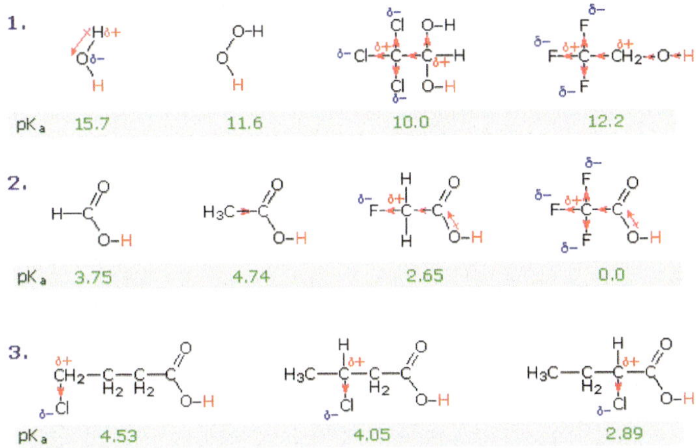

Figure 5.5: Effect of electron removal on acidity of carboxylic acids.

[Source: https://chem.libretexts.org/Bookshelves/Organic_Chemistry/Map%3A_Organic_Chemistry_(Bruice)/17%3A_Carbonyl_Compounds_I%3A_Reactions_of_Carboxylic_Acids_and_Carboxylic_Derivatives/17.2____The_Structures_of_Carboxylic_Acids_and_Carboxylic_Derivatives]

As hydrogen is less electronegative than oxygen so water is less acidic than hydrogen peroxide, and the covalent bond linking these atoms is polarized in the manner displayed. Alcohols are marginally less acidic than water, due to the low electronegativity of carbon, but 2,2,2,-trifluoroethanol, and chloral hydrate, $Cl_3CCH(OH)_2$ are significantly more acidic than water, because of inductive electron removal by the electronegative halogens (and the 2nd oxygen in chloral hydrate). In the case of carboxylic acids, if the electrophilic behavior of the carbonyl carbon is reduced the acidity of the carboxylic acid will also decrease (Passerini, 1921; Passerini & Simone, 1921). Likewise, a rise in its electrophilicity will increase the acidity of the acid. Acetic acid is 10 times weaker acid than the formic acid (first two records in the 2nd row), checking the electron giving the character of an alkyl group comparative to hydrogen, as noted previously in a discussion of carbocation constancy. Electronegative substituents increase acidity by inaugural electron removal (Diorazio et al., 2006; Jahani et al., 2010). As anticipated, the greater the electronegativity of the substituent the larger the increase in acidity (F > Cl > Br > I), and the nearer the substituent is to the carboxyl group the more is its effect (isomers in the 3rd row). Substituents also affect the acidity of benzoic acid products, but resonance effects contend with inductive effects (Samanta & Das, 2015; Dadhania et al., 2017; Pai et al., 2017). The methoxy group is electron giving and the nitro group is electron removing (last three entries in the table of pK_a values).

5.7. CARBOXYLIC ACID DERIVATIVES AND ACYL GROUPS

The functional groups that experience nucleophilic acyl replacements are called carboxylic acid products: they contain carboxylic acids themselves, carboxylates (deprotonated carboxylic acids), thioesters, esters, acyl phosphates, and amides. Two more instances of carboxylic acid products which are less biologically significant but vital in laboratory synthesis are anhydrides acid, chlorides, and carboxylic acid (Abbott et al., 2003; Panigrahi et al., 2006; Teles et al., 2017).

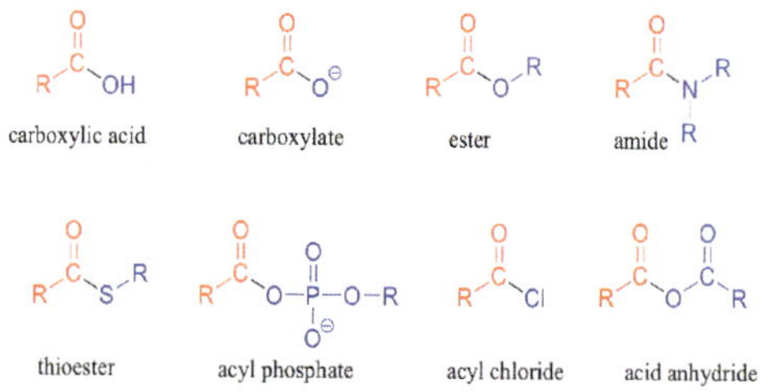

Figure 5.6: Carboxylic acid and its derivatives.

[Source: https://chem.libretexts.org/Bookshelves/Organic_Chemistry/Book%3A_Organic_Chemistry_with_a_Biological_Emphasis_(Soderberg)/12%3A_Acyl_substitution_reactions/12.1%3A_Introduction_to_carboxylic_acid_derivatives_and_the_nucleophilic_acyl_substitution_reaction]

The carboxylic acid products can be illustrious from ketones and aldehydes by the existence of a group having an electronegative heteroatom – typically, nitrogen sulfur, or oxygen-bonded straight to the carbonyl carbon (Sarkar et al., 2005; Sun & Luo, 2005). We can think of a carboxylic acid derived as containing two sides. One side is the carbonyl group and the joined alkyl group: this is known as an *acyl group* (in the particular case where R is a methyl group, the term *acetyl group* is used). Another side is the heteroatom-containing group: in this edition, we will occasionally refer to this constituent as the 'acyl X' group (this, nevertheless, is *not* a regular term in organic chemistry).

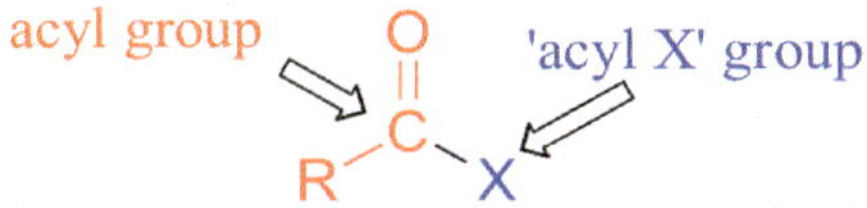

Figure 5.7: Description of acyl-X group.

[Source: https://chem.libretexts.org/Bookshelves/Organic_Chemistry/Book%3A_Organic_Chemistry_with_a_Biological_Emphasis_(Soderberg)/12%3A_Acyl_substitution_reactions/12.1%3A_Introduction_to_carboxylic_acid_derivatives_and_the_nucleophilic_acyl_substitution_reaction]

Notice that the acyl X groups are basically deprotonated (removal of the proton) forms of other functional groups: in an amide, for instance, the acyl X group is an amine, whereas in an ester the acyl X group is alcohol (Yang et al., 2008; Hosseini-Monfared et al., 2015).

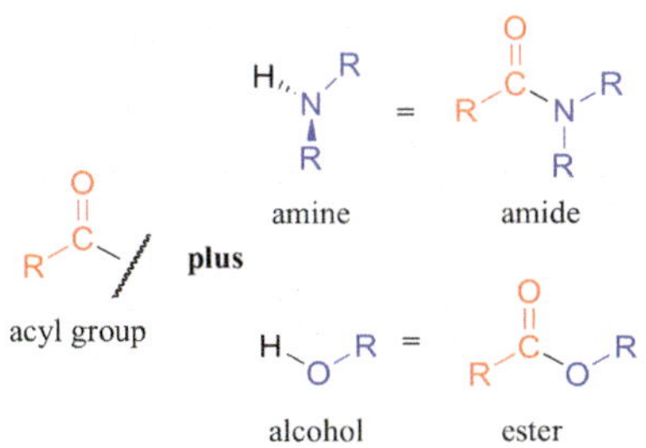

Figure 5.8: Different compounds containing acyl group.

[Source: https://chem.libretexts.org/Bookshelves/Organic_Chemistry/Book%3A_Organic_Chemistry_with_a_Biological_Emphasis_(Soderberg)/12%3A_Acyl_substitution_reactions/12.1%3A_Introduction_to_carboxylic_acid_derivatives_and_the_nucleophilic_acyl_substitution_reaction]

Here are a few instances of molecules with carboxylic acid derived groups. Acetylsalicylic acid, the anodyne medication commonly called aspirin, comprises an ester and a carboxylic acid. The peptide bonds that bond amino acids together in proteins are amides (Armelao et al., 2006; Hojjati et al., 2007; Cabello et al., 2015).

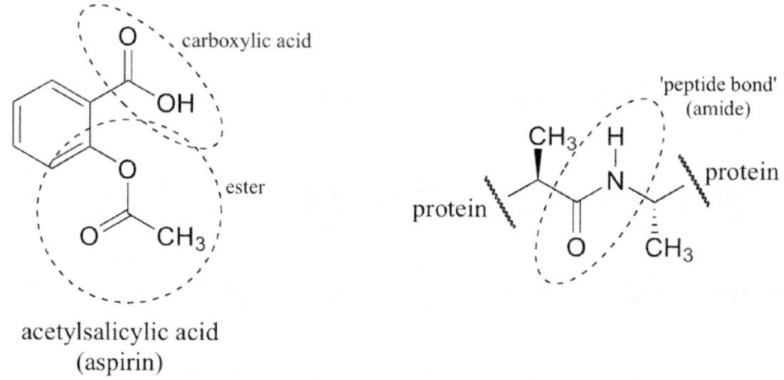

Figure 5.9: Molecules containing carboxylic acid derived groups.

[Source: https://chem.libretexts.org/Bookshelves/Organic_Chemistry/Book%3A_Organic_Chemistry_with_a_Biological_Emphasis_(Soderberg)/12%3A_Acyl_substitution_reactions/12.1%3A_Introduction_to_carboxylic_acid_derivatives_and_the_nucleophilic_acyl_substitution_reaction]

Soap is serene mainly of ‘fatty acid’ molecules, which are in fact carboxylates with long carbon chains. Stearate is the main component in several types of soap.

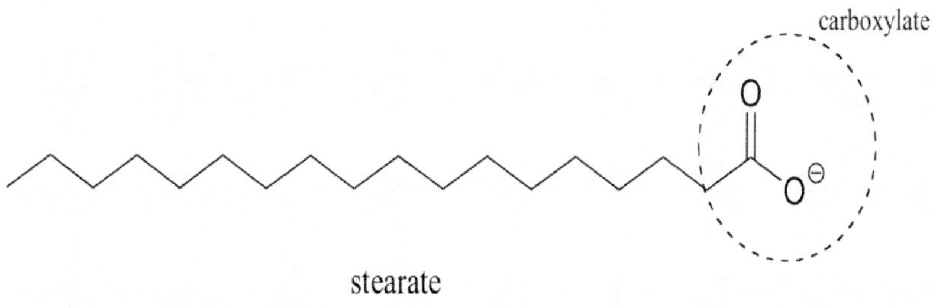

Figure 5.10: Organic structure of soap molecule.

[Source: https://chem.libretexts.org/Bookshelves/Organic_Chemistry/Book%3A_Organic_Chemistry_with_a_Biological_Emphasis_(Soderberg)/12%3A_Acyl_substitution_reactions/12.1%3A_Introduction_to_carboxylic_acid_derivatives_and_the_nucleophilic_acyl_substitution_reaction]

Acetyl-CoA, a very essential molecule in metabolism, is described by its thioester group (it also covers two amide groups – can you find them?).

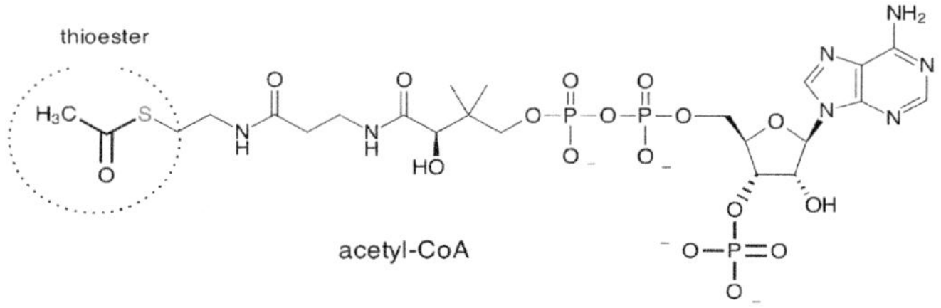

Figure 5.11: Acetyl molecule with thioester group.

[Source: https://chem.libretexts.org/Bookshelves/Organic_Chemistry/Book%3A_Organic_Chemistry_with_a_Biological_Emphasis_(Soderberg)/12%3A_Acyl_substitution_reactions/12.1%3A_Introduction_to_carboxylic_acid_derivatives_and_the_nucleophilic_acyl_substitution_reaction]

5.8. THE NUCLEOPHILIC ACYL SUBSTITUTION REACTION

The point that the atom nearby to the carbonyl carbon in carboxylic acid products is an electronegative heteroatom – rather than a carbon alike in ketones or hydrogen alike in aldehydes – is essential to understand the reactivity of these functional groups. Alike in ketones and aldehydes, carboxylic acid products are attacked from a side of their trigonal planar carbonyl carbon by a nucleophile, changing this carbon to tetrahedral (sp^3) geometry. In carboxylic acid products, the acyl X group is a potential withdrawal group (Sarkany et al., 2005; Narayanan & Sakthivel, 2008). What this means is that the tetrahedral produce formed from an attack of the nucleophile on the carbonyl carbon is not regarded as the product: it is a *reactive intermediary*. The tetrahedral intermediary quickly collapses: the carbon-oxygen double bond re-forms, and the acyl X group is eliminated (Qu et al., 2010; Cash et al., 2012).

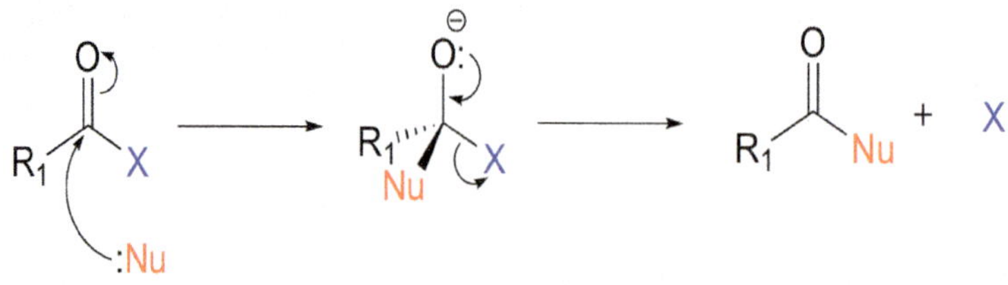

Figure 5.12: Nucleophilic acyl substitution reaction.

[Source: https://chem.libretexts.org/Bookshelves/Organic_Chemistry/Book%3A_Organic_Chemistry_with_a_Biological_Emphasis_(Soderberg)/12%3A_Acyl_substitution_reactions/12.1%3A_Introduction_to_carboxylic_acid_derivatives_and_the_nucleophilic_acyl_substitution_reaction]

Notice that in the production, the nucleophile turn out to be the new acyl X group. Due to which this reaction type is known as a nucleophilic acyl *substitution*: one acyl X group is replaced for another. For instance, in the reaction beneath, one alcohol X group (3-methyl-1-butanol) is swapped by another alcohol X group (methanol), as one ester is changed to another.

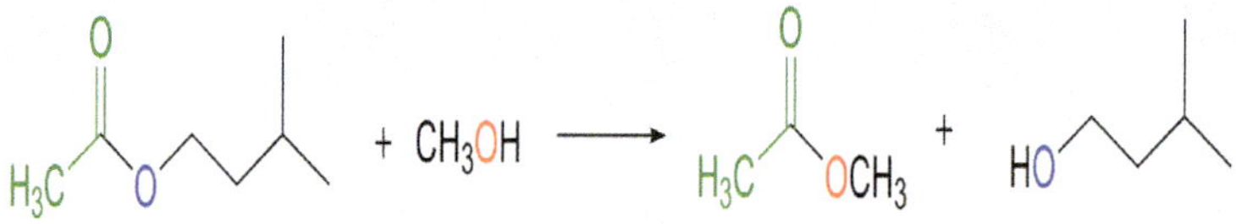

Figure 5.13: Acyl transfer reaction.

[Source: https://chem.libretexts.org/Bookshelves/Organic_Chemistry/Book%3A_Organic_Chemistry_with_a_Biological_Emphasis_(Soderberg)/12%3A_Acyl_substitution_reactions/12.1%3A_Introduction_to_carboxylic_acid_derivatives_and_the_nucleophilic_acyl_substitution_reaction]

Another approach of observing at this reaction is to picture the *acyl* group being shifted from one acyl X group to another: in the sample above, the acetyl group is being shifted from 3-methyl-1-butanol to methanol. For this reason, nucleophilic acyl substitutions are also normally known as *acyl transfer* reactions (Masala & Seshadri, 2004; Poole et al., 2004).

When the entering nucleophile in an acyl replacement is a water molecule, the reaction is also raised to as acyl hydrolysis. For instance, the next reaction can be defined as the hydrolysis of an ester (to form carboxylic acid and an alcohol).

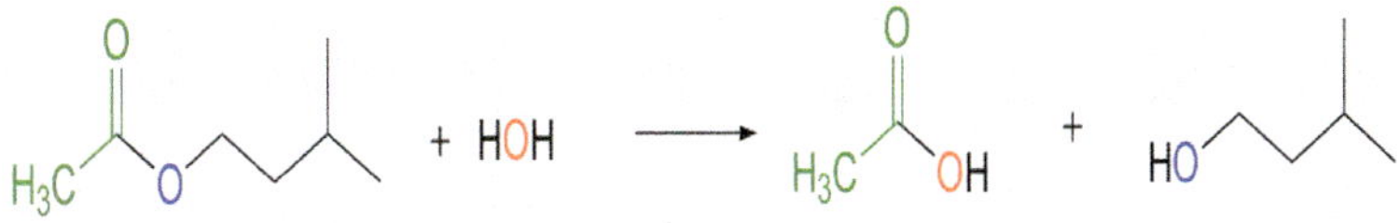

Figure 5.14: Acyl hydrolysis reaction.

[Source: https://chem.libretexts.org/Bookshelves/Organic_Chemistry/Book%3A_Organic_Chemistry_with_a_Biological_Emphasis_(Soderberg)/12%3A_Acyl_substitution_reactions/12.1%3A_Introduction_to_carboxylic_acid_derivatives_and_the_nucleophilic_acyl_substitution_reaction]

We could also explain this reaction as the transmission of an acyl group from alcohol to a water molecule. In a similar trace, the hydrolysis of an amide to form a carboxylic acid could be explained as the transmission of an acyl group from ammonia (NH3) to water.

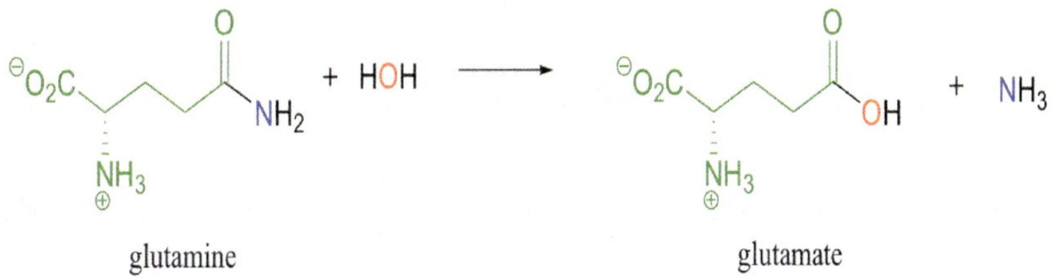

Figure 5.15: Conversion of an acyl group from ammonia (NH3) to water.

[Source: https://chem.libretexts.org/Bookshelves/Organic_Chemistry/Book%3A_Organic_Chemistry_with_a_Biological_Emphasis_(Soderberg)/12%3A_Acyl_substitution_reactions/12.1%3A_Introduction_to_carboxylic_acid_derivatives_and_the_nucleophilic_acyl_substitution_reaction]

As we will understand in later divisions of this chapter the hydrolysis of amides and esters are mainly significant reaction types in biochemical ways. When your body assimilates the fat in a hamburger, for instance, enzymes in your pancreas known as lipases first catalyze ester hydrolysis reactions to the fatty acids (Raveendran et al., 2006; Schmid, 2008; Slistan-Grijalva et al., 2008).

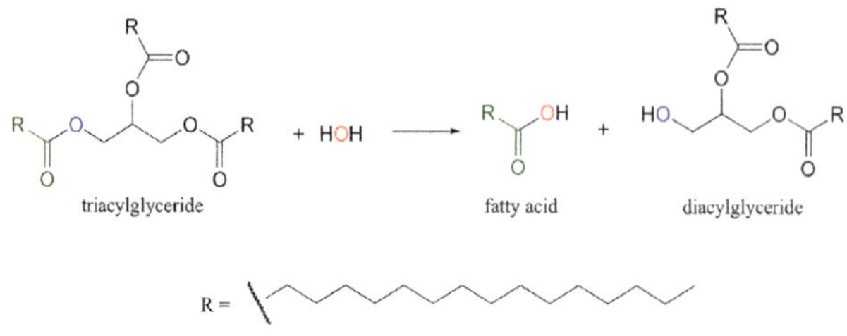

Figure 5.16: Catalysis of ester hydrolysis reactions to the fatty acids.

[Source: https://chem.libretexts.org/Bookshelves/Organic_Chemistry/Book%3A_Organic_Chemistry_with_a_Biological_Emphasis_(Soderberg)/12%3A_Acyl_substitution_reactions/12.1%3A_Introduction_to_carboxylic_acid_derivatives_and_the_nucleophilic_acyl_substitution_reaction]

5.9. THE RELATIVE REACTIVITY OF CARBOXYLIC ACID DERIVATIVES

The comparative reactivity of the carboxylic acid products is an essential concept to understand first before entering into a full analysis of nucleophilic acyl substitutions. As a common rule, the carbonyl carbon in an acyl group is fewer electrophilic than that in a ketone or aldehyde. This is because, in carboxylic acid products, the fractional positive charge on the carbon is stabilized slightly by resonance effects from the nearby heteroatom (Luo et al., 2005; Abdel-Aziz et al., 2014).

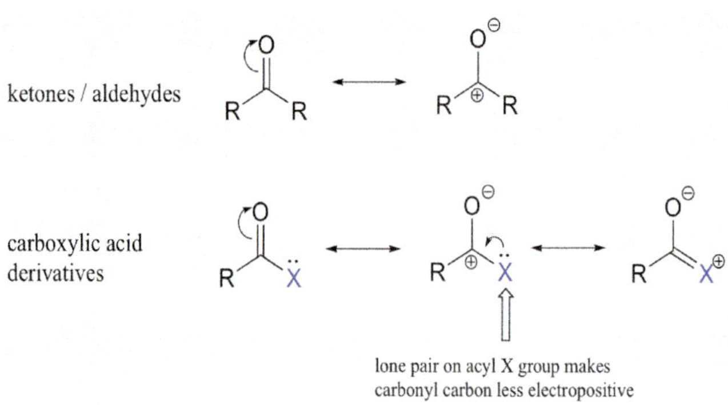

Figure 5.17: Nucleophilic acyl substitution reactions.

[Source: https://chem.libretexts.org/Bookshelves/Organic_Chemistry/Book%3A_Organic_Chemistry_with_a_Biological_Emphasis_(Soderberg)/12%3A_Acyl_substitution_reactions/12.1%3A_Introduction_to_carboxylic_acid_derivatives_and_the_nucleophilic_acyl_substitution_reaction]

Among the carboxylic acid products, carboxylate groups are the minimum reactive towards nucleophilic acyl substitution, after that amides, then esters and (protonated) carboxylic acids, thioesters, and lastly acyl phosphates, which are the utmost reactive among the biologically related acyl groups.

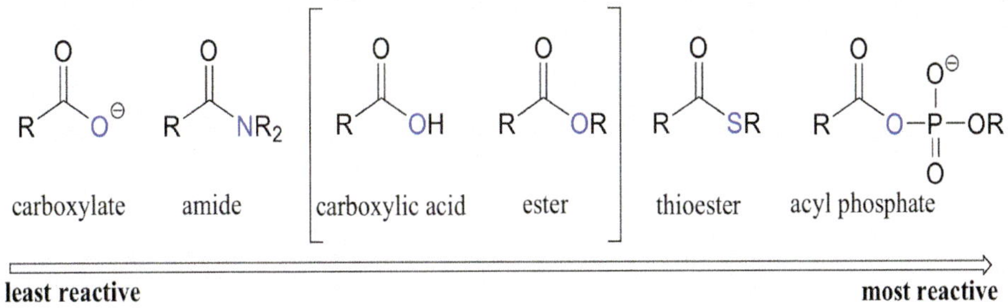

Figure 5.18: Reactivity of carboxylic acid products.

[Source: https://chem.libretexts.org/Bookshelves/Organic_Chemistry/Book%3A_Organic_Chemistry_with_a_Biological_Emphasis_(Soderberg)/12%3A_Acyl_substitution_reactions/12.1%3A_Introduction_to_carboxylic_acid_derivatives_and_the_nucleophilic_acyl_substitution_reaction]

The dissimilar reactivities of the functional groups can be understood by valuing the basicity of the withdrawal group in each case – it is established fact that weaker bases are good leaving groups! A thioester has reactivity than an ester, for instance, because a thiolate (RS^-) is a weaker base than an alkoxide (RO^-). In common, if the entering nucleophile is a weaker base than the 'acyl X' group that is present there, the first nucleophilic step will basically reverse itself and we'll get the initial materials back:

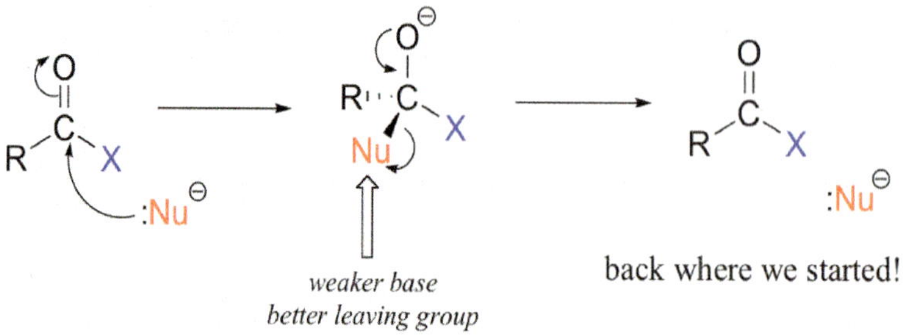

Figure 5.19: Reverse nucleophilic reaction.

[Source: https://chem.libretexts.org/Bookshelves/Organic_Chemistry/Book%3A_Organic_Chemistry_with_a_Biological_Emphasis_(Soderberg)/12%3A_Acyl_substitution_reactions/12.1%3A_Introduction_to_carboxylic_acid_derivatives_and_the_nucleophilic_acyl_substitution_reaction]

This is why it is not possible to *straight* change an ester, for instance, into a thioester by an acyl substitution reaction – this would be a difficult reaction.

Here's another approach to ponder the comparative reactivities of the different carboxylic acid derivatives: consider the comparative electrophilicity, or degree of partial positive charge, on the carbonyl carbon in each type. This depends on how much electron density the neighboring heteroatom on the acyl X group is capable to donate: higher the electron donation by the heteroatom indicates lower partial positive charge on the carbonyl carbon, which sequentially indicates lower electrophilicity (Sun & Luo, 2005; Kango et al., 2013; Hethnawi et al., 2017).

The negatively charged oxygen on the carboxylate group has heaps of electron density to contribute, therefore the carbonyl carbon is not very electrophilic. In amides, the nitrogen atom is a dominant electron contributing group by resonance – remember that the carbon-nitrogen link in peptides has substantial double-bond character – therefore amides are relatively unreactive. Amides *do* experience acyl substitution reactions in biochemical ways, but these reactions are fundamentally slow and the enzymes catalyzing them have advanced effective strategies to lower the activation energy hurdle (He et al., 2017; Sato et al., 2017).

Esters and carboxylic acids are in the intermediate range of reactivity, whereas thioesters are slightly more reactive. The most reactive of the carboxylic acid products often found in biomolecules are the acyl phosphates. These are most frequently present in two forms: the acyl-adenosine monophosphate and the simple acyl monophosphate (Shi et al., 2001; 2003; Sperling & Parak, 2010).

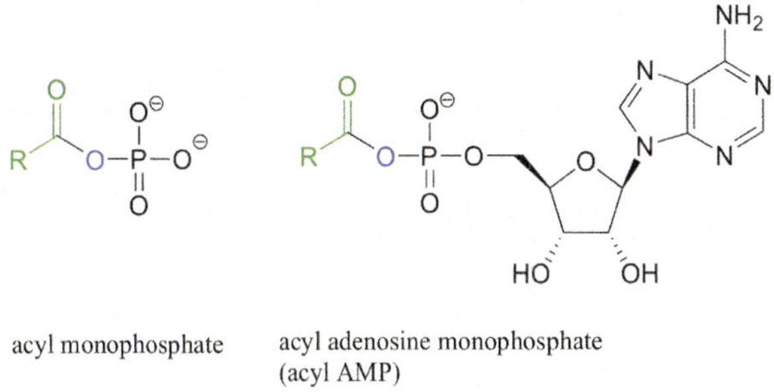

Figure 5.20: Acyl monophosphate molecules.

[Source: https://chem.libretexts.org/Bookshelves/Organic_Chemistry/Book%3A_Organic_Chemistry_with_a_Biological_Emphasis_(Soderberg)/12%3A_Acyl_substitution_reactions/12.1%3A_Introduction_to_carboxylic_acid_derivatives_and_the_nucleophilic_acyl_substitution_reaction]

Both are extremely reactive to acyl substitution reactions, and are frequently referred to as 'activated carboxylic acids' or 'activated acyl groups.' The great reactivity of acyl phosphates is due to the capability to form complexes with magnesium ions.

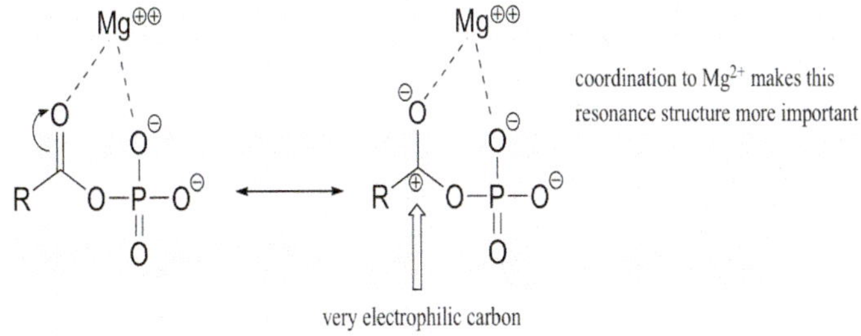

Figure 5.21: Capability of acyl phosphates to form complexes with magnesium ions.

[Source: https://chem.libretexts.org/Bookshelves/Organic_Chemistry/Book%3A_Organic_Chemistry_with_a_Biological_Emphasis_(Soderberg)/12%3A_Acyl_substitution_reactions/12.1%3A_Introduction_to_ carboxylic_acid_derivatives_and_the_nucleophilic_acyl_substitution_reaction]

The magnesium ion acts as a Lewis acid to take electron density from the oxygen end of the acyl carbonyl bond, therefore significantly increasing the degree of partial positive charge – and therefore the electrophilicity – of the carbonyl carbon. The magnesium ion also equilibriums negative charge on the phosphate, making it an exceptional leaving group (Schneiderman & Hillmyer, 2017; Sezen-Edmonds & Loo, 2017).

In our analysis of acyl substitution reactions, we will start with the foundation and reactions of the extremely reactive acyl phosphates. We will then argue how thioesters play an important part in the acyl substitution reactions of lipid metabolism. Lastly, we will take a look at some significant acyl substitution reactions containing esters, as well as the foundation and cleavage of the amide connections in the peptide bonds of proteins (Shi et al., 2015; Mohammadifar et al., 2016; Zhang et al., 2017).

5.10. USES OF CARBOXYLIC ACIDS

Some common usages of carboxylic acids are given beneath:

- Fatty acids that are important to human beings are containing carboxylic acids. Examples contain omega-3 fatty acids and omega-6.
- Advanced fatty acids are also used in the production of soaps.

The manufacturing of many other food products and soft drinks include the use of many carboxylic acids.

- The creation of rubber includes the practice of acetic acid as a coagulant.
- Hexanedioic acid is used in the production of nylon-6,6.

Carboxylic acids have several uses in the textile, leather, and rubber industries. Ethylenediaminetetra acetic acid is a generally used chelating proxy.

- The production of many drugs includes the use of these compounds. Consequently, carboxylic acids are very vital in pharmaceuticals.
- The manufacture of many polymers includes the use of compounds comprising the carboxyl functional group.

Three of the main uses of carboxylic acids are exemplified in the following subsectios.

5.10.1. Organic Synthesis

The use of carboxylic acids in organic production is a very extensive area and the chemical alterations of this group to another have made it a very handy functional group. These chemical transformations have got development when they carry out over Green chemistry processes (Lazzarato et al., 2008; Bian et al., 2017).

One of the approaches to aim for energy productivity (one of the principles of Green Chemistry) is to make reactions below microwave irradiation. The first report by this approach was an esterification reaction with carboxylic acids and alcohols attaining high yields in a little reaction time (Laaziz et al., 2017).

Besides the esterification, the amidation reaction by a transformation of the carboxylic acids is also significant because of the new covalent bond shaped. This bond is of great significance because it can be found in a varied variety of molecules both in small molecules and normal products with pharmacological activity (Laadila et al., 2017). So, the improvement of new straight amidation reactions is significant. One of those straight approaches of amide formation is by reacting amines with carboxylic acids by using toluene as a solvent or by using radiofrequency heating in neat environments. Nevertheless, this reagent-free pathway has restrictions in the substrate scope.

Conversely, Lanigan et al. (2013) reported a method in which they used simple borate esters that are effective reagents for the straight synthesis of amides, by using a variation of amines and carboxylic acids. This reaction can be carried out willingly to the air in acetonitrile reflux. The amidation product can be refined in a very simple method, in most circumstances, it only requires a simple clarification procedure by using commercially open resins giving tremendous yields (Figure 2.22).

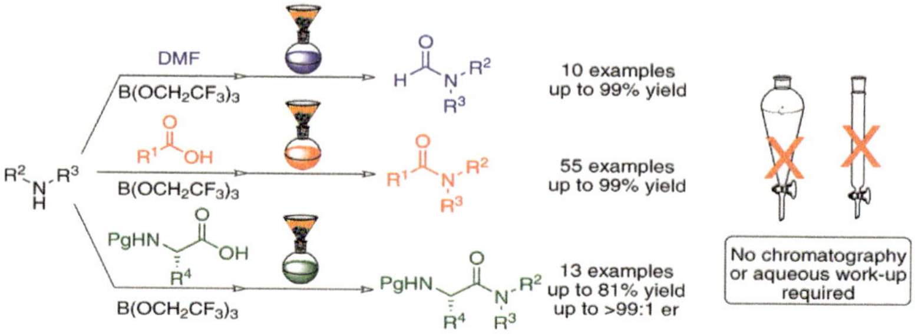

Figure 5.22: Carboxylic acids were used to obtain a variety of amides.

[Source: <http://pubs.acs.org/doi/abs/10.1021%2Fjo400509n>]

Another approach in which microwave purification was used in the amidation reaction has proven to be an effective production of amides (Jašo et al., 2015). For instance, Ojeda-Porras et al. (2015) explained a green approach for the direct amidation (**3**) of carboxylic acids (**1**) and amines (**2**) by using silica gel as a solid backing (Figure 2.23).

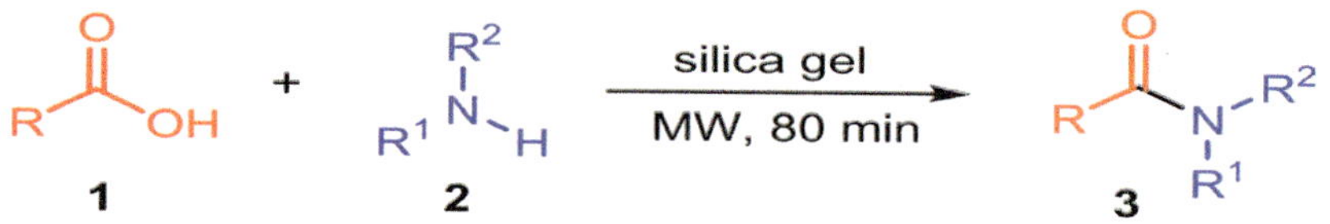

Figure 5.23: Mediation reaction via carboxylic acids and amines.

[Source: https://www.intechopen.com/books/carboxylic-acid-key-role-in-life-sciences/applications-of-carboxylic-acids-in-organic-synthesis-nanotechnology-and-polymers]

Besides the distinctive use of carboxylic acids in alterations other carbonyl groups, they can be used as a substrate in multicomponent reactions like the renowned Ugi (1959) and Passerini (1921) reactions. An example is the reaction between an, an isocyanide 4, oxazolidine 5 and a carboxylic acid 6 to offers *N*-acyloxy ethylamino acid products (7) (Figure 2.24) so that it can be completed with those structures that are formed by the well-established Passerini and Ugi reactions and they permit the generation of a chemical libraries.

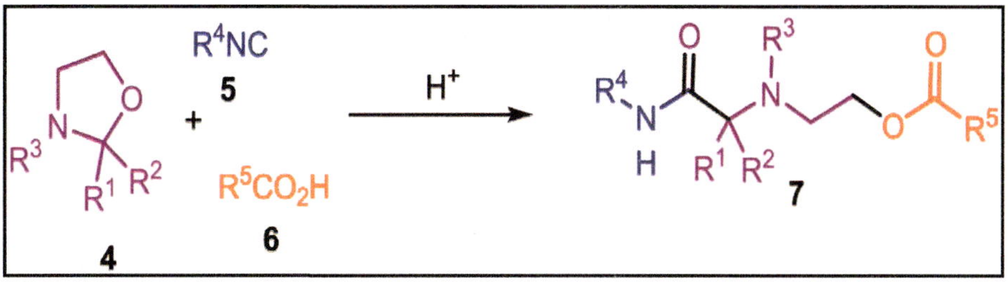

Figure 5.24: Use of carboxylic acid in a multicomponent reaction.

[Source: https://www.intechopen.com/books/carboxylic-acid-key-role-in-life-sciences/applications-of-carboxylic-acids-in-organic-synthesis-nanotechnology-and-polymers]

Also, the carboxylic acids can be used in organocatalytic reactions. As is in the case, carboxylic acid 11 can be used straight to gain the α-hydroxy phosphonates **12**. The reaction was carried on in a simple approach with a range of ketones 8 and aldehydes 9 with trimethylphosphite 10 in the company of catalytic extents of pyridine 2,6-dicarboxylic acid 11 in water as a solvent. This produces a low cost and environmentally friendly approach (Figure 2.25).

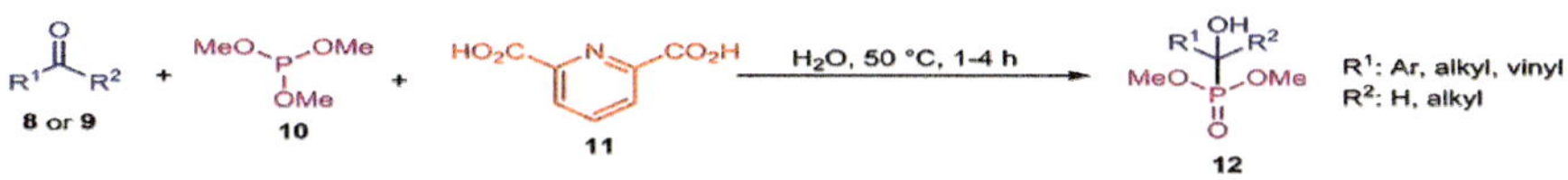

Figure 5.25: Carboxylic acid as a catalyst.

[Source: https://www.intechopen.com/books/carboxylic-acid-key-role-in-life-sciences/applications-of-carboxylic-acids-in-organic-synthesis-nanotechnology-and-polymers]

An example of using carboxylic acids as reagents is by integrating them into the structure of ionic liquids (IL). Jahani et al. explained the condensation of 1,8-dioxo-octa hydro xanthene 16 with 5,5-dimethyl-1,3-cyclohexanedione (13) and aldehyde products 14 by using carboxylic acid-functionalized IL (15) beneath microwave radiation (Figure 2.26). offers and offers the advantages of using microwave radiation and IL are making the reaction an effective and eco-friendly process.

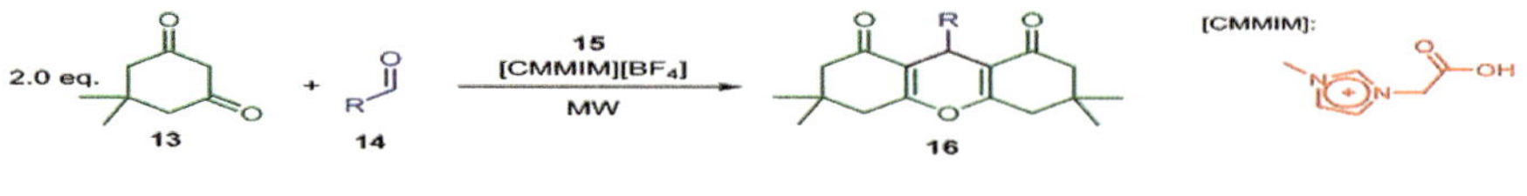

Figure 5.26: Carboxylic acid in ionic liquid in the multicomponent reaction.

[Source: https://creativecommons.org/licenses/by-nc-nd/4.0/]

Carboxylic acids offer significant applications in the pharmaceutical area, because of their chemical structure. Different approaches have been settled for their detection, in cosmetics, in food additives, in medicines, etc. In 2015, Samanta et al. (2015) described a selective chromogenic structure, which not only can distinguish fumaric acid vs. maleic acid but can also distinguish maleic acid among diverse carboxylic acids. This technique uses sharp colorimetric, along with fluorogenic responses in both food additives and physiological conditions. The detection of this kind of organic acids is very significant. For instance, maleic acid plays a vital part in the organism because is a Krebs cycle inhibitor while fumarate is formed in the Krebs cycle. Too much consumption of maleic acid can origin different kidney diseases (Robin et al., 2017).

5.10.2. Nanotechnology

Lately, carboxylic acids have been studied widely, because of their vital applications in the food industry, petrochemical, dyes, stabilizers and now in nanotechnology. This kind of acid has become very significant because it has been reflected as a green solvent as part of the Eutectic Deep Solvents (DES), studied in 2003 by Abbott et al. (2003) which are attained by a blend of hydrogen bond acceptor (such as a quaternary ammonium salt) and a hydrogen bond contributor species (proton contributing species). One of the most general acids is carboxylic acids, belonging to green solvents since they are not very cheap and toxic (Wu et al., 2017).

One of the most significant applications now of the carboxylic acids is the surface amendment of the nanoparticles, this because during the production of the nanoparticles by any approach these tend to agglomerate because of the van der Waals forces and the lack of repulsive forces. In addition, oxidation at the surface of the nanoparticles creates uncertainty which guides to accumulation. One of the approaches to avoiding this problem is to shield the colloidal particles with a stabilizing or passivating agent, which connects with the surface of the nanoparticles to retain them suspended, and therefore to avoid their accumulation. Besides acting as stabilizers, carboxylic acids also affect the reactivity, solubility, size, and shape of nanoparticles. Amongst the other surface modifiers, the most consumed agents to stabilize the surface of the nanoparticles are polyvinylpyrrolidone (PVP), starch, chitosan, gelatin, and cellulose. Carboxylic acids work as stabilizers for the preparation of nanoparticles because the carboxyl group offers coordination to the surface of the nanoparticles and thus they stabilize. For example the oleic acid, which is broadly used for the stabilization of nanoparticles, along with for controlling the morphology and size of nanoparticles. In this regard in 2008 Yang et al. (2008) attained nanoparticles of Fe3O4 with sizes of 6–30 nm by using oleic acid as surfactant, this group of examiners functionalized the nanoparticles with carboxylic acids attaining acid catalysts for the hydrolysis of carbohydrates, being capable to detect that the acid functionalization can have great advantages in generating more active catalysts and therefore an application in green procedures. In contrast, Cabello et al. (2015) were capable to surface functionalize multiwall carbon nanotubes by using aniline and acetic acid, helped by ultrasound, indicating the hydrophilic behavior of carbon nanotubes.

Hojjati et al. (2007) improved TiO_2 nanoparticles with carboxylic acid monitored by polymerization with acrylic acid to get a well-dispersed nanoparticle in polyacrylic acid. Armenalo uses the carboxylic acid as a solvent to get CuS particles, finding numerous advantages with the use of this solvent because it favors the hydrolysis of the C-S bond thus producing a firm CuS supersaturation and a great speed of nucleation, preferring the growth of the particles and avoids agglomeration of the particles. Qu et al. (2010) studied the chemical adjustment of nanoparticles of TiO_2 with carboxylic acids by the solvothermal technique finding development in the photovoltaic performance of the TiO_2 nanoparticles in spite of being covered with carboxylic acid.

In other readings, it has been presented that the use of carboxylic acid as a surface convertor of nanoparticles can be simply redispersed in varied matrix or solvents and increases properties as antibacterial action. In current years, inorganic nanoparticles have been generally studied because of the exceptional properties that they offer, because of their large surface area, highlighting applications such as catalytic, optical, sensing, electrical, magnetic, transport, and thermal conductivity, electromagnetic. These properties are the product of the large surface area that they have. Metals like silver, gold, copper, and palladium have been used to create inorganic nanoparticles of several sizes and shapes. The process and the conditions of production of the nanoparticles directly impact its size and shape.

An extensive variety of approaches have been developed to produce metallic nanoparticles with different morphologies such as nanodisks, nanotubes, nanofibers, and others. In common, these processes are categorized into three groups: biological methods, physical and chemical methods.

Chemical means are the best used, because of their easiness of climbing. The chemical decline of metal salts in solution is the best frequently used method. As stated by Slistan-Grijalva et al. (2008), the chemical reduction lets adequate control of the size distribution, size, and the shape of nanoparticles.

In aqueous structures, the reducing agent is produced or added in situ, amongst the most frequently used reducing agents are sodium hydrazine, dimethylformamide, and borohydride, though in recent year's nontoxic and alike effective substances have been used, like glucose and sodium citrate. In contrast, in non-aqueous schemes, the solvent can also act like a reducing agent, these solvents can be alcohols for example glycerol, polyethylene glycol, and ethylene glycol, from which colloidal nanoparticles are attained. The benefit of these systems is that adding reducing agents is not compulsory, even producing at room temperature. Physical methods contain electrochemical approaches, laser ablation, thermolysis, sonochemistry, and microwave radiation. For instance, the thermolysis method includes the decay of solids at high temperature; over this process, it is likely to obtain particles < 5 nm.

Biologicals means are advanced by using a metal salt and a reducing agent, which may be enzymes plant extracts, or microorganisms.

Nevertheless, in common, the nanoparticles gained by any approaches tend to agglomerate because of the absence of the forces of attraction of van der Waals and repulsive forces. Another common distinctive of metallic nanoparticles (Ag, Cu, etc.) is the oxidation of the surface initiating instability of the nanoparticles and growing in the aggregation of alike. As a plan to avoid these two main problems, the nanoparticles are improved superficially, to make them stable in an aqueous solution and thus to avoid their aggregation. Along with acting as stabilizers, surface modernizers also affect the reactivity, solubility, shape, and size of nanoparticles. The surface modification of inorganic nanoparticles has nowadays attracted consideration, since they can be incorporated impeccably into some ceramic, polymer or metal medium, by improving the relations between the two phases (the discrete phase refers to the inorganic nanoparticles and the constant phase is referred to the ceramic, metallic or polymer phase) formed hybrid materials or nanocomposite, joining the properties of the two phases.

There are numerous types of surface variation like synthetic polymer grafts, chemical treatments, the ligand exchange method, amongst the others; modification by chemical behaviors may contain reactions with carboxylic acid epoxides, metal alkoxides, silane combination agents, among others; this being a suitable technique to avoid the agglomeration of the nanoparticles and therefore increase the dispersion, to attain the surface alteration of nanoparticles has been used ultrasonic radiation as a method that follows the principles of green chemistry, since it decreases the use of solvents along with energy. Another advantage of this method is that the reaction mixture can be heated swiftly and consistently, resulting in a little reaction time and a complete interface avoiding the manufacture of by-products.

Nanoparticles can be surface adjusted with adsorption of polymers on the apparent, is one of the greenest techniques to advance the dispersal of nanoparticles in aqueous systems. Nanoparticles with hydrophilic manners can be isolated in polar solvents by amending them with cationic or anionic polymeric dispersants. These dispersants produce repulsive forces among the particles and rise the dispersibility of the nanoparticles. In this regard, the trioctylphosphine (TOPO), polycarboxylic acids, oleic acid, and amines are cases of surfactants that adjust nanoparticles of Fe_2O_3, Al_2O_3, TiO_2 and that can even be added throughout the production of the nanoparticles.

The nanoparticles can be functionalized with organic particles with biological functions, for example, sugars, lipids, and peptides, along with other macromolecules like enzymes, proteins, RNA and DNA. The mixture of biomolecules and inorganic nanoparticles permits the usage of these in biological systems because they contain distinctive properties for applications like molecular recognition.

Another very significant approach found within the principles of green chemistry is the usage of plasma to change the surface of nanoparticles. Plasma surface modification is an operative and inexpensive surface treatment skill for several nanoparticles.

This is a comparatively simple, dry and fast technique which has been used to adjust the surface interaction of different nanoparticles. It has also verified to be a useful method since it permits the use of organic molecules and/or gases, whereby the surface modification can be accustomed to precise necessities, in other words, a specific functional group can be selected.

This method was initially applied to modify the polymer surfaces, but in the latest years has been used for the dealing of different nanoparticles. The principle of the plasma polymerization method is the formation of radical and ionized molecules by the bombing of an electromagnetic field. These radicals and molecules can react with the surface of the substrate by removal, erosion, and deposition. Thus, the surface kinds of stuff of the substrate will be modified.

Numerous organic molecules like pyrrole, acrylic acid and styrene have been used in plasma to amend nanoparticles of carbon nanofibers, alumina and zinc oxide, individually. By this procedure, it is likely to form a thin layer reaching from 1 to 3 nm in depth.

The analytical techniques used for describing the obtained nanoparticles are transmission electron microscopy (TEM) for shape and size determination, fluorescence spectroscopy, dynamic light scattering. Moreover, the bond of ligand molecules to metal nanoparticles can check out by surface-enhanced Raman scattering.

Silver nanoparticles are known with antibacterial and antifungal properties and have used in biosensing, antivirals in opposition to HIV-I, in paint products water and refinement systems. Just in case of Au nanoparticles, these are applicable in therapy and cancer diagnosis, antiviral and antibacterial, biosensing, MRI applications. Magnetic nanoparticles have antimicrobial stuff and are applicable in biomedicine like magnetic resonance, drug delivery, and cancer treatment. The usage of TiO_2 nanoparticles that are also antimicrobial is in skincare stuff, photocatalysis, nanomedicine, wastewater, gas sensor to remove inorganic and organic pollutants. ZnO nanoparticles also have anti-cancer and antimicrobial properties, used in medical fillers, cosmetics, optoelectronic and electronic devices, and gas sensors. The main uses of the nanoparticles of Al_2O_3 are in drug delivery, in membranes to remove pathogenic microorganisms, antimicrobial applications, and removal of dense metals from wastewater, in gas separation and in catalysis processes. On the other hand, the uses of SiO_2 nanoparticles are in biosensing, also work as a transporter for antimicrobial applications, tissue engineering and drug delivery.

Carboxylic acids are now used as surface modifiers in carbon-placed nanostructure: carbon nanotubes, multiple walls, and single wall, nanofibers, graphene, etc., with the purpose of refining the dispersion in polar solvents like ethanol, water, ethyl acetate, methanol, etc. This kind of apparent modifications can be carried out through diverse options making use of green chemistry, which indorse the use of maintainable activation energies like microwaves, ultrasound, and plasma, that supports to decrease energy consumption and reduction the time period of the reaction in the surface modification.

5.10.3. Polymer

The carboxylic acids current applications in the gaining of polymers, acting as additives, monomers, catalysts, initiators, dopants, etc. Now, an area of excessive interest in the manufacture of acidic polymers, with dissimilar applications, for instance, in an electronic area essential that present characteristics like high solubility, electron donors in aqueous solvents, etc.

As improver, the carboxylic acids have been studied. In 2017, it was stated the analysis of a series of direct carboxylic acids whit dissimilar chain lengths of trans-2-decanoic acid (CA-10), 9-tetradecanoic acid (CA-14), 6 trans-2-hexanoic acid carbon atoms (CA-6), used as halogen-free additives-solvent, reflected a maintainable and feasible process useful for the manufacture of polymeric films whit optical properties, whit potential use in solar cells. The completion of the study displayed that growing the length of the carboxylic acid chain alterations the cartography of the polymeric film.

Current studies described the application of acids polymers as dopants, to stabilize and approach the electrically conducting states of accompanying. The acids polymers are used for change small-molecules acids as: poly(styrene sulfonate) (PSS), poly(acrylic acid), and (2-acrylamido-2-methyl-1-propanesulfonic acid) (PAAMPSA), poly(2-acrylamido-2-methyl-1-propanesulfonic acid), because its chemical nature, aids to stabilize the passage of the polymers, in end, the existence of an extra of carboxylic acid in the chemical structure indorses good dispersion, thus soothing the electric nature of the doped polymer.

Shi et al. (2012) described the gaining of two fresh blue transmissive donor-acceptor electrochromic polymers: a polymeric material produced by interchanging copolymerization route and another arbitrary copolymerization, indicating that this kind of polymeric materials having distinctive as advanced water solubility. The outcomes of this study, display that these polymeric acids show electrochromic properties.

Lately Mohammadifar et al. (2016) described the reaction of cationic polymerization at normal temperature and solvent-free to get polyglycerol, the polymerization is categorized as green polymerization, in this kind of polymerization, citric acid contributes as a donor of proton, encouraging polymerization, it was described that this kind of polymerization does not need a purification procedure which shows that the process is very maintainable, these types of polymers may be potential nominees for biomedical applications.

There are diverse kinds of polymer acids, synthetic or natural, for instance, polylactic acid presents uses like polymeric acid antibacterial, because of its chemical and physical features, the polylactic acid can be combined or modified whit other polymer mediums and form composites, it also shows great biocompatibility and biodegradability, which makes it a maintainable material.

Poly (methacrylic acid), is an irregular material, its chemical structure contains a polar backbone where the carboxylic acid is and the remaining structure is no polar. In 2017, Robin et al., (2017) described the study of character rheological of focus of the poly (methacrylic acid) in aqueous solutions and effect in viscosity, the intent of this study is identify the physic-chemical of the polymeric material and to be capable to formulate dissimilar composites, closing that the poly (methacrylic acid) in aqueous solutions offer a rheological behavior restrained by the equilibrium between in the different connections intramolecular such as hydrogen-bonding. Similarly, the poly (methacrylic acid) is measured a biodegradable and maintainable polymeric material, used for the gain of biomaterials. Poly (methacrylic acid) brushes are extremely vulnerable to swelling in aqueous solution because of ionization of carboxylic groups offerings in their chemical structure.

In common, the uses of carboxylic acid in the polymer are because of the occurrence of a polar group that aids solvate in aqueous systems, easing its processing.

REFERENCES

1. Abbott, A. P., Capper, G., Davies, D. L., Rasheed, R. K., & Tambyrajah, V. (2003). Novel solvent properties of choline chloride/urea mixtures. *Chemical Communications*, (1), 70–71.
2. Abdel-Aziz, M. S., Shaheen, M. S., El-Nekeety, A. A., & Abdel-Wahhab, M. A. (2014). Antioxidant and antibacterial activity of silver nanoparticles biosynthesized using *Chenopodium murale* leaf extract. *Journal of Saudi Chemical Society*, *18*(4), 356–363.
3. Allen, C. L., Chhatwal, A. R., & Williams, J. M. (2012). Direct amide formation from unactivated carboxylic acids and amines. *Chemical Communications*, *48*(5), 666–668.
4. Amarnath, L., Andrews, I., Bandichhor, R., Bhattacharya, A., Dunn, P., Hayler, J., & Kaptein, B. (2012). Green chemistry articles of interest to the pharmaceutical industry. *Organic Process Research & Development*, *16*(4), 535–544.
5. Armelao, L., Camozzo, D., Gross, S., & Tondello, E. (2006). Synthesis of copper sulphide nanoparticles in carboxylic acids as solvent. *Journal of Nanoscience and Nanotechnology*, *6*(2), 401–408.
6. Bian, Z., Xin, Z., Meng, X., Tao, M., Lv, Y., & Gu, J. (2017). Effect of citric acid on the synthesis of CO methanation catalysts with high activity and excellent stability. *Industrial & Engineering Chemistry Research*, *56*(9), 2383–2392.
7. Cabello, CJ, Sáenz, A., Pérez, C., López, L., Barajas, L., Cantú Sifuentes, L., & Ávila, C. (2015). Modification of multiple-walled carbon nanotubes (MWNTC's) using acetic acid and aniline by ultrasonic radiation. *Latin American Journal of Metallurgy and Materials*, *35* (1), 27–33.
8. Carey, J. S., Laffan, D., Thomson, C., & Williams, M. T. (2006). Analysis of the reactions used for the preparation of drug candidate molecules. *Organic & biomolecular chemistry*, *4*(12), 2337–2347.
9. Cash, B. M., Wang, L., & Benicewicz, B. C. (2012). The preparation and characterization of carboxylic acid-coated silica nanoparticles. *Journal of Polymer Science Part A: Polymer Chemistry, 50*(13), 2533–2540.
10. Constable, D. J., Dunn, P. J., Hayler, J. D., Humphrey, G. R., Leazer Jr, J. L., Linderman, R. J., ... & Zaks, A. (2007). Key green chemistry

research areas—a perspective from pharmaceutical manufacturers. *Green Chemistry*, *9*(5), 411–420.

11. Dadhania, A. N., Patel, V. K., & Raval, D. K. (2017). Ionic liquid promoted facile and green synthesis of 1, 8-dioxo-octahydroxanthene derivatives under microwave irradiation. *Journal of Saudi Chemical Society*, *21*, S163-S169.
12. Diorazio, L. J., Motherwell, W. B., Sheppard, T. D., & Waller, R. W. (2006). Observations on the reaction of N-alkyloxazolidines, isocyanides and carboxylic acids: A novel three-component reaction leading to N-acyloxyethylamino acid amides. *Synlett*, *2006*(14), 2281–2283.
13. Gedye, R., Smith, F., Westaway, K., Ali, H., Baldisera, L., Laberge, L., & Rousell, J. (1986). The use of microwave ovens for rapid organic synthesis. *Tetrahedron Letters*, *27*(3), 279–282.
14. Ghose, A. K., Viswanadhan, V. N., & Wendoloski, J. J. (1999). A knowledge-based approach in designing combinatorial or medicinal chemistry libraries for drug discovery. 1. A qualitative and quantitative characterization of known drug databases. *Journal of Combinatorial Chemistry*, *1*(1), 55–68.
15. He, J., Zhou, L., Liu, J., Yang, L., Zou, L., Xiang, J., ... & Yang, X. (2017). Modulation of surface structure and catalytic properties of cerium oxide nanoparticles by thermal and microwave synthesis techniques. *Applied Surface Science*, *402*, 469–477.
16. Hethnawi, A., Nassar, N. N., Manasrah, A. D., & Vitale, G. (2017). Polyethylenimine-functionalized pyroxene nanoparticles embedded on Diatomite for adsorptive removal of dye from textile wastewater in a fixed-bed column. *Chemical Engineering Journal*, *320*, 389–404.
17. Hojjati, B., Sui, R., & Charpentier, P. A. (2007). Synthesis of TiO2/PAA nanocomposite by RAFT polymerization. *Polymer*, *48*(20), 5850–5858.
18. Hosseini-Monfared, H., Parchegani, F., & Alavi, S. (2015). Carboxylic acid effects on the size and catalytic activity of magnetite nanoparticles. *Journal of Colloid and Interface Science*, *437*, 1–9.
19. Houlding, T. K., Tchabanenko, K., Rahman, M. T., & Rebrov, E. V. (2013). Direct amide formation using radiofrequency heating. *Organic & Biomolecular Chemistry, 11*(25), 4171–4177.
20. Jahani, F., Zamenian, B., Khaksar, S., & Tajbakhsh, M. (2010).

Pyridine 2, 6-dicarboxylic acid as a bifunctional organocatalyst for hydrophosphonylation of aldehydes and ketones in water. *Synthesis, 2010*(19), 3315–3318.

21. Jašo, V., Glenn, G., Klamczynski, A., & Petrović, Z. S. (2015). Biodegradability study of polylactic acid/thermoplastic polyurethane blends. *Polymer Testing*, *47*, 1–3.
22. Kango, S., Kalia, S., Celli, A., Njuguna, J., Habibi, Y., & Kumar, R. (2013). Surface modification of inorganic nanoparticles for development of organic–inorganic nanocomposites—A review. *Progress in Polymer Science*, *38*(8), 1232–1261.
23. Laadila, M. A., Hegde, K., Rouissi, T., Brar, S. K., Galvez, R., Sorelli, L., ... & Abokitse, K. (2017). Green synthesis of novel biocomposites from treated cellulosic fibers and recycled bio-plastic polylactic acid. *Journal of Cleaner Production*, *164*, 575–586.
24. Laaziz, S. A., Raji, M., Hilali, E., Essabir, H., Rodrigue, D., Bouhfid, R., & el kacem Qaiss, A. (2017). Bio-composites based on polylactic acid and argan nut shell: Production and properties. *International Journal of Biological Macromolecules*, *104*, 30–42.
25. Lanigan, R. M., Starkov, P., & Sheppard, T. D. (2013). Direct synthesis of amides from carboxylic acids and amines using B (OCH2CF3) 3. *The Journal of Organic Chemistry, 78*(9), 4512–4523.
26. Lazzarato, L., Donnola, M., Rolando, B., Marini, E., Cena, C., Coruzzi, G., & Biondi, S. (2008). Searching for new NO-donor aspirin-like molecules: a new class of nitrooxy-acyl derivatives of salicylic acid. *Journal of medicinal chemistry*, *51*(6), 1894–1903.
27. Lukasik, N., & Wagner-Wysiecka, E. (2014). A review of amide bond formation in microwave organic synthesis. *Current Organic Synthesis*, *11*(4), 592–604.
28. Luo, C., Zhang, Y., Zeng, X., Zeng, Y., & Wang, Y. (2005). The role of poly (ethylene glycol) in the formation of silver nanoparticles. *Journal of colloid and interface science*, *288*(2), 444–448.
29. Masala, O., & Seshadri, R. (2004). Synthesis routes for large volumes of nanoparticles. *Annu. Rev. Mater. Res.*, *34*, 41–81.
30. Mohammadifar, E., Bodaghi, A., Dadkhahtehrani, A., Nemati Kharat, A., Adeli, M., & Haag, R. (2016). Green synthesis of hyperbranched polyglycerol at room temperature. *ACS Macro Letters*, *6*(1), 35–40.
31. Narayanan, K. B., & Sakthivel, N. (2008). Coriander leaf mediated

biosynthesis of gold nanoparticles. *Materials Letters*, *62*(30), 4588–4590.

32. Oguz, O., Bilge, K., Simsek, E., Citak, M. K., Wis, A. A., Ozkoc, G., & Menceloglu, Y. Z. (2017). High-performance green composites of poly (lactic acid) and waste cellulose fibers prepared by high-shear thermokinetic mixing. *Industrial & Engineering Chemistry Research*, *56*(30), 8568–8579.
33. Ojeda-Porras, A., Hernández-Santana, A., & Gamba-Sánchez, D. (2015). Direct amidation of carboxylic acids with amines under microwave irradiation using silica gel as a solid support. *Green Chemistry*, *17*(5), 3157–3163.
34. Pai, Z. P., Selivanova, N. V., Oleneva, P. V., Berdnikova, P. V., & Beskopyl'nyi, A. M. (2017). Catalytic oxidation of α-alkenes with hydrogen peroxide to carboxylic acids in the presence of peroxopolyoxotungstate complexes. *Catalysis Communications*, *88*, 45–49.
35. Panigrahi, S., Praharaj, S., Basu, S., Ghosh, S. K., Jana, S., Pande, S., & Pal, T. (2006). Self-assembly of silver nanoparticles: synthesis, stabilization, optical properties, and application in surface-enhanced Raman scattering. *The Journal of Physical Chemistry B*, *110*(27), 13436–13444.
36. Passerini, M. (1921). Above the isonitriles (II). Compounds with aldehydes or ketones and monobasic organic acids. *Gazz. Chim. Ital*, *51*, 181–189.
37. Passerini, M., & Simone, L. (1921). Above the isonitriles (I). P-isonitril-azobenzolo compound with acetone and acetic acid. *Gazz. Chim. Ital*, *51*, 126–129.
38. Perreux, L., Loupy, A., & Volatron, F. (2002). Solvent-free preparation of amides from acids and primary amines under microwave irradiation. *Tetrahedron*, *58*(11), 2155–2162.
39. Poole, Jr. C. P., Owens, F. J., & Averback, R. S. (2004). Books-Introduction to Nanotechnology. *Physics Today*, *57*(9), 57.
40. Qu, Q., Geng, H., Peng, R., Cui, Q., Gu, X., Li, F., & Wang, M. (2010). Chemically binding carboxylic acids onto TiO2 nanoparticles with adjustable coverage by solvothermal strategy. *Langmuir*, *26*(12), 9539–9546.
41. Raveendran, P., Fu, J., & Wallen, S. L. (2006). A simple and "green"

method for the synthesis of Au, Ag, and Au–Ag alloy nanoparticles. *Green Chemistry, 8*(1), 34–38.

42. Robin, C., Lorthioir, C., Amiel, C., Fall, A., Ovarlez, G., & Le Cœur, C. (2017). Unexpected rheological behavior of concentrated poly (methacrylic acid) aqueous solutions. *Macromolecules*, *50*(2), 700–710.
43. Roughley, S. D., & Jordan, A. M. (2011). The medicinal chemist's toolbox: an analysis of reactions used in the pursuit of drug candidates. *Journal of Medicinal Chemistry*, *54*(10), 3451–3479.
44. Sáenz-Galindo, A., López-López, L. I., Fabiola, N., Castañeda-Facio, A. O., Rámirez-Mendoza, L. A., Córdova-Cisneros, K. C., & de Loera-Carrera, D. (2018). Applications of Carboxylic Acids in Organic Synthesis, Nanotechnology and Polymers. *Carboxylic Acid: Key Role in Life Sciences*, 35, 1–10.
45. Samanta, S., Kar, C., & Das, G. (2015). Colorimetric and fluorometric discrimination of geometrical isomers (Maleic Acid vs Fumaric Acid) with real-time detection of maleic acid in solution and food additives. *Analytical Chemistry*, *87*(17), 9002–9008.
46. Sarkany, A., Papp, Z., Sajo, I., & Schay, Z. (2005). Unsupported Pd nanoparticles prepared by γ-radiolysis of PdCl2. *Solid State Ionics*, *176*(1–2), 209–215.
47. Sarkar, A., Kapoor, S., & Mukherjee, T. (2005). Synthesis of silver nanoprisms in formamide. *Journal of Colloid and Interface Science*, *287*(2), 496–500.
48. Sato, K., Kondo, S., Tsukada, M., Ishigaki, T., & Kamiya, H. (2007). Influence of solid fraction on the optimum molecular weight of polymer dispersants in aqueous TiO_2 nanoparticle suspensions. *Journal of the American Ceramic Society*, *90*(11), 3401–3406.
49. Schmid, G. (2008). *Clusters and Colloids: From Theory to Applications* (vol. 1, pp. 1–15). John Wiley & Sons.
50. Schneiderman, D. K., & Hillmyer, M. A. (2017). 50th anniversary perspective: There is a great future in sustainable polymers. *Macromolecules*, *50*(10), 3733–3749.
51. Sezen-Edmonds, M., & Loo, Y. L. (2017). Beyond doping and charge balancing: How polymer acid templates impact the properties of conducting polymer complexes. *The journal of physical chemistry letters*, *8*(18), 4530–4539.

52. Shi, D., Lian, J., He, P., Wang, L. M., Xiao, F., Yang, L., ... & Mast, D. B. (2003). Plasma coating of carbon nanofibers for enhanced dispersion and interfacial bonding in polymer composites. *Applied Physics Letters*, *83*(25), 5301–5303.
53. Shi, D., Wang, S. X., Van Ooij, W. J., Wang, L. M., Zhao, J., & Yu, Z. (2001). Uniform deposition of ultrathin polymer films on the surfaces of Al 2 O 3 nanoparticles by a plasma treatment. *Applied Physics Letters*, *78*(9), 1243–1245.
54. Shi, P., Amb, C. M., Dyer, A. L., & Reynolds, J. R. (2012). Fast switching water processable electrochromic polymers. *ACS Applied Materials & Interfaces*, *4*(12), 6512–6521.
55. Slistan-Grijalva, A., Herrera-Urbina, R., Rivas-Silva, J. F., Avalos-Borja, M., Castillón-Barraza, F. F., & Posada-Amarillas, A. (2008). Synthesis of silver nanoparticles in a polyvinylpyrrolidone (PVP) paste, and their optical properties in a film and in ethylene glycol. *Materials Research Bulletin*, *43*(1), 90–96.
56. Sperling, R. A., & Parak, W. J. (2010). Surface modification, functionalization and bioconjugation of colloidal inorganic nanoparticles. *Philosophical Transactions of the Royal Society A: Mathematical, Physical and Engineering Sciences*, *368*(1915), 1333–1383.
57. Sun, X., & Luo, Y. (2005). Preparation and size control of silver nanoparticles by a thermal method. *Materials Letters*, *59*(29–30), 3847–3850.
58. Teles, A. R. R., Capela, E. V., Carmo, R. S., Coutinho, J. A., Silvestre, A. J., & Freire, M. G. (2017). Solvatochromic parameters of deep eutectic solvents formed by ammonium-based salts and carboxylic acids. *Fluid Phase Equilibria*, *448*, 15–21.
59. Ugi, I. (1959, January). Versuche mit isonitrilen. In *Angewandte Chemie-International Edition* (Vol. 71, No. 11, pp. 386–386). Muhlenstrasse 33–34, D-13187 Berlin, Germany: Wiley-V Ch Verlag GMBH.
60. Ugi, I. (1962). The α-addition of immonium ions and anions to isonitriles accompanied by secondary reactions. *Angewandte Chemie International Edition in English*, *1*(1), 8–21.
61. Varma, R. S., & Naicker, K. P. (1999). Solvent-free synthesis of amides from non-enolizable esters and amines using microwave irradiation.

Tetrahedron Letters, *40*(34), 6177–6180.

62. Wu, Y., Nizam, M. N., Ding, X., & Xu, F. J. (2017). Rational Design of Peptide-Functionalized Poly (Methacrylic Acid) Brushes for On-Chip Detection of Protease Biomarkers. *ACS Biomaterials Science & Engineering*, *4*(6), 2018–2025.
63. Yang, H., Ogawa, T., Hasegawa, D., & Takahashi, M. (2008). Synthesis and magnetic properties of monodisperse magnetite nanocubes. *Journal of Applied Physics*, *103*(7), 07D526.
64. Yaragorla, S., Singh, G., Saini, P. L., & Reddy, M. K. (2014). Microwave assisted, Ca (II)-catalyzed Ritter reaction for the green synthesis of amides. *Tetrahedron Letters, 55*(33), 4657–4660.
65. Yasa, S. R., Krishnasamy, S., Singh, R. K., & Penumarthy, V. (2017). Synthesis and characterization of iso-undecenoic and iso-undecanoic acids based polyol esters. *Industrial & Engineering Chemistry Research*, *56*(26), 7423–7433.
66. Zhang, Y., Yuan, J., Sun, J., Ding, G., Han, L., Ling, X., & Ma, W. (2017). Alkenyl carboxylic acid: engineering the nanomorphology in polymer–polymer solar cells as solvent additive. *ACS Applied Materials & Interfaces*, *9*(15), 13396–13405.

6

Chemical Characteristics and Uses of Acetaldehyde

CONTENTS

6.1. INTRODUCTION

Acetaldehyde was perceived by Scheele in 1774 during the reaction of sulfuric acid and black manganese dioxide with alcohol. Its composition was explained by Liebig in 1835 who made pure acetaldehyde by the oxidation of ethanol (C_2H_5OH) with chromic acid (H_2CrO_4) and named this product aldehyde, a reduced form of the term "alcohol dehydrogenatus."

Acetaldehyde is a low-boiling, mobile, highly flammable liquid having a pungent smell. Due to its high chemical reactivity, this product is a significant intermediate in the making of ethyl acetate, acetic acid, peracetic acid, acetic anhydride, butanol, pentaerythritol, ethylhexanol, chlorinated acetaldehydes, alkyl amines, glyoxal, pyridines, and several other chemicals. The first commercial utilization was the generation of acetone through acetic acid in the middle of 1914 and 1918 in Germany and in Canada (Sopyan et al., 1996; Dumitriu et al., 1999).

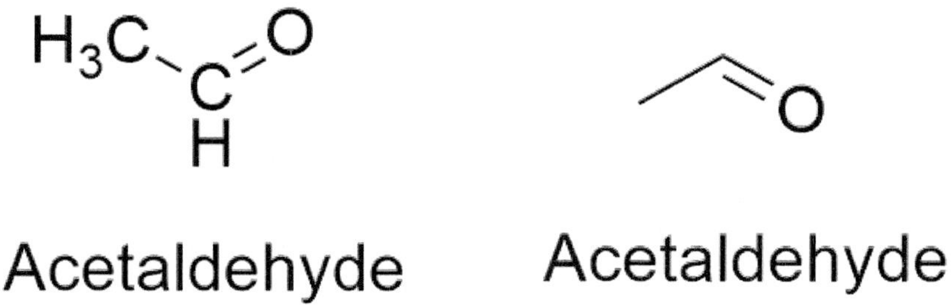

Figure 6.1: Chemical structure of acetaldehyde.

[Source: http://www.softschools.com/formulas/chemistry/acetaldehyde_formula/415/]

Existence: Acetaldehyde is a median in the digestion of animal and plant organisms, in which acetaldehyde can be perceived in small amounts. Larger amounts of this product hinder with the biological processes. As a median in the processes of alcoholic fermentation, it is existent in small quantities in all of the alcoholic beverages, like wine, beer, and spirits. Acetaldehyde has been identified in the plant juices and vital oils, tobacco smoke, and roasted coffee (Payne et al., 1966; Amit & Smith, 1985)

Commercial production processes comprise of oxidation or dehydrogenation of ethanol (C_2H_5OH), the addition of the water to acetylene, incomplete oxidation of the hydrocarbons, and oxidation of ethylene. In the era of 1970s, the world's capacity of this last method, Wacker-Hoechst direct oxidation, escalated to around 2×10^6 t/a. Conversely, the significance of acetaldehyde (C_2H_4O) as an organic median is now decreasing steadily, as new methods for some of the acetaldehyde byproducts have been established, like the oxo process for 2-ethyl hexanol and butanol and the Monsanto procedure for acetic acid. In future, new procedures for the acetic anhydride (Halcon, Eastman, Hoechst), for the vinyl acetate (Halcon), and for the alkyl amines will diminish the utilization of acetaldehyde as the starting material (Ullrich & Mannschreck, 1967; Tateishi et al., 1993; Chu et al., 2011).

6.2. PHYSICAL PROPERTIES

Acetaldehyde, M_r 44.054, is the colorless liquid having a pungent, suffocating smell which is to some extent fruity when diluted.

Acetaldehyde is totally miscible with most of the organic solvents and water. It doesn't form an azeotrope with methanol, water, ethanol, acetic acid, acetone, or benzene. Binary azeotropes are made with butane and diethyl ether (Matveev, 1977; Farnworth et al., 2001).

Physical Data: viscosity and compressibility at higher pressure are specified by Chaudhuri et al. (1968), the vapor pressure of the aqueous acetaldehyde solutions are studied by Mielcarski & Puchalska, (2002). Freezing points of the aqueous acetaldehyde solutions are specified by Soffritti et al. (2002); vapor-liquid equilibria of the binary systems of acetaldehyde (C_2H_4O) with ethanol, water, ethylene oxide, and acetic acid, are specified by Heitz (1974), with the vinyl acetate (Smith & Bonner, 1951).

Safety Data: Flashpoint (Abel–Pensky; DIN 51755; ASTM 56–70) −20°C (−40°C according to safety regulations of the Berufsgenossenschaftder Chemischen Industrie, Germany). Ignition temperature (DIN 51794; ASTM D 2155–66) 140°C, ignition retardation when introduced into the hot air stream is specified by Mullins (1953). Explosive limits in the air: 4–57 vol%; for effect of pressure on the explosive limits (Gaines et al., 1977).

6.3. CHEMICAL PROPERTIES

Acetaldehyde is a very highly reactive compound displaying all of the characteristic aldehyde reactions along with those of the alkyl groups in which the hydrogen atoms are triggered by the carbonyl groups in the *alpha* position. When heated beyond 420°C acetaldehyde decomposes into carbon monoxide and methane (Bangle, Jr., 1954; Ohta & Baba, 1986).

6.3.1. Addition Reactions

With water, acetaldehyde makes an uneven hydrate; separate solid hydrates are identified only with the chlorinated acetaldehydes. *Alcohols* add to the acetaldehyde producing hemiacetals, which make acetals with surplus alcohol in the existence of acids by the elimination of water (Payne et al., 1966; Chu et al., 2011). Diols form cyclic acetals; for instance, 2-methyl-1,3-dioxolane is attained from ethylene acetaldehyde and glycol, and 2-methyl-1,3-dioxane from the 1,3-propanediol (Sopyan et al., 1996; Shimizu et al., 1999).

The aqueous solution of *sodium bisulfite* and acetaldehyde forms the crystalline chemical compound from which the acetaldehyde can be set free. Dry *ammonia produces* crystalline acetaldehyde ammonia. H*ydrocyanic acid and* acetaldehyde react to generate Lacto-nitrile (α-hydroxy propionitrile), a probable intermediate in the acrylonitrile production (Mori et al., 1991; Shandura et al., 2009).

Acetaldehyde reacts with the *acetic anhydride* to produce ethylidene diacetate, a median in the vinyl acetate procedure of Celanese Corp (Miller, 1966; Trofimov, 1990).

6.3.2. Products of Aldol Addition

Two molecules of C_2H_4O combine in the existence of dilute acids or alkaline catalysts at the room temperature or with adequate heating to produce acetaldol, CH_3CH (OH) CH_2CHO. At higher temperatures, water is easily cleaved from this acetaldol, making crotonaldehyde. Further condensation under severe conditions to produce aldehyde resins has no industrial significance.

Acetaldehyde and *Urea* condense in the existence of sulfuric acid to produce crotonylidenediurea, which is utilized as the long term nitrogen fertilizer (Braun et al., 1986; Stoner Jr, & Miller, 2006).

Acetaldehyde is also a median in the synthesis of butadiene beginning from acetylene and continuing through acetaldol and hydrogenation product of acetaldol, 1,3-butanediol (Sheldon & Kochi, 1976; Al-Shammary et al., 1991). This procedure was presented around 1918 and is carried out on the commercial scale in a few Eastern European countries (Cornils & Kuntz, 1995; Stoner Jr & Miller, 2006).

Acrolein is attained by the aldol condensation of formaldehyde and acetaldehyde and succeeding water eradication, analogous to the creation of crotonaldehyde. This technique is also without any commercial significance today, while the formation of pentaerythritol from C_2H_4O and the fourfold amount of CH_2O in the existence of NaOH or $Ca(OH)_2$ is very vital industrially (Miller et al., 1970; Hodge et al., 1986).

6.3.3. Reaction with Compounds of Nitrogen

With Schiff bases, major amines, $CH_3CH{=}NR$, are made. Nitrogen compounds like hydrazine, hydroxylamine, semicarbazide, and phenylhydrazine react with acetaldehyde (C_2H_4O) to produce easily crystallizable compounds which are utilized for the analytical characterization and determination of aldehydes. Several other ketones and aldehydes can usually be characterized in a similar way as their analogous derivatives normally have sharp melting points (Fischer & Jira, 1954; Lee et al., 2000).

The amalgamation of pyridine and its derivatives is of great importance. 5-Ethyl-2-methylpyridine is attained by a reaction of the aqueous ammonia with C_2H_4O in the existence of fluoride ions. In the added existence of acrolein or formaldehyde, the mixtures of alkyl pyridines and pyridine form (Farias et al., 2013; Lin et al., 2017).

6.3.4. Oxidation

The main part of the acetaldehyde C_2H_4O produced commercially is utilized for making acetic acid by the oxidation with air or oxygen. Acetaldehyde monomer acetate is created as a median and decomposes into acetaldehyde and peracetic acid at the elevated temperatures and in existence of the catalytic amounts of Fe or Co salts. In the existence of Mn^{2+} salts, acetic acid (CH_3COOH) is attained from the acetaldehyde monomeracetate, and in the existence of Cu^{2+} and Co^{2+} salts, acetic anhydride could be formed (Sieber et al., 1983a, 1983b).

Oxidation with the nitric acid produces glyoxal. Halogenated acetaldehydes are generally prepared by halogenation.

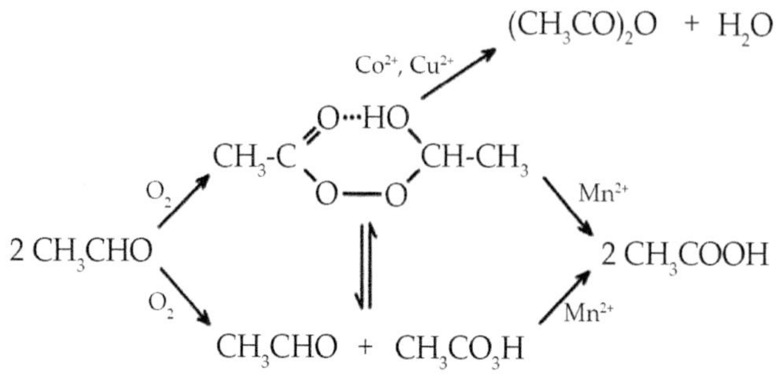

Figure 6.2: Production of glyoxal from acetaldehyde.

[Source: https://onlinelibrary.wiley.com/doi/abs/10.1002/14356007.a01_031]

Mono-, di-, and trichloro acetaldehydes and tribromo acetaldehyde are beneficial for manufacturing pharmaceuticals, insecticides, and dyes.

6.3.5. Reduction

Acetaldehyde is readily hydrogenated to ethanol. Before 1939, which is, before the petrochemically made ethylene became accessible in Europe, the reaction was utilized industrially to form ethanol from acetaldehyde (C_2H_4O) and, thus, from acetylene.

Mono, di-, and tri-ethylamine can be made from acetaldehyde, ammonia (NH_3), and hydrogen in the existence of the hydrogenation catalyst (Flores-Soto et al., 2012; Wiemann et al., 2012).

6.3.6. Miscellaneous Reactions

Tishchenko reaction of C_2H_4O gives the commercially significant solvent ethyl acetate ($C_4H_8O_2$); it is readily catalyzed by the aluminum alcoholate.

Table 6.1: Consumption of Acetaldehyde (C_2H_4O) (103 t) in 1993 (Eckert et al., 2000)

Product	U.S.A	Mexico	Western Europe	Japan	Total
Acetate esters	32	5	133	158	328
Acetic acid/acetic anhydride		139	307	97	543
Pyridine and pyridine bases	45		44	17	106
Pentaerythritol	22		37	12	71
1,3-Butylene glycol	13			*	13
Peracetic acid	19			*	19
Others	16	105	85	65	271
Total	147	249	606	349	1351

* Included in others (glyoxalic acid/ glyoxal, crotonaldehyde, *n*-butanol, lactic acid, 2- ethylhexanol).

As the radical trapping agent, acetaldehyde is utilized to govern the length of chain in the polymerization of the vinyl compounds.

6.3.7. Consumption

The consumption of C_2H_4O has varied during the past few years. Since 1993 in the United States of America, acetaldehyde is now not utilized for the manufacturing of acetic acid, 2-ethyl hexanol, or butanol which are manufactured by other routes (Table 6.1). The consumption of C_2H_4O for other chemicals such as pyridine bases or peracetic acid is increasing.

6.4. PRODUCTION

Raw materials which have been utilized for the making of C_2H_4O are:

- Ethylene
- Carbon monoxide and hydrogen
- Acetylene
- Lower hydrocarbons
- Methanol
- Ethanol obtained from the hydration of the ethylene or from the fermentation of the carbohydrates.

The economy of commercial procedures depends fundamentally upon the availability of raw materials and prices. In the highly industrialized countries sustaining high costs for ethanol by economic measures or where the petrochemical ethanol wasn't accessible, as in Japan or Germany prior to 1939, acetylene (C_2H_2) was the preferred opening material for acetaldehyde. The process of acetylene is still functioned in a few Eastern European countries and by the companies where inexpensive acetylene is available. Petro-chemically made ethanol, conversely, was the preferred raw material in the rest of the countries, while ethanol produced by fermentation was and is still utilized on the small scale in countries having less chemical industry (Zhang et al., 2014; Baranyi et al., 2019).

In Western countries, all these procedures have been almost entirely substituted by the process of direct oxidation established by Wacker-Chemie and Hoechst in the era of late 1950s. This is as ethylene is accessible at a lower price than C_2H_2 (Acker et al., 2014; Jordan et al., 2016).

Even the 2-stage processes utilizing ethanol from ethylene (C_2H_4) as the initiating material aren't competitive due to the decreasing significance of acetaldehyde (C_2H_4O) as an organic median.

Usually, all procedures based on acetylene (C_2H_2), ethylene (C_2H_4) and ethanol's (C_2H_5OH) are more discriminating than oxidation of the saturated hydrocarbons. This is due to the fact that in the latter circumstance, other products of oxidation are manufactured in addition to the acetaldehyde. Due to the great cost of isolating the product mixture, these processes are cost-effective only in the large units and when all secondary and main products attained in the procedure are utilized (Funder et al., 1998; Rustamov, 2007).

6.4.1. Production from Ethanol (C2H5OH)

For the making of acetaldehyde, ethanol can normally be dehydrogenated or oxidized in the existence of oxygen. Amongst 1918 and 1939, dehydrogenation gained superiority over oxidation due to the instantaneous production of hydrogen. Later, though, catalytic vapor-phase oxidation of the ethanol became the favored process, possibly due to the long life of catalyst and the probability of recovering energy (Hendl et al., 1994; Scalabrin et al., 2002).

6.4.1.1. Dehydrogenation of Ethanol

In the first investigation on the dehydrogenation of ethanol, published in 1886, C2H5OH was passed through the glass tubes at around 260°C.

$CH_3CH_2OH(l) \rightarrow CH_3CHO(l) + H_2\ (g)$

$\Delta H = +82.5$ kJ/mol

Enhanced yields are attained in the existence of catalysts like copper, platinum, or oxides of nickel, zinc, or cobalt. In later patents, chromium and zinc catalysts, oxides of the uncommon earth metals, and mixtures of chromium and copper oxides have been reported (Smith & Bonner, 1951; Das & Kuloor, 2013). The least quantity of decomposition products is attained using copper catalysts and repeated redevelopment of the catalysts is needed (Weast & Astle, 1979; Nötzold et al., 1994).

Process Description: Vapor of ethanol is passed at around 260 to 290°C over the catalyst comprising of Cu sponge or Cu triggered with chromium oxide in the tubular reactor [37]. The conversion of 25 to 50% per run is attained. By washing with water and alcohol, ethanol and acetaldehyde are alienated from the exhaust gas, which is mostly hydrogen. Pure acetaldehyde is attained by distillation; ethanol is isolated from the water and higher-boiling produces by distillation and usually flows back to the reactor. The final acetaldehyde produce is ca. 90%. Byproducts include crotonaldehyde, butyric acid, and ethyl acetate (Craven & Lambert, 1951; Yaws, 1990).

6.4.1.2. Oxidation of Ethanol

Ethanol Oxidation is the best and oldest laboratory process for making acetaldehyde. In the commercial procedure, ethanol is generally oxidized catalytically with the oxygen in the vapor phase.

$CH_3CH_2OH(g) + 1/2O_2\ (g) \rightarrow CH_3CHO(l) + H_2O(l)$

$\Delta H = -242.0$ kJ/mol

Silver, Copper, and their alloys or oxides are the most regularly utilized catalysts. For an instance of the simultaneous oxidation–dehydrogenation procedure (Lambert et al., 1950; Hubbell, 1971).

Veba-Chemie Process (Figure 6.3). Ethanol is assorted with air and then passed over the silver catalyst at around 500 to 650°C (c). The temperature is dependent on the alcohol to air ratio and flow rate of the gas via a catalyst. Alcohol conversion differs among 50 and 70% and the produce is usually between 97 and 99% dependent on the conditions of the reaction. Acetaldehyde and unreacted alcohols are eliminated from the remaining gas usually by washing with the cold alcohol (e) and isolated by fractional distillation (h); after the concentration, alcohol returns back to the reactor. The heat produced in this reaction is used for steam production utilizing the waste-heat recovery systems instantaneously after the reaction zones (Chaudhuri et al., 1968; Heitz, 1974; Soffritti et al., 2002).

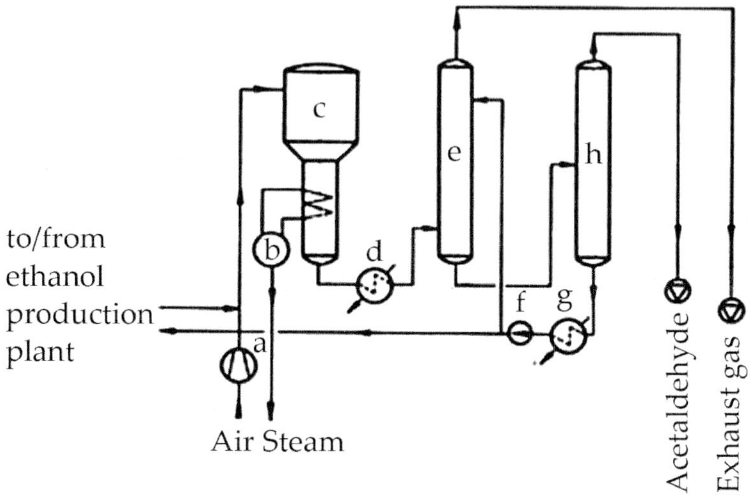

Figure 6.3: Production of acetaldehyde by Veba-Chemie process.

[Source: https://onlinelibrary.wiley.com/doi/abs/10.1002/14356007.a01_031]

- Air compressor
- Heat recovery system
- Reactor
- Cooler
- Waste-gas scrubber
- Washing alcohol and returning pump;
- Cooler
- Acetaldehyde rectification

The waste gas comprises chiefly of hydrogen, nitrogen, methane, carbon dioxide and carbon monoxide; it is then burned as the lean gas with low fatty value in the steam generators. Small amounts of CH_3COOH (acetic acid) are obtained as the byproduct (Hartmann & Leitinger, 1984; Kutzelnigg et al., 1990).

6.4.2. Production from Acetylene (C_2H_2)

The most vital catalysts for the industrial hydration (water addition) are mercury compounds:

$$C_2H_2\,(g) + H_2O\,(l) \xrightarrow[H_2SO_4]{Hg^{2+}} CH_3CHO\,(l)\ \Delta H = -138.2\ kJ/mol$$

This process only prospers industrially when the condensation and polymerization products of the acetaldehyde produced in the medium of acid are eliminated. To accomplish this, German consortium for electrochemical industry in (1912) proposed the process utilizing excess acetylene (C_2H_2) at the elevated temperature and eliminating the acetaldehyde product instantaneously from the reaction liquid. Simultaneously, the heat of the reaction is eliminated by distilling a suitable amount of water. Secondary reactions, like the oxidation of C_2H_4O to produce carbon dioxide and acetic acid, the outcome in reduction of the Hg^{2+} to the metallic mercury (Morrison & Shaw, 1933; Sarkowski & Götze, 1996).

6.4.2.1. Wet Oxidation Process

The process of wet oxidation circumvented direct usage of the poisonous mercury compounds. It was functioned, among others, by the Wacker-Chemie till they switch to ethylene as the beginning material in 1962.

In this method, Fe (III) sulfate is normally added to re-oxidized the Hg metal to Hg (II) salt, therefore ensuring adequate concentrations of the active catalyst. The acetylene then reacts at 90 to 95°C with the aqueous solution of catalyst; amongst 30 and 50% of the introduced acetylene reacts normally in one run (Mullins, 1953; Rauch-Puntigam & Völker, 2013). The gas evolving from the reactor is then cooled; mostly water and traces of Hg are separated and reverted to the reactor. Water and Acetaldehyde are condensed in the additional coolers and acetaldehyde is finally washed out with H2O from the cycle gas that has normally been cooled to 25 to 30°C. An 8 to 10% aqueous solution of acetaldehyde is obtained. Nitrogen is released with the feed gas whereas carbon dioxide is produced as the byproduct; to evade excessive buildup, these gases are eradicated by withdrawing the small stream of cycle gas. Fe(II) sulfate is made in this reaction and is then oxidized in the separate reactor with 30% nitric acid at around 95°C. Pure acetaldehyde is attained by the fractional distillation of aqueous solution at around 200kPa (Yelle, 1979; Robin et al., 2002).

6.4.2.2. Chisso Process

The Chisso process utilizes sulphuric acid/mercury sulfate solution as the catalyst. The acetylene reacts entirely with the solution of catalyst at 68 to 78°C and the gauge pressure of around 140kPa. The combination of vacuum and pressure process stage at the low temperature and without surplus acetylene is utilized; pure C_2H_4O can be separated and distilled by using the heat of reaction.

6.4.2.3. Production through Vinyl Ether

Reppe at BASF established the process utilizing vinyl ether; it was functioned in the pilot plant amongst 1939 and 1945. The utilization of poisonous mercury compounds is evaded altogether. Methanol (CH3OH) is added to the acetylene at 150 to 160°C and around 1600 kPa in the existence of potassium hydroxide to produce methyl vinyl ether (Stamfer et al., 1992; Krapcho & Waterhouse, 1998). The methyl vinyl ether is hydrolyzed with the dilute acid:

$$C_2H_2 + CH_3OH \xrightarrow{KOH} CH_3OCH = CH_2 \xrightarrow{H_2O,\ H^+} CH_3OH + CH_3CHO$$

6.4.2.4. Production through Ethylidene Diacetate

Addition of CH3COOH to C2H2 in the existence of Hg (II) salts produces ethylidene diacetate $CH_3CH\ (OCOCH_3)_2$, which disintegrate into acetic anhydride and acetaldehyde at 130 to 145°C in the existence of acid catalysts. This process was made by Societe Chimique des Usines du Rhone in 1914 on the industrial scale but is without any importance now.

6.4.3. Production from Ethylene (C2H4)

Ethylene is the most vital starting material now for the making of acetaldehyde. Most of the current capability operates by direct oxidation of the ethylene (Wacker process).

6.4.3.1. Direct Oxidation of C_2H_4

This process was established by Wacker-Chemie and Hoechst between 1957 and 1959. Officially, the reaction continues as follows:

$C_2H_4 + 1/2O_2 \rightarrow CH_3CHO \Delta H = -244$ kJ/mol

The aqueous solution of $CuCl_2$ and $pdCl_2$ is utilized as a catalyst.

Acetaldehyde creation had already been noticed in the reaction amongst ethylene (C_2H_4) and aqueous palladium chloride. This reaction is usually quantitative:

$$C_2H_4 + PdCl_2 + H_2O \rightarrow CH_3CHO + Pd + 2HCl$$

In Wacker-Hoechst process, the metallic palladium (Pd) is re-oxidized by copper chloride ($CuCl_2$), which is regenerated with oxygen:

Pd + 2CuC12 →PdC12 + 2 CuCl

$$2CuCl + 1/2O_2 + 2HCl \rightarrow 2CuCl_2 + H_2O$$

Thus only the small amount of palladium chloride ($PdCl_2$) is needed for the transformation of ethylene. The reaction of C_2H_4 with $PdCl_2$ is the rate-determining step.

1- and 2-stage versions of the procedure are on stream. In the *1-stage method*, the mixture of ethylene–oxygen reacts with the solution of catalyst. The stationary state is developed during the reaction in which reaction and oxidation continue at the same rate. The stationary state is controlled by the extent of catalyst's oxidation, as given by the ratio of $c\text{Cu}_2$+/($c\text{Cu}_2$+ +cCu+). In the 2-*stage process, the* reaction is generally carried out with C_2H_4 and then with O_2 in two distinct reactors. The solution of catalyst is consecutively reduced and oxidized. Simultaneously the extent of catalyst's oxidation changes alternately. Air is utilized rather than pure oxygen for the oxidation of catalyst (Ullmann, 1985; Takata et al., 2003).

The reaction rate is reduced by the acid made in the depletion of palladium chloride. This can be prohibited by protecting the acids with basic Cu salts. Restoration of the basic Cu salts occurs during catalyst oxidation (Chernyuk, 1971; Landolt & Kerr, 1973).

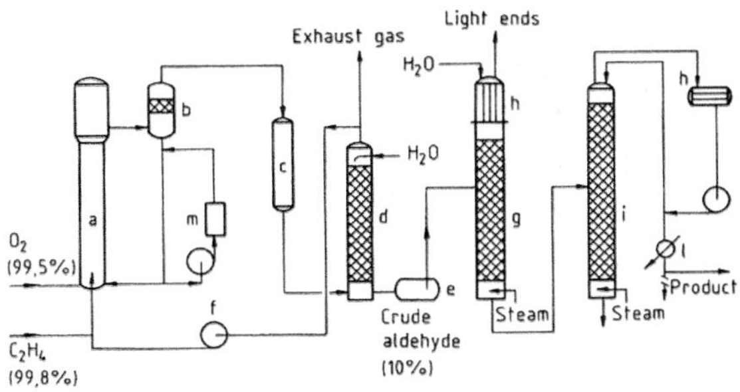

Figure 6.4: One-stage process.

[Source: https://onlinelibrary.wiley.com/doi/abs/10.1002/14356007.a01_031]

- Reactor
- Separating vessel
- Cooler
- Scrubber
- Crude aldehyde tank
- Cycle-gas compressor
- Light-ends distillation
- Condensers
- Purification column
- Product cooler
- Regeneration

In 1-stage process (Figure 6.4) oxygen and ethylene are normally charged into the lower portion of the reaction tower (a); the catalyst is circulated through the separating vessel (b) by the principle of airlift and carefully assorted with the gas. Reaction conditions are 130°C and 400 kilos Pascal. The mixture of acetaldehyde–water vapor, jointly with the unreacted gas, is reserved from the separating vessel; the reaction products from the mixture are isolated by cooling (c) and washing with the water (d); the unreacted gas is reverted to the reactor. The small portion is released from cycle gas as the exhaust gas to stop the buildup of inactive gases; these inactive gases are either announced as pollution of the feed gas or created as byproducts. The partial stream of the catalyst is heated to around 160°C to decompose derivatives which have gathered in the catalyst (Kandiner & Brinkley, 1950; Simoneit & Mazurek, 1982).

Crude acetaldehyde attained during washing of reaction products is purified in two stages. The first stage is an extractive purification with H_2O in which the light ends with lower boiling points than the acetaldehyde are divided at the top, whereas water and the higher-boiling derivatives, like chlorinated acetaldehydes, acetic acid, or crotonaldehyde, are reserved together with acetaldehyde (C_2H_4O) at the bottom. In the second column (i) the acetaldehyde is distilled by fractional distillation (James & Parks, 1982; Bringmann et al., 2001).

2-Stage Process (Figure 6.3) Tubular reactors (a), (d) are utilized for both reaction and oxidation. The gases react totally in the existence of the catalyst. The reaction of ethylene occurs at around 105 to 110°C and 900 to 1000kPa. Catalyst solution comprising of acetaldehyde (C_2H_4O) is then prolonged in the flash tower (b) by decreasing the pressure to the atmospheric level. The mixture of acetaldehyde–water vapor distills overhead whereas catalyst is sent through the pump (c) to oxidation reactor (d), where it reacts with O_2 at around 1000kPa. As reaction and oxidation are carried out distinctly, no high-purity initiating gas is needed. Normally, the air is utilized instead of oxygen. Oxygen conversion is nearly complete; exhaust air from (e) can be utilized as inactive gas for plant utilization (Elke et al., 1998; Lesch et al., 2011). The oxidized solution of catalyst separated from the exhaust air in the separator (e) is reutilized for reaction with the ethylene in (a).

The mixture of acetaldehyde–water vapor from a flash tower (b) is pre-concentrated in column (f) to about 60 to 90% acetaldehyde by using the heat of the reaction. Process water released at the bottom (f) is reverted to flash tower in order to maintain the constant catalyst concentration. The portion of process water is utilized for scrubbing the exhaust air in (i) and the exhaust gas in (k) free of acetaldehyde (C_2H_4O). Scrubber water then drifts to the rough aldehyde column (f).

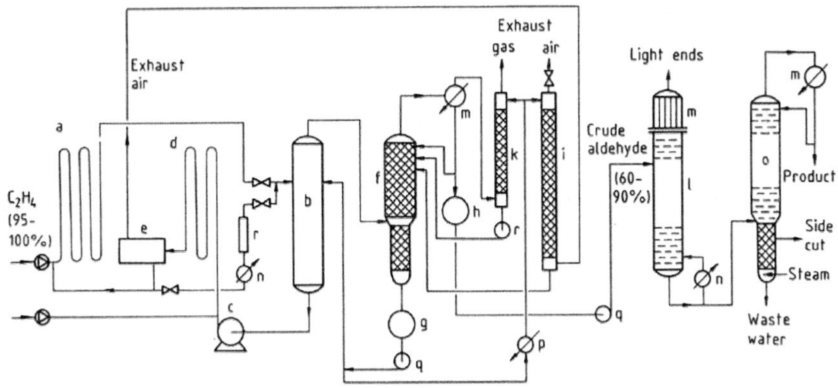

Figure 6.5: Two-stage process.

[Source: https://onlinelibrary.wiley.com/doi/abs/10.1002/14356007.a01_031]

- Reactor
- Flash tower
- Catalyst pump
- Oxidation reactor
- Exhaust-air separator
- Crude-aldehyde column

- Process-water tank
- Crude-aldehyde container
- Exhaust-air scrubber
- Exhaust-gas scrubber
- Light-ends distillation
- Condensers
- Heater
- Purification column
- Cooler
- Pumps
- Regeneration

The 2-stage purification of the rough acetaldehyde follows. In the 1st stage (l), the low-boiling substances, like carbon dioxide, chloromethane, and chloroethane are separated. In the 2nd stage (o), water and the higher-boiling derivatives, like acetic acid and chlorinated acetaldehydes, are eradicated from acetaldehyde, and the concluding is attained in pure form. Chlorinated acetaldehydes generally become concentrated in the columns as medium-boiling elements and are discharged horizontally. From this mixture, monochloro acetaldehyde can be attained as the hemihydrate. Remaining byproducts can be reverted to the catalyst for the oxidative decomposition (Hanke et al., 1991; Feng et al., 2007). This oxidative self-distillation is assisted by thermal treatment of the incomplete stream of catalyst at around 160 to 165°C (regeneration, r).

Balance of Reactions and the Side Products

The produce in both kinds of the process is almost the same. The balance of the 2-stage process is given below:

100 parts of C_2H_4 (ethylene) gives:

- 95 parts C_2H_4O (acetaldehyde)
- 1.9 parts chlorinated C_2H_4O (aldehydes)
- 1.1 parts unreacted C_2H_4 (ethylene)
- 0.8 parts CO_2 (carbon dioxide)
- 0.7 parts CH_3COOH (acetic acid)
- 0.1 parts CH_3Cl (chloromethane)
- 0.1 parts C_2H_5Cl (ethyl chloride)

- 0.3 parts crotonaldehyde, methane, ethane, and other trivial side products

Process Variant: An interesting alternative of the process, even though of no technical significance, utilizes glycol as the medium of reaction. Cyclic acetal of C_2H_4O, namely the hydrolyzable 2-methyl-1,3-dioloxane, is attained. The benefit of this process is the high reaction rate (Angino & Andrews, 1968; Garcıa-Sánchez et al., 1999).

Construction Materials: During process improvement, serious difficulties have occurred by the tremendously acidic aqueous $CuCl_2$–$PdCl_2$ solution. These issues have been resolved in the 2-stage process by making parts in interaction with the solution of catalyst completely from titanium (Ti) or by lining the parts with the metal (Sennewald et al., 1963; Hall et al., 1973; Fiserova-Bergerova et al., 1990).

Waste Air: The leftover air from the process of oxidation comprises of small amounts of the unreacted ethylene, acetaldehyde, and the side products of the reaction like chloromethane, ethane, methane, and chloroethane. In Germany, the leftover air should be filtered of the side products in order to meet the standards of TA-Luft (Veith et al., 1983; Grandjean & Landrigan, 2006).

6.4.3.2. Acetaldehyde as the Byproduct

Acetaldehyde is also created in the making of vinyl acetate from acetylene or ethylene. It is isolated by distillation and is usually transformed into acetic acid for reuse.

6.4.3.3. Isomerization of the Ethylene Oxide

The research was done on this method before approving the direct oxidation of C_2H_4 (ethylene). Catalysts were SiO_2, Al_2O_3, and acid salts of the mineral acids like molybdic acid, sulfuric acid, or phosphoric acid. Yields of 90 to 95% have been described, but the method hasn't gained industrial significance.

6.4.4. Production from Sources of C_1

Since the upsurges in oil cost in 1973 and 1977, C_1 has gained concern as the feedstock for the organic chemicals and as the alternative for petrochemicals. Though, for the production of acetaldehyde, C_1 material appears to be of trivial significance as most of the typical acetaldehyde byproducts can be prepared from C_1 sources.

6.4.4.1. Production from the Synthesis Gas

Acetaldehyde is made with low selectivity and the yield of ca. 30% from synthesis gas, jointly with ethanol, acetic acid, and saturated hydrocarbons, chiefly methane. Catalysts are Co and Rh compounds triggered by magnesium chloride or iodine compounds and sustained on the silicate carrier. There is no industrial utilization of this process so far.

6.4.4.2.Production through Methanol, Acetic Anhydride, or Methyl Acetate

Hydroformylation of CH_3OH (methanol) with CO/H_2 has quite well been known since the innovation of the oxo process. It occurs in the existence of hydroformylation catalysts, like nickel, cobalt, and iron salts or the conforming metal carbonyls, at the increased temperature of 180 to 200°C and high pressure of 30 to 40 MPa. The acetaldehyde selectivity of about 80% has been claimed to utilize a Fe–Co carbonyl or instead of a Co-Ni catalyst in the existence of nitriles, tertiary amines, or phosphines as the catalyst.

Likewise, high selectivity has been declared for *hydrocarbonylation of the methyl acetate* with Pd or Rh catalysts in the existence of iodine compounds and tertiary phosphines along with with Co–Ru catalysts in the existence of sodium and methyl iodides.

REFERENCES

1. Acker, J. P., Hansen, A. L., Kurach, J. D., Turner, T. R., Croteau, I., & Jenkins, C. (2014). A quality monitoring program for red blood cell components: in vitro quality indicators before and after implementation of semiautomated processing. *Transfusion*, *54*(10), 2534–2543.
2. Al-Shammary, A. F. Y., Caga, I. T., Winterbottom, J. M., Tata, A. Y., & Harris, I. R. (1991). Palladium-based diffusion membranes as catalysts in ethylene hydrogenation. *Journal of Chemical Technology & Biotechnology*, *52*(4), 571–585.
3. Amit, Z., & Smith, B. R. (1985). A multi-dimensional examination of the positive reinforcing properties of acetaldehyde. *Alcohol*, *2*(2), 367–370.
4. Angino, E. E., & Andrews, R. S. (1968). Trace element chemistry, heavy minerals, and sediment statistics of Weddell Sea sediments. *Journal of Sedimentary Research*, *38*(2), 634–642.
5. Bangle, Jr., R. (1954). Gomori's paraldehyde-fuchsin stain. I. Physico-chemical and staining properties of the dye. *Journal of Histochemistry & Cytochemistry*, *2*(4), 291–299.
6. Baranyi, G., Sieber, S., Pearce, J., Cheval, B., Dibben, C., Kliegel, M., & Cullati, S. (2019). A longitudinal study of neighbourhood conditions and depression in ageing European adults: Do the associations vary by exposure to childhood stressors?. *Preventive Medicine*, *126*, 105764.
7. Braun, D., Böhringer, B., Ivan, B., Kelen, T., & Tüdös, F. (1986). Structural defects in poly (vinyl chloride)—IV. Thermal degradation of vinyl chloride/acetylene copolymers. *European Polymer Journal*, *22*(1), 1–4.
8. Bringmann, G., Tasler, S., Endress, H., Kraus, J., Messer, K., Wohlfarth, M., & Lobin, W. (2001). Murrastifoline-F: first total synthesis, atropo-enantiomer resolution, and stereoanalysis of an axially chiral N, C-coupled biaryl alkaloid. *Journal of the American Chemical Society*, *123*(12), 2703–2711.
9. Chaudhuri, P. M., Stager, R. A., & Mathur, G. P. (1968). Properties of aliphatic and aromatic aldehydes under high pressure. Compressibility and viscosity determination. *Journal of Chemical & Engineering Data*, *13*(1), 9–11.
10. Chernyuk, G. P. (1971). Thorough Elimination of Acetylenic

Hydrocarbon Impurities from Various Mixtures. *Russian Chemical Reviews*, *40*(12), 1047–1060.

11. Chu, D., Mo, J., Peng, Q., Zhang, Y., Wei, Y., Zhuang, Z., & Li, Y. (2011). Enhanced Photocatalytic Properties of SnO2 Nanocrystals with Decreased Size for ppb-level Acetaldehyde Decomposition. *ChemCatChem*, *3*(2), 371–377.
12. Cornils, B., & Kuntz, E. G. (1995). Introducing TPPTS and related ligands for industrial biphasic processes. *Journal of organometallic chemistry*, *502*(1–2), 177–186.
13. Craven, P. M., & Lambert, J. D. (1951). The viscosities of organic vapours. *Proceedings of the Royal Society of London. Series A. Mathematical and Physical Sciences*, *205*(1082), 439–449.
14. Das, T. R., & Kuloor, N. R. (2013). Thermodynamic Properties of Acetaldehyde. *Journal of the Indian Institute of Science*, *50*(2), 45.
15. Dumitriu, E., Hulea, V., Chelaru, C., Catrinescu, C., Tichit, D., & Durand, R. (1999). Influence of the acid–base properties of solid catalysts derived from hydrotalcite-like compounds on the condensation of formaldehyde and acetaldehyde. *Applied Catalysis A: General*, *178*(2), 145–157.
16. Eckert, M., Fleischmann, G., Jira, R., Bolt, H. M., & Golka, K. (2000). Acetaldehyde. *Ullmann's Encyclopedia of Industrial Chemistry, 1, 1–10.*
17. Elke, K., Jermann, E., Begerow, J., & Dunemann, L. (1998). Determination of benzene, toluene, ethylbenzene and xylenes in indoor air at environmental levels using diffusive samplers in combination with headspace solid-phase microextraction and high-resolution gas chromatography–flame ionization detection. *Journal of Chromatography A*, *826*(2), 191–200.
18. Farias, K., Garcia, A., Whittle, J., & Lucena, C. (2013, September). Analyzing the effort of composing design models of large-scale software in industrial case studies. In *International Conference on Model Driven Engineering Languages and Systems* (vol.1, pp. 639–655). Springer, Berlin, Heidelberg.
19. Farnworth, E. R., Lagace, M., Couture, R., Yaylayan, V., & Stewart, B. (2001). Thermal processing, storage conditions, and the composition and physical properties of orange juice. *Food Research International*, *34*(1), 25–30.

20. Feng, X. H., Zhai, L. M., Tan, W. F., Liu, F., & He, J. Z. (2007). Adsorption and redox reactions of heavy metals on synthesized Mn oxide minerals. *Environmental Pollution*, *147*(2), 366–373.
21. Fischer, EO, & Jira, R. (1954). Cyclopentadienyl manganese tricarbonyl. *Journal of Natural Sciences B*, *9* (9), 618–619.
22. Fiserova-Bergerova, V., Pierce, J. T., & Droz, P. O. (1990). Dermal absorption potential of industrial chemicals: criteria for skin notation. *American Journal of Industrial Medicine*, *17*(5), 617–635.
23. Flores-Soto, M. E., Chaparro-Huerta, V., Escoto-Delgadillo, M., Vazquez-Valls, E., Gonzalez-Castaneda, R. E., & Beas-Zarate, C. (2012). Structure and function of NMDA-type glutamate receptor subunits. *Neurología (English Edition)*, *27*(5), 301–310.
24. Funder, J. W., Pearce, P. T., Smith, R., & Smith, A. I. (1988). Mineralocorticoid action: target tissue specificity is enzyme, not receptor, mediated. *Science*, *242*(4878), 583–585.
25. Gaines, K. C., Salhany, J. M., Tuma, D. J., & Sorrell, M. F. (1977). Reaction of acetaldehyde with human erythrocyte membrane proteins. *FEBS letters*, *75*(1–2), 115–119.
26. Garcıa-Sánchez, A., Alastuey, A., & Querol, X. (1999). Heavy metal adsorption by different minerals: application to the remediation of polluted soils. *Science of the total environment*, *242*(1–3), 179–188.
27. Grandjean, P., & Landrigan, P. J. (2006). Developmental neurotoxicity of industrial chemicals. *The Lancet*, *368*(9553), 2167–2178.
28. Hall, Jr, H. K., Smith, C. D., & Plorde, D. E. (1973). Cycloaddition of ethylene to acrylonitrile. *The Journal of Organic Chemistry, 38*(11), 2084–2087.
29. Hanko, R., Rabe, K., Dally, R., & Hoppe, D. (1991). Stereoselective Synthesis of Cyclic Dipeptide Hydroxyalkyl Isosteres via Metalated N, N-Dialkylcarbamate 2-Alkenyl Esters. *Angewandte Chemie International Edition in English*, *30*(12), 1690–1693.
30. Hartmann, G. K., & Leitinger, R. (1984). Range errors due to ionospheric and tropospheric effects for signal frequencies above 100 MHz. *Bulletin géodésique*, *58*(2), 109–136.
31. Heitz, E. (1974). Corrosion of metals in organic solvents. In *Advances in Corrosion Science and Technology* (vol. 1, pp. 149–243). Springer, Boston, MA.
32. Hendl, S., Millat, J., Vogel, E., Vesovic, V., Wakeham, W. A., Luettmer-

Strathmann, J., ... & Assael, M. J. (1994). The transport properties of ethane. I. Viscosity. *International journal of thermophysics*, *15*(1), 1–31.

33. Hodge, V., Stallard, M., Koide, M., & Goldberg, E. D. (1986). Determination of platinum and iridium in marine waters, sediments, and organisms. *Analytical Chemistry*, *58*(3), 616–620.
34. Hubbell, J. H. (1971). Survey of photon-attenuation-coefficient measurements 10 eV to 100 GeV. *Atomic Data and Nuclear Data Tables*, *3*, 241–297.
35. James, R. O., & Parks, G. A. (1982). Characterization of aqueous colloids by their electrical double-layer and intrinsic surface chemical properties. In *Surface and colloid science*(vol. 1, pp. 119–216). Springer, Boston, MA.
36. Jordan, A., Chen, D., Yi, Q. L., Kanias, T., Gladwin, M. T., & Acker, J. P. (2016). Assessing the influence of component processing and donor characteristics on quality of red cell concentrates using quality control data. *Vox sanguinis*, *111*(1), 8–15.
37. Kandiner, H., & Brinkley, S. (1950). Correction: Calculation of Complex Equilibrium Relations. *Industrial & Engineering Chemistry*, *42*(8), 1526–1526.
38. Krapcho, A. P., & Waterhouse, D. (1998). Sodium Trimethyl Silanoate. A Hydroxyl Synthon for Fluoride SNAr Type Displacements from Anthracene-9, 10-diones, Benz [g] isoquinoline-5–10-diones and Nitrobenzenes. *Synthetic communications*, *28*(18), 3415–3422.
39. Kutzelnigg, W., Fleischer, U., & Schindler, M. (1990). The IGLO-Method: ab-initio calculation and interpretation of NMR chemical shifts and magnetic susceptibilities. In *Deuterium and shift calculation* (vol.1, pp. 165–262). Springer, Berlin, Heidelberg.
40. Lambert, J. D., Staines, E. N., & Woods, S. D. (1950). Thermal conductivities of organic vapours. *Proceedings of the Royal Society of London. Series A. Mathematical and Physical Sciences*, *200*(1061), 262–271.
41. Landolt, G. R., & Kerr, G. T. (1973). Methods of Separation and Purification Using the Molecular Sieving Properties of Zeolites. *Separation and Purification Methods*, *2*(1), 283–359.
42. Lee, H. J., Yeo, Y. K., & Changn, J. Y. (2000). Modeling of industrial high pressure autoclave polyethylene reactor including decomposition

phenomena. *Korean Journal of Chemical Engineering*, *17*(2), 223–229.

43. Lesch, K. P., Selch, S., Renner, T. J., Jacob, C., Nguyen, T. T., Hahn, T., ... & Heine, M. (2011). Genome-wide copy number variation analysis in attention-deficit/hyperactivity disorder: association with neuropeptide Y gene dosage in an extended pedigree. *Molecular psychiatry*, *16*(5), 491.
44. Lin, YP, Lee, YL, Hung, CY, Huang, WJ, & Lin, SC (2017). Determination of multiresidue analysis of β-agonists in muscle and viscera using liquid chromatograph/tandem mass spectrometry with Quick, Easy, Cheap, Effective, Rugged, and Safe methodologies. *Journal of Food and Drug Analysis*.
45. Matveev, K. I. (1977). Development of new homogeneous catalysts for the oxidation of ethylene to acetaldehyde. *Kinetika i Kataliz*, *18*(4), 862–877.
46. Mielcarski, M., & Puchalska, I. (2002). Deposition of Ru-106 and I-125 on silver by internal electrolysis. *Nukleonika*, *47*(2), 83–86.
47. Miller, S. A. (1966). *Acetylene; Its Properties, Manufacture and Uses* (Vol. 1, pp. 1–10). Benn.
48. Miller, S. A., Rast, H. E., & Caspers, H. H. (1970). Lattice vibrations of LiYF4. *The Journal of Chemical Physics*, *52*(8), 4172–4175.
49. Mori, H. O., Mizuno, N., & Misono, M. (1991). Factors controlling the selectivity of the oxidation of acetaldehyde over heteropoly compounds. *Journal of Catalysis*, *131*(1), 133–142.
50. Morrison, G. O., & Shaw, T. G. (1933). By-Products of the Carbide Industry: The Manufacture of Ethylidene Diacetate and Vinyl Acetate. *Transactions of The Electrochemical Society*, *63*(1), 425–447.
51. Mullins, B. P. (1953). Studies on the spontaneous ignition of fuels injected into a hot air stream. *Fuel*, *32*(4), 451.
52. Nötzold, D., Wulff, H., & Herzog, G. (1994). Differenzthermoanalyse der Bildung des Pentastrontiumchloridphosphats und röntgenographische Untersuchung seiner Struktur. *Journal of alloys and compounds*, *215*(1–2), 281–288.
53. Ohta, N., & Baba, H. (1986). Emission properties to acetaldehyde vapor in relation to photodissociation. *The Journal of Physical Chemistry*, *90*(12), 2654–2661.
54. Payne Jr, L. K., Stansbury Jr, H. A., & Weiden, M. H. J. (1966). Synthesis

and Insecticidal Properties of Some Cholinergic Trisubstituted Acetaldehyde O-(Methylcarbamoyl) oximes. *Journal of Agricultural and Food Chemistry*, *14*(4), 356–365.

55. Rauch-Puntigam, H., & Völker, T. (2013). *Acryl-und Methacrylverbindungen* (Vol. 9, pp. 1–19). Springer-Verlag.
56. Robin, M., Pique, V., Faure, R., & Galy, J. P. (2002). Ultrasonic irradiation of the Ullmann condensation: Application to the preparation of dioxolo, dioxino, cyclopent, and imidazolo anthranilic acid derivatives. *Journal of Heterocyclic Chemistry*, *39*(5), 1083–1085.
57. Rustamov, R. M. (2007, July). Laplace-Beltrami eigenfunctions for deformation invariant shape representation. In *Proceedings of the fifth Eurographics Symposium on Geometry Processing* (pp. 225–233). Eurographics Association.
58. Sarkowski, H., & Götze, H. (1996). *Springer-Verlag: History of a Scientific Publishing House: Part 2: 1945–1992. Rebuilding-Opening Frontiers-Securing the Future* (Vol. 2, pp. 1–20). Springer Science & Business Media.
59. Scalabrin, G., Cristofoli, G., & Richon, D. (2002). Viscosity equations of pure fluids in an innovative extended corresponding states framework: II. Application to four fluids. *Fluid Phase Equilibria*, *199*(1–2), 281–294.
60. Sennewald, K., Schallus, E., & Pohl, F. (1963). Erzeugung von Acetylen durch thermische Spaltung von Kohlenwasserstoffen mittels hocherhitzten Wasserstoffes.(WLP-Verfahren der Knapsack-Griesheim AG). *Chemie Ingenieur Technik*, *35*(1), 1–6.
61. Shandura, M. P., Yakubovskyi, V. P., & Kovtun, Y. P. (2009). (4, 4-Difluoro-4-bora-3a, 4a,-diaza-s-indacen-3-yl) acetaldehyde: Synthesis and chemical properties. *Journal of Heterocyclic Chemistry*, *46*(6), 1386–1391.
62. Sheldon, R. A., & Kochi, J. K. (1976). Metal-catalyzed oxidations of organic compounds in the liquid phase: A mechanistic approach. In *Advances in Catalysis* (Vol. 25, pp. 272–413). Academic Press.
63. Shimizu, Y., Yamaguchi, K., Fukunaga, K., Takao, Y., Hyodo, T., & Egashira, M. (1999). Acetaldehyde gas-sensing properties and surface chemistry of SnO2-based sensor materials. *Journal of the Electrochemical Society*, *146*(3), 1222–1226.
64. Sieber, W., Wolfgruber, M., Neugebauer, D., Orama, O., & Kreissl,

FR (1983). Transition-Metal-Keten Compounds, XII [1] Synthesis and Structure of η5-C5H5 (1) $CO)_2$ [P (CH3) 3] WC≡C-C3H5. *Journal of Natural Sciences B*, *38* (1), 67–70.

65. Sieber, WJ, Eberl, K., Wolfgruber, M., & Kreissl, FR (1983). Transition-Metal-Keten Compounds, XVII [1] Preparation of an Anionic η2-Ketenyl Complex/Compounds of Transition Metal Ketene, XVII [1] Synthesis of an Anionic η2-Ketenyl Complex *Journal of Natural Sciences B*, *38* (9), 1159–1160.
66. Simoneit, B. R., & Mazurek, M. A. (1982). Organic matter of the troposphere—II. Natural background of biogenic lipid matter in aerosols over the rural western United States. *Atmospheric Environment (1967)*, *16*(9), 2139–2159.
67. Smith, T. E., & Bonner, R. F. (1951). Acetaldehyde, Propionaldehyde, and n-Butyraldehyde. *Industrial & Engineering Chemistry*, *43*(5), 1169–1173.
68. Soffritti, M., Belpoggi, F., Lambertin, L., Lauriola, M., Padovani, M., & Maltoni, C. (2002). Results of long-term experimental studies on the carcinogenicity of formaldehyde and acetaldehyde in rats. *Annals of the New York Academy of Sciences*, *982*(1), 87–105.
69. Sopyan, I., Watanabe, M., Murasawa, S., Hashimoto, K., & Fujishima, A. (1996). An efficient TiO_2 thin-film photocatalyst: photocatalytic properties in gas-phase acetaldehyde degradation. *Journal of Photochemistry and Photobiology A: Chemistry*, *98*(1–2), 79–86.
70. Stampfer, M. J., Malinow, M. R., Willett, W. C., Newcomer, L. M., Upson, B., Ullmann, D., ... & Hennekens, C. H. (1992). A prospective study of plasma homocyst(e)ine and risk of myocardial infarction in US physicians. *Jama*, *268*(7), 877–881.
71. Stoner, Jr. J. O., & Miller, S. A. (2006). Isotopic ^{13}C target foils 0.4–2.2 mg/cm^2 by pyrolysis of methane, with alternative methods of production. *Nuclear Instruments and Methods in Physics Research Section A: Accelerators, Spectrometers, Detectors and Associated Equipment*, *561*(1), 24–37.
72. Takata, T., Saeki, D., Makita, Y., Yamada, N., & Kihara, N. (2003). Aromatic hydrocarbon-catalyzed direct reaction of sulfur and sodium in a heterogeneous system: selective and facile synthesis of sodium monosulfide and disulfide. *Inorganic Chemistry*, *42*(12), 3712–3714.
73. Tateishi, N., Nishimura, K., Yahikozawa, K., Nakagawa, M., Yamada,

M., & Takasu, Y. (1993). Electrocatalytic properties of ultrafine gold particles towards oxidation of acetaldehyde and ethanol. *Journal of Electroanalytical Chemistry, 352*(1–2), 243–252.

74. Trofimov, B. A. (1990). Preparation of pyrroles from ketoximes and acetylenes. In *Advances in Heterocyclic Chemistry* (Vol. 51, pp. 177–301). Academic Press.
75. Ullmann, A. A. (1985). Data in search of a theory: A critical examination of the relationships among social performance, social disclosure, and economic performance of US firms. *Academy of Management Review, 10*(3), 540–557.
76. Ullrich, J., & Mannschreck, A. (1967). Studies on the Properties of (—)-2-α-Hydroxyethyl-Thiamine Pyrophosphate ("Active Acetaldehyde"). *European Journal of Biochemistry, 1*(1), 110–116.
77. Veith, G. D., Call, D. J., & Brooke, L. T. (1983). Structure–toxicity relationships for the fathead minnow, Pimephales promelas: Narcotic industrial chemicals. *Canadian Journal of Fisheries and Aquatic Sciences, 40*(6), 743–748.
78. Weast, R. C., & Astle, M. J. (1979). *CRC Handbook of Chemistry and Physics: A Ready-Reference Book of Chemical and Physical Data* (Vol. 60, pp. 1–18) (1979–1980). CRC Press.
79. Wiemann, L. O., Faltl, C., & Sieber, V. (2012). Lipase-mediated epoxidation of the cyclic monoterpene limonene to limonene oxide and limonene dioxide. *Zeitschrift für Naturforschung B, 67*(10), 1056–1060.
80. Yaws, C. L. (1990). Physical Properties: Part 11. Specific Heats of Organic Compounds. *Encyclopedia of Chemical Processing and Design: Volume 36-Phosphorus to Pipeline Failure: Subsidence Strains*, 36, 157.
81. Yelle, L. E. (1979). The learning curve: Historical review and comprehensive survey. *Decision Sciences, 10*(2), 302–328.
82. Zhang, D., Wang, G., Han, D., Zhang, Y., Xu, J., Lu, J., ... & Li, M. (2014). Activation of PPAR-γ ameliorates pulmonary arterial hypertension via inducing heme oxygenase-1 and p21WAF1: An in vivo study in rats. *Life Sciences, 98*(1), 39–43.

7

Clemmensen Reduction Mechanism

CONTENTS

7.1. INTRODUCTION

The Clemmensen reduction of carbonyl complexes to alkane parallels by the virtue of combined hydrochloric and zinc has been investigated with many substrates. Aromatic and aliphatic compounds have been engaged and have revealed different chemical manners. Hence, the initial reaction providing methylene equivalents is not firmly followed by the aryl products since other products are also produced (Suh et al., 1991; Quintero et al., 2000).

A further point of notice has been the reaction mechanism because with the aliphatic complexes the reduction does not drive over the alcohol. There is not a full presentation on this topic. We give a clear and thorough reaction mechanism, with the particular comment in every step (Weber et al., 1969; Bae et al., 1991).

Our order is a combined theory of the carbanionic mechanism and the carbenoid suggestion. There should not be two very diverse mechanisms in order to describe the same reaction, but a clear and complete one. This mode, the two theories are pacified. This will also occupy a space in Chemical Education (Hewthorne, 1956; Bartlett & Storey, 1958).

Clemmensen reduction is a chemical reaction defined as a decrease of ketones (or aldehydes) to alkanes by using concentrated hydrochloric acid and zinc amalgam (Flammia et al., 1999; Sánchez-Viesca et al., 2018). This reaction is termed after Erik Christian Clemmensen, a Danish chemist (Cumming, 1950; Denney, 1969).

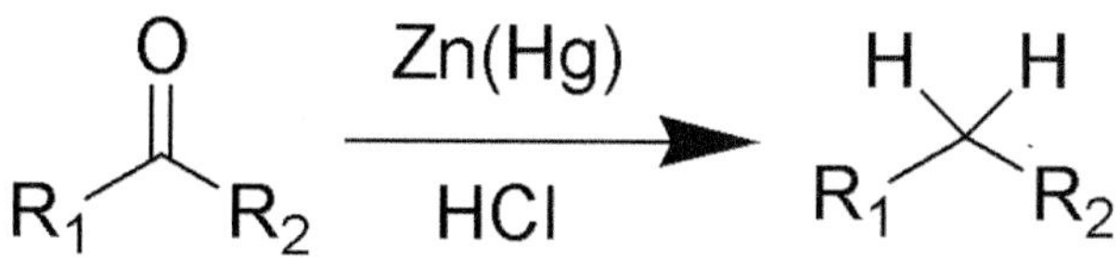

Figure 7.1: Clemmensen reduction reaction.

The initial Clemmensen reduction circumstances are mainly operative at reducing aryl-alkyl ketones, like those produced in a Friedel-Crafts acylation. The two-step order of Friedel-Crafts acylation trailed by Clemmensen reduction constitutes a standard strategy for the main alkylation of arenes. With cyclical ketones or aliphatic, adjusted Clemmensen conditions using initiated zinc dust in an anhydrous mixture of hydrogen chloride in acetic anhydride or diethyl ether is very effective (Baldinger & Nieuwland, 1933; Boone & Ashby, 1979).

The substrate must be forbearing of the muscularly acidic conditions of the Clemmensen reduction (37% HCl). Numerous alternatives are presented. Acid-sensitive substrates that are constant to the solid base can be condensed by using the Wolff-Kishner reduction; an extra, milder technique for substrates constant to hydrogenolysis in the existence of Raney nickel is a two-step Mozingo reduction (Lipp, 1922; Stoll & Rouvé, 1935).

Despite the distant past of this reaction, the mechanism of the Clemmensen reduction remains ambiguous. Because of the heterogeneous structure of the reaction, mechanistic investigations are tough, and only a few studies have been released. Mechanistic suggestions usually invoke organozinc intermediates, at times with zinc carbenoids, either as distinct types or as organic fragments bound to the zinc metal surface. Nevertheless, the analogous alcohol is supposed not to be an intermediary, since subjection of the alcohol to Clemmensen circumstances usually does not extend the alkane product (Franke & Kroupa, 1930; Ruzicka et al., 1945; Roylance & Choi, 2016).

Clemmensen reduction is one of the supreme dominant reduction approaches in organic production, and many analyses have been printed. Normally, zinc amalgam and greatly rigorous hydrochloric acid under reflux are engaged to suppress the production of by-products like alcohols, dimerization yields containing pinacols, and associated compounds. Though, among numerous modifications of this synthetic process, the process that using zinc powder in ether saturated or acetic anhydride with hydrogen chloride is suggested for its minor conditions (0°C, 1–2 h), and is presented for reductions of ketones and aldehydes in molecules carrying these functional groups as alkoxycarbonyl, phenol ether, acetoxy and cyano (Train et al., 2008; Sofer et al., 2015; Bhattacharyya, 2016).

7.2. EARLY APPROACHES AND MODERN THEORIES

The Clemmensen reaction lies in the reduction of a carbonyl compound to a methylene or methyl group by the means of hydrochloric acid and zinc amalgam. There are two analyses and general data (Gowan & Wheeler, 1962; Surrey, 2013).

Brewster (1972) explained the Clemmensen reduction. He reflects the solid metal as the "electron pump," this mechanical simile can be confusing since zinc has been taken by other writers as a foundation of free electrons in a reaction medium. He characterizes the metal as – $(:M)_x^+$. This is excluded because there is no purpose to diverge the reagent. There is no such kind of dipole in Zn^0 (Royals, 1954; House, 1972; Brewster, 1954).

In another statement, Brewster considered the reduction of some sterically stuck ketones and proposes as intermediate the structure in Figure 7.2 (Stashevsky, 1959; Brewster et al., 1963). This is in conflict with his former paper in which he reflects there is no carbonium ion. If so, how the suggested intermediate can be produced?

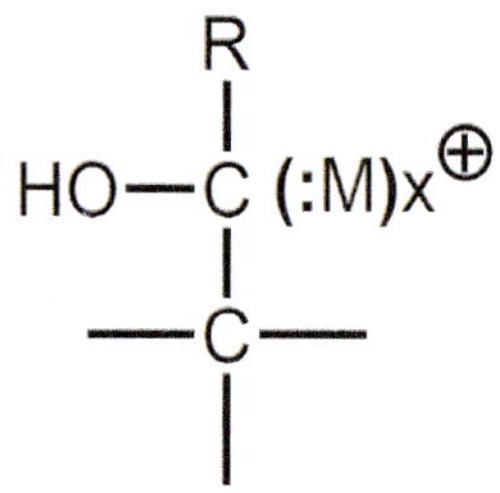

Figure 7.2: Brewster's outmoded representation of a reaction intermediate.

[Source: https://www.researchgate.net/publication/322530918_A_Complete_and_Sustained_Clemmensen_Reduction_Mechanism]

Staschewski (1959) repeated Brewster's incorrect illustration of zinc, but he encompassed the desired carbocation preceding to the relations with the metal. The next two steps are unclear because of the infrequent distribution of the electrical charges and electrons which is completely inadequate nowadays.

Nakabayashi (1960) gave the signal for the formation of an intermediary carbonium ion in the Clemmensen reduction. He investigated the primary stages with t-butyl phenyl ketone and acetophenone (Nakabayashi, 1960).

After the creation of a cationic intermediary, he suggests more reaction with another zinc atom producing a dipole with negative zinc.

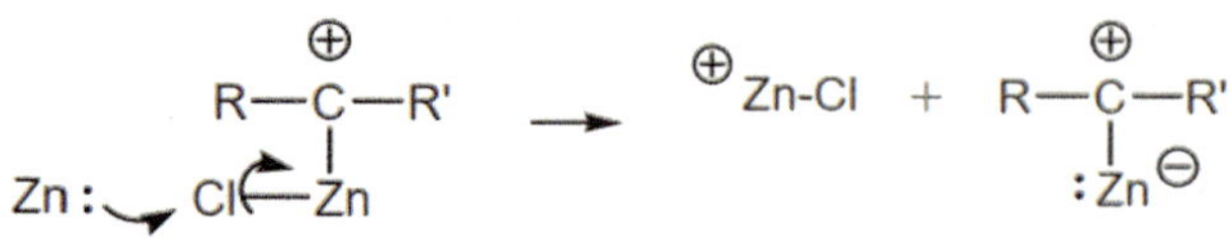

Figure 7.3: Unfeasible electrophilic reaction.

[Source: https://www.researchgate.net/publication/322530918_A_Complete_and_Sustained_Clemmensen_Reduction_Mechanism]

This is entirely unlikely, in the chloro-zinc bond chlorine contains δ_- and therefore it is not electrophilic. Therefore, there is not any negative zinc. Furthermore, in the zincate anions, the negative charge rises from the ligands, i.e., chlorine, oxygen, etc., not from the zinc.

The reduction of definite esters and 2-benzoyl benzoic acids was discussed by Thomson and Risinger (1963). This study discovered neighboring group involvement and the development of bimolecular reduction products. Nevertheless, this theme is distant from the unique Clemmensen reaction.

Reusch (1968), in Augustine's book 'Reduction,' remarks the Clemmensen reduction in the division on 'Deoxygenation of Carbonyl Mixtures.' He grants a modified type of Brewster's mechanism. A projected intermediary is in Figure 7.4, which is intolerable since active metals form just positive ions, not negative ones. Moreover, there are misplaced steps in order to have a clear and articulate reaction mechanism.

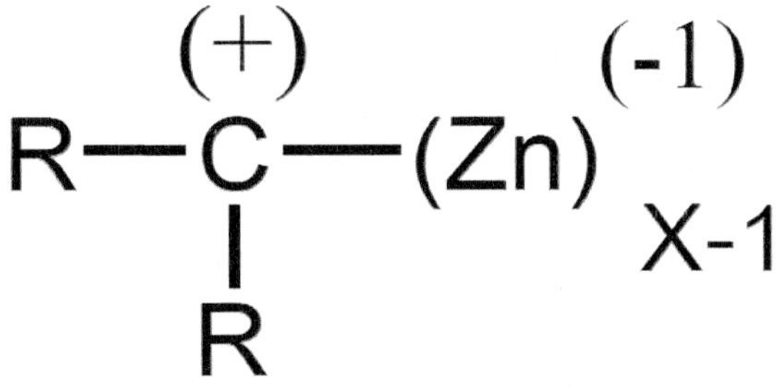

Figure 7.4: A variant of the previously rejected intermediate.

[Source: https://www.researchgate.net/publication/322530918_A_Complete_and_Sustained_Clemmensen_Reduction_Mechanism]

Ten years late, Schmitt and Horner (1978) suggested a reaction mechanism for the Clemmensen reduction. Two zinc atoms were distinguished like Zn' and Zn." In Figure 7.5, the electric charges are absent, the zinc and the halogen must be positive and negative, individually, as specified below. Then the zinc atom must leave the metal apparent and make a covalent bond with the chloride ion.

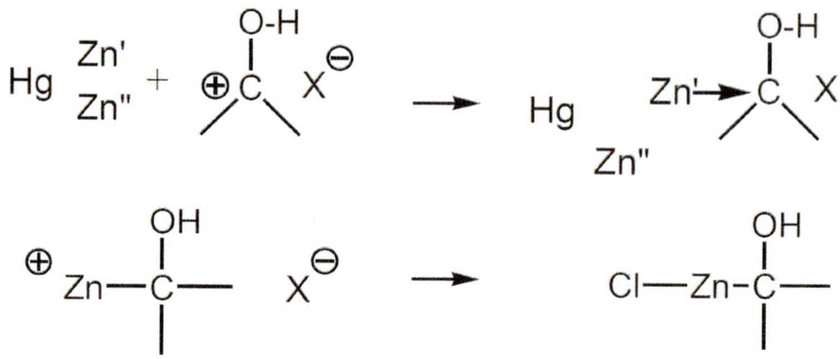

Figure 7.5: Incorrect and corrected structure after reaction with zinc.

[Source: http://article.sapub.org/10.5923.j.chemistry.20180801.02.html]

The offered the following step is in Figure 7.6, a paired 1,2 migration. This is very impracticable because there is no purpose for the separation of two sigma bonds.

Figure 7.6: Arbitrary twofold atom transfer.

[Source: http://article.sapub.org/10.5923.j.chemistry.20180801.02.html]

Further paper is that of Davis and Crump (1970), from New Zealand. They investigated the Clemmensen reduction offer 1,4-diketones and achieved alcohols which are not normally produced in this reaction. For example, hexane-2,5-dione gave hexane-2-ol, as well as cis- and trans-hex-4-en-2-ol, Figure 7.7: Clearly, neighboring group contribution deviates entirely the normal reactivity.

The Clemmensen reduction of 2-acetylflourene is irregular because it produces not only carbinol but also the 2-ethylflourene, 2-(α-hydroxyethyl) fluorene, and eight products of this alcohol (Minabe et al., 1976).

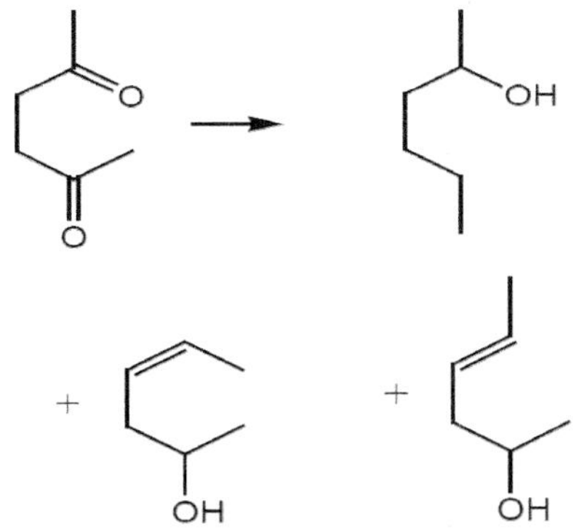

Figure 7.7: An example of I,4 diketone reduction by Clemmensen reaction.

[Source: http://article.sapub.org/10.5923.j.chemistry.20180801.02.html]

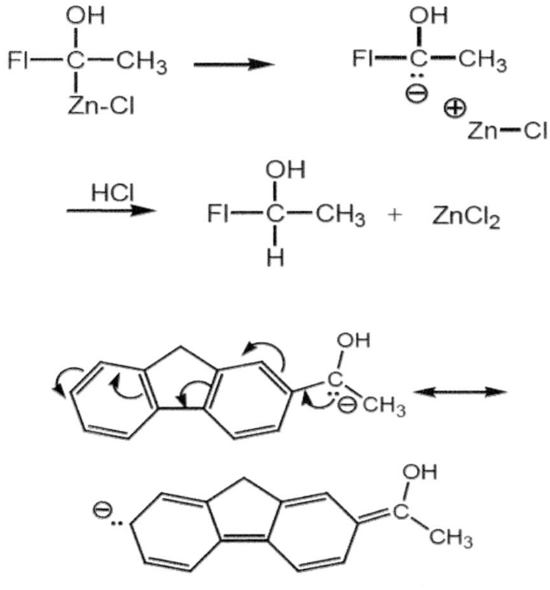

Figure 7.8: Formation of methyl-2-fluorylcarbinol and carbanion stabilization.

[Source: http://article.sapub.org/10.5923.j.chemistry.20180801.02.html]

The constancy of this carbanion can be described by the prolonged cross conjugated double bond structure that fallouts from electron delocalization by resonance. This is not the circumstance in normal Clemmensen reductions and outcomes in a different chemical manner. Price and Burdon (1986) showed the existence of carbene intermediates because they were seized by an alkene to provide a cyclopropane. The writers studied the reaction of replaced acetophenones with combined zinc in regular (50% aq. EtOH) solution at 20°C and HCl. They offer direct development of a zinc-carbene from the ketone, Figure 7.9.

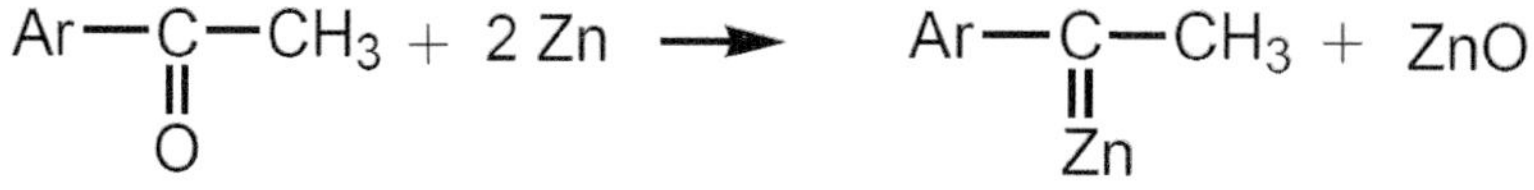

Figure 7.9: Main reaction of the so-called 'carbenoid mechanism'.

[Source: http://article.sapub.org/10.5923.j.chemistry.20180801.02.html]

The feasibility or not of this suggestion is discoursed later in 'The Carbenoid mechanism.' We reflect that the carbene is made over preceding intermediates as we will understand in the Discussion.

The free carbene was seized by reaction with extra styrene as shown in Figure 7.10.

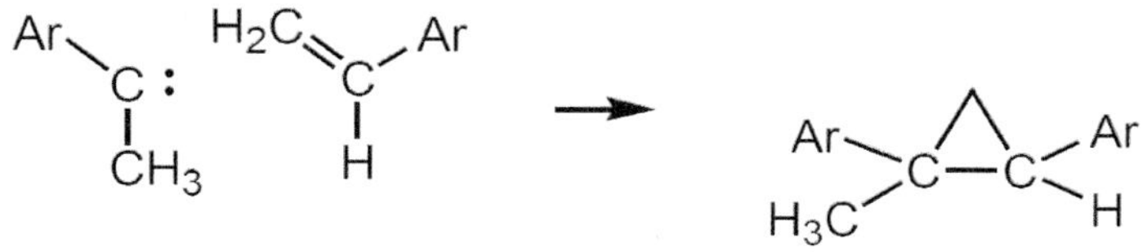

Figure 7.10: Formation of a cyclopropane derivative with capturing styrene.

[Source: http://article.sapub.org/10.5923.j.chemistry.20180801.02.html]

Real info online on the Clemmensen Reduction explains 'the mechanism is not however completely understood' and there are two major proposals: the 'Carbenoid mechanism' and the 'Carbanionic mechanism' (Buchanan & Woodgate, 1969).

In one exhibition of the 'Carbanionic mechanism,' each carbocation responds with two free electrons, purportedly released by the zinc, making a carbanion that later is protonated. The resultant short order is apparently faultless but it ignores the development of a carbene, the main intermediary that has been noticed experimentally. But if an impartial organometallic intermediary is formed in the order, as we will understand later, there is an opportunity to form a carbene, else no. The misreading of free electrons in the reaction intermediate develops from considering exactly the zinc as an 'electron pump.' Moreover, this would upsurge hydrogen development in the acidic medium (Sentz&Leonard, 1952).

In the further mechanism existing, the 'Carbenoid,' the first formed intermediary is a zinc-carbenoid, deprived of hydrogen chloride contribution, Figure 7.11.

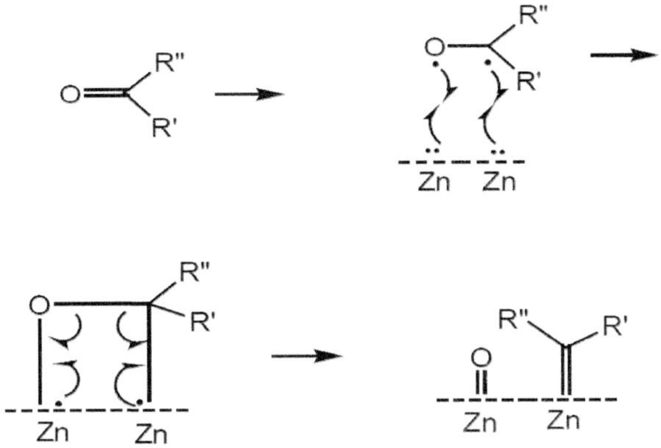

Figure 7.11: Supposed formation of zinc oxide and a zinc-carbene.

[Source: http://article.sapub.org/10.5923.j.chemistry.20180801.02.html]

The concurrent interaction of the carbonyl complex with two vicinal zinc atoms in the zinc surface is not feasible since the Zn-Zn distance is 2.3 times lengthier (2.84 Å) than a C=O bond length (1.21 Å). Further, it is very uncertain what would be the reaction among a carbonyl compound and zinc only, i.e., deprived of the interference of hydrochloric acid.

In the following section, we give the creation of a zinc-carbene and a free carbene after more steps and by the reaction with a second zinc atom.

In this analysis, devoted completely to the mechanical theme, we have pointed out the inaccuracies, the incorrect reactivities and the absent links that we have faced in the communications on the Clemmensen reduction system.

7.3. DESCRIPTION OF COMPLETE CLEMMENSEN REDUCTION MECHANISM

In place of two dissimilar reaction mechanisms, the 'Carbenoid' and the 'Carbanionic,' we present a whole mechanism that contains the formation of both types. This has been probably since the formation of sigma bonded organometallic intermediary, Figure 7.12.

The first stage is an interface of the aldehyde or ketone with a solvated hydroxonium ion approaching from the aqueous HCL (hydrochloric acid) (White, 1960; Banigan, 1966). The resultant oxonium chloride is in significance with the carbonium ion electrometer. This carbocation, in a chemisorption stage, responds with elemental zinc, an electrocoating component, leading to a two-electron reduction. The achieved organometallic intermediary reacts with hydrochloric acid (HCL): a carbene is produced with zinc chloride elimination and concomitant water (Luder & Zuffanti, 1961). The electron short species counters with another a dipolar ion and zinc atom is formed. This metal carbene compound is of Schrock kind, a carbanion and an electron lacking metal. It can be signified double bonded to zinc, however, *ab initio* quantum mechanical approaches show a zinc-carbon single bond for $ZnCH_2$ (Hamilton & Schaefer, 1991; Long, 1995).

The reaction of the zwitterion with HCL (hydrochloric acid) produces a deoxy organometallic derived. Lastly, ionization provides a carbanion which by protonation gives the reduction product, like with methylene rather than the unique carbonyl group. The ionization can be aided by through formation of ZnCl (zinc chloride) because of interaction with a chloride ion. This is reinforced by the fact that zinc chloride in the existence of HCL (hydrochloric acid) forms the tetrachlorozincate anion, $ZnCl_4$=, show the zinc accessibility to chloride ions (Génin et al., 1998).

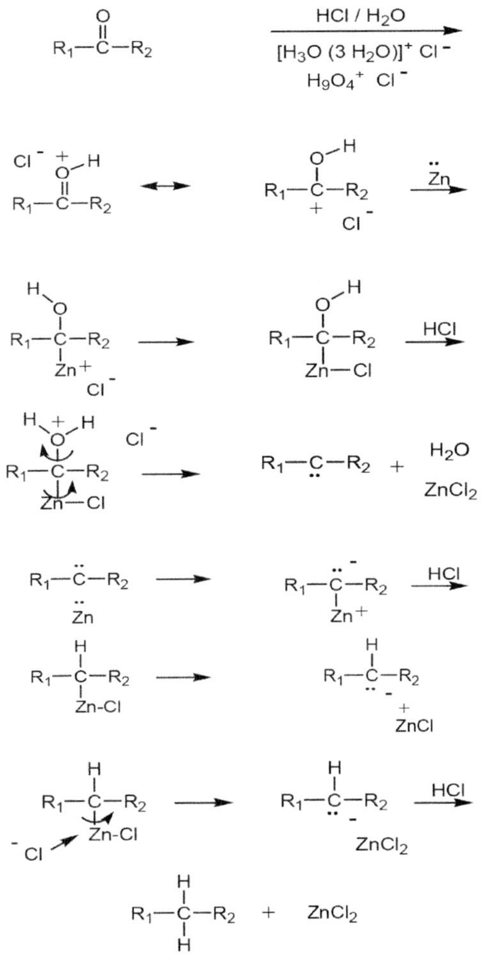

Figure 7.12: Complete reaction mechanism of the Clemmensen reduction.

[Source: http://article.sapub.org/10.5923.j.chemistry.20180801.02.html]

The Clemmensen reduction can be summarized in the following equation:

$R_1\text{-CO-}R_2 + 2\ Zn + 4\ HCl \longrightarrow R_1\text{-}CH_2\text{-}R_2 + H_2O + 2\ ZnCl_2$

In the mechanism offered by Vedejs (2004), the hydroxy organometallic intermediary, subsequently water removal and protonation produces a bis-chlorozinc structure, Figure 7.13, and no carbene is produced. Nevertheless, carbenes are recognized intermediates whose formation has been proved experimentally.

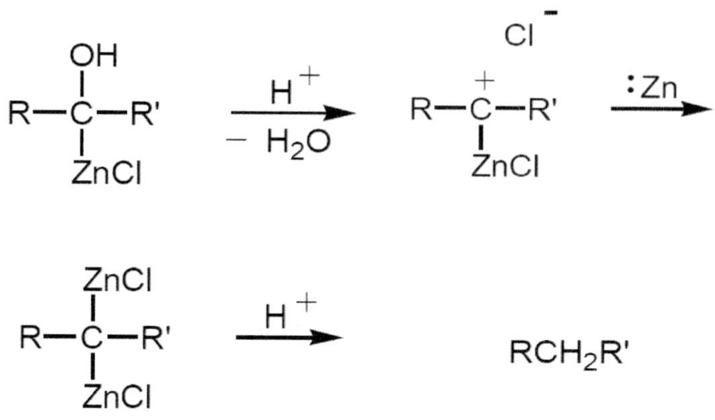

Figure 7.13: Improbable protolysis to the hydrocarbon.

[Source: http://article.sapub.org/10.5923.j.chemistry.20180801.02.html]

Furthermore, the bis-chlorozinc structure is not subjected to the necessary ionization so that it should be protonated and produce the reduction product since a chlorozinc ligand origins a δ_- at the carbon atom and therefore a carbanion from the second chlorozinc clutch would be at the semi negative carbon atom, which is extremely unlikely. Contrariwise, the carbanions in our reaction order are distant likely.

Figure 7.14 displays our alternate to the former reactions. The protonated hydroxy organometallic intermediary goes ahead with practically synchronic water removal, carbene creation, and $^+$ZnCl disarticulation.

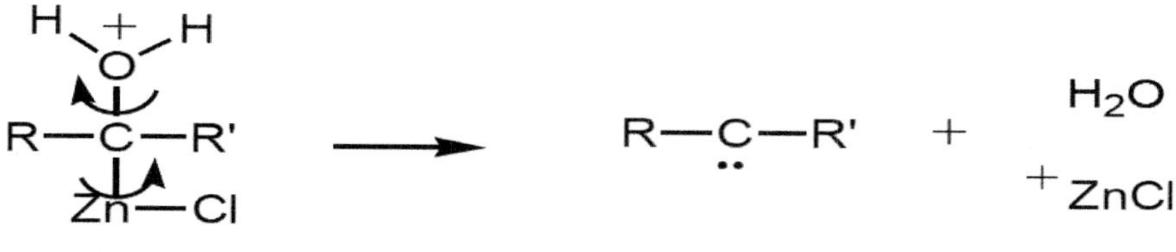

Figure 7.14: The required carbene formation via a concerted reaction mechanism.

[Source: http://article.sapub.org/10.5923.j.chemistry.20180801.02.html]

7.4. CLEMMENSEN REDUCTION EXAMPLES

Clemmensen reduction and its yields of some ketones and aldehydes are described below (Sorokin et al., 1995).

7.4.1. Clemmensen Reduction of Ethanal

Ethanal is an aldehyde. Through the Clemmensen reduction, Ethanal can be reduced to ethane. Concentrated HCl and Zinc amalgam is used as the Clemmensen reducing substance. Ethane is an alkane complex.

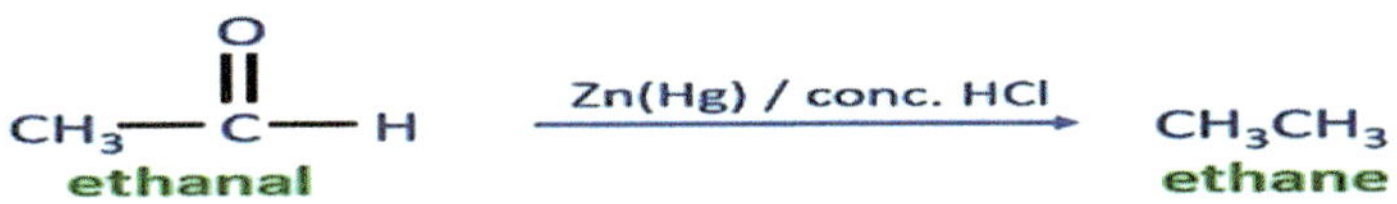

Figure 7.15: Clemmensen reduction of ethanol.

[Source: https://www.chemistryscl.com/reactions/Clemmensen-reduction/main.html]

7.4.2. Clemmensen Reduction of Propanal

The Propanal is an aldehyde. When Zn (Hg)/concentrated HCl and propanal counter, **propane** is given as the result. Propane also an alkane complex.

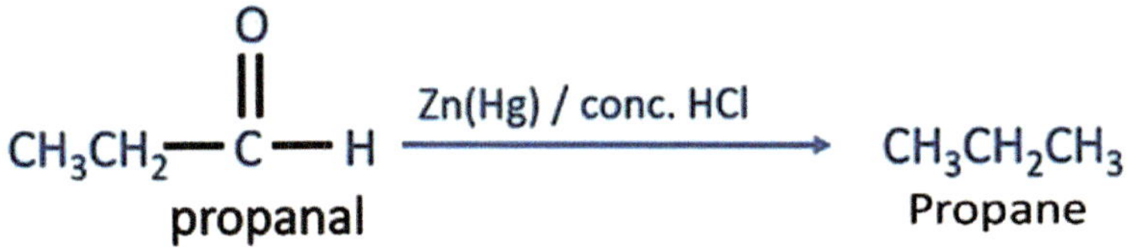

Figure 7.16: Clemmensen reduction of propanal.

[Source: https://www.chemistryscl.com/reactions/Clemmensen-reduction/main.html]

7.4.3. Clemmensen Reduction of Propanone

Propanone is a ketone complex. Propanone provides **propane** as the result with Zn(Hg)/concentrated HCl. Propane is an alkane complex.

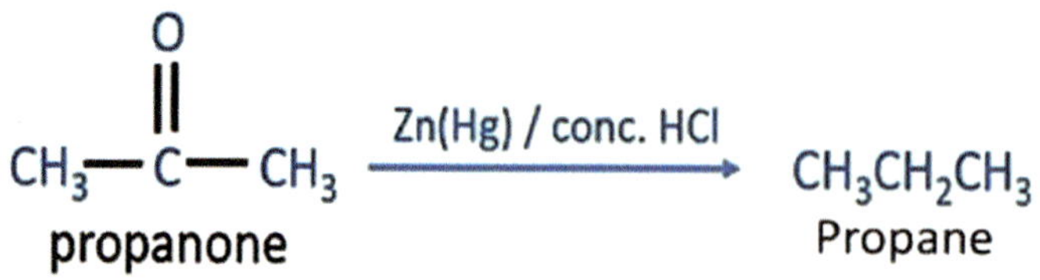

Figure 7.17: Clemmensen reduction of propanone.

[Source: https://www.chemistryscl.com/reactions/Clemmensen-reduction/main.html]

7.4.4. Benzaldehyde and Clemmensen Reduction

Methylbenzene (toluene) is specified when benzaldehyde reacts with Clemmensen mixture (Zn(Hg)/concentrated hydrochloric acid (HCl)).

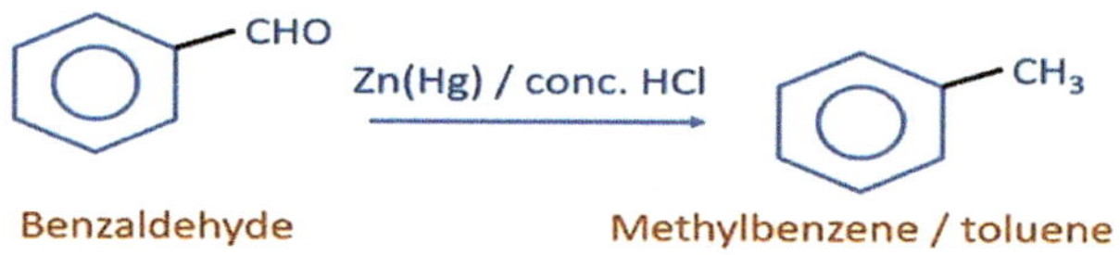

Figure 7.18: Clemmensen reduction of propanone.

[Source: https://www.chemistryscl.com/reactions/Clemmensen-reduction/main.html]

7.4.5. Zinc Amalgam and Concentrated DCl with Ethanal

Concentrated DCl and Zinc amalgam also work as Clemmensen reagent. This DCl and Zn(Hg) decrease ketones and aldehydes to hydrocarbons. The hydrogen atom in the hydrochloric acid (HCl) is exchanged with a deuterium atom. Deuterium and Hydrogen are the Isotopes of hydrogen.

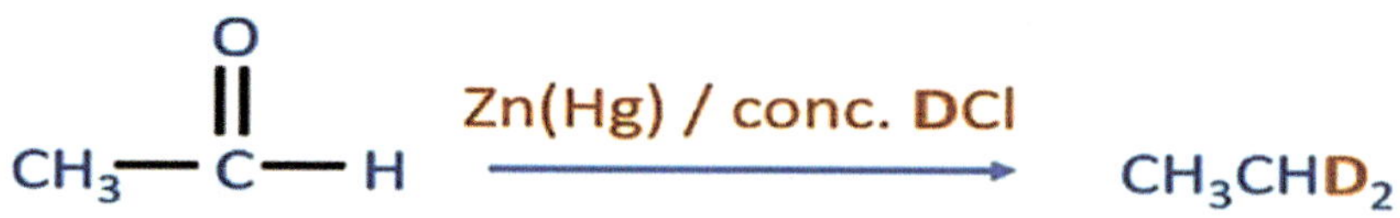

Figure 7.19: Zinc amalgam and concentrated DCl reaction with ethanol.

[Source: https://www.chemistryscl.com/reactions/Clemmensen-reduction/main.html]

7.4.6. Clemmensen Reduction Mechanism

Several mechanisms have been proposed for the Clemmensen reduction. One mechanism is explained below.

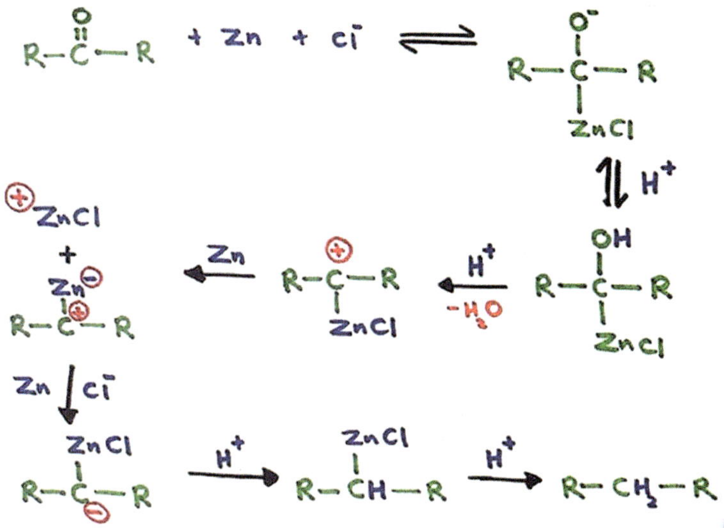

Figure 7.20: Clemmensen reduction mechanism.

[Source: https://www.chemistryscl.com/reactions/Clemmensen-reduction/main.html]

7.4.7. Separation of Aldehydes and Ketones from Clemmensen Reduction

You can't isolate ketone and aldehyde from this test since both ketone and aldehyde counter with Zn[Hg]/concentrated HCl and produce hydrocarbons as the result. Thus you are incapable to understand any difference throughout the reaction or after the reaction. As an instance, you cannot distinct propanone and propanal from the Clemmensen reduction reaction (Amouri et al., 2001; Andrés et al., 2001).

7.4.8. Phase Change in Clemmensen Reduction

Clemmensen reduction produces alkane from ketones and aldehydes. Maximum of ketones and aldehydes be present in the solid or liquid phase. Regardless of ketones and aldehydes, alkanes are in a gaseous phase. At times there may be a phase transformation in products and reactants. Few examples are listed beneath.

- Ethanal to ethane: Ethanal is a gas or liquid. At room temperature, Ethane is a gas.
- Benzaldehyde to toluene: liquid-to-liquid – there is no phase change.

7.4.9. Clemmensen Reduction and LiAlH4/ dry ether

Both ketones and aldehydes can be condensed by $LiAlH_4$/dry ether or Clemmensen reagent.

- Ketones and aldehydes are condensed to **hydrocarbons** by Clemmensen reagent.
- Ketones and aldehydes are condensed to **alcohols** by $LiAlH_4$/dry ether.

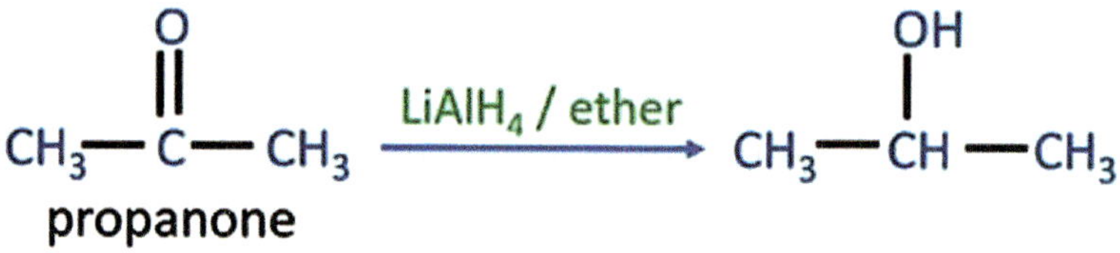

Figure 7.21: Clemmensen reduction and LiAlH4/ dry ether.

[Source: https://www.chemistryscl.com/reactions/Clemmensen-reduction/main.html]

7.4.10. Reduction of Acetaldehyde in the Presence of Zinc Amalgam (Zn(Hg)) with Concentrated HCl

Concentrated HCl and Zinc amalgam is used to condense acetaldehyde to ethane. Ethane is CH_3CH_3 and acetaldehyde is CH_3CHO which is an alkane complex.

REFERENCES

1. Amouri, H., Thouvenot, R., Gruselle, M., Malézieux, B., & Vaissermann, J. (2001). Synthesis, NMR, and X-ray Molecular Structure of a Unique Chiral Propeller-Like Cobalt Complex [(Co_2 (CO) 4) μ, η2, η2-(− H2CC: CCH2−)(− dppm) 2][BF4] 2 and Differentiation of Its Enantiomers by the Trisphat Anion. *Organometallics*, *20*(10), 1904–1906.
2. Andrés, R., Brissard, M., Gruselle, M., Train, C., Vaissermann, J., Malézieux, B., ... & Verdaguer, M. (2001). Rational Design of Three-Dimensional (3d) Optically Active Molecule-Based Magnets: Synthesis, Structure, Optical and Magnetic Properties of {[Ru (Bpy) 3] 2+, Clo4-,[Mniicriii (Ox) 3]-} N and {[Ru (Bpy) 2ppy]+,[Miicriii (Ox) 3]-} N, with Mii= Mnii, Niii. X-Ray Structure of {[Δru (Bpy) 3] 2+, Clo4-,[Δmniiδcriii (Ox) 3]-} N and {[Λru (Bpy) 2ppy]+,[Λmniiλcriii (Ox) 3]-} N. *Inorganic chemistry*, *40*(18), 4633–4640.
3. Bae, K. H., Koo, S. H., & Seo, W. J. (1991). Synthesis and antibacterial activities of 4-hydroxy-o-phenylphenol and 3, 6-diallyl-4-hydroxy-o-phenylphenol against a cariogenic bacteriumStreptococcus mutans OMZ 176. *Archives of Pharmacal Research*, *14*(1), 41.
4. Baldinger, LH, & Nieuwland, JA (1933). The Preparation of Some Alkyl Substituted Phenylacetonitriles in Liquid Ammonia. *Journal of the American Chemical Society*, *55* (7), 2851–2853.
5. Banigan, Sh. Ed. (1966). *The Binding Force,* By Scientists of the Westinghouse Research Laboratories, New York, USA: Walker and Co., 1966, vol.1, p. 70, The Hydrated Proton.
6. Bartlett, P. D., & Storey, B. T. (1958). Ionic Fission of the OO Bond in t-Butyl Arylpersulfonates1. *Journal of the American Chemical Society*, *80*(18), 4954–4961.
7. Bhattacharyya, B. (2016). Synthetic and Mechanistic Approach to Reduction of, β-Unsaturated Ketones via Zinc-mediated Single Electron Transfer (SET) Process: A Preliminary Review. *International Journal of Environmental Sciences*, *5*(2), 110–116.
8. Boone, J. R., & Ashby, E. C. (1979). Reduction of cyclic and bicyclic ketones by complex metal hydrides. In *Topics in Stereo Chemistry* (vol. 1, pp. 53–95).
9. Brewster, J. H. (1954). Reductions at Metal Surfaces. II. A Mechanism for the Clemmensen Reduction1. *Journal of the American Chemical*

Society, 76(24), 6364–6368.

10. Brewster, J. H., Fusco, A. M., Carosino, L. E., & Corman, B. G. (1963). Reduction of Phthalic Acid and Its Derivatives by Zinc1. *The Journal of Organic Chemistry, 28*(2), 498–501.
11. Buchanan, J. S. C., & Woodgate, P. D. (1969). The Clemmensen reduction of difunctional ketones. *Quarterly Reviews, Chemical Society, 23*(4), 522–536.
12. Burdon, J. and Price, R. C., (1986). The mechanism of the Clemmensen reduction: the substrates, *J. Chem. Soc., Chem. Comm., 12,* 893–894.
13. Crump, D. R. and Davis, B. R., (1970). The Clemmensen reduction of 1,4-diketones, *Chemical Communications,* pp.768–769.
14. Cumming, W. M. (1950). *Systematic Organic Chemistry, 1*(1), 1–10.
15. Denney, R. C. (1969). *Named Organic Reactions* (vol.1, pp. 118–118). London: Butterworths.
16. Flammia, D., Dukat, M., Damaj, M. I., Martin, B., & Glennon, R. A. (1999). Lobeline: Structure-Affinity Investigation of Nicotinic Acetylcholinergic Receptor Binding. *Journal of Medicinal Chemistry, 42*(18), 3726–3731.
17. Franke, A., & Kroupa, A. (1930). About "Ring Narrowing" in the formation of internal ethers (oxides) from glycols. *Monatshefte for Chemistry and Related Parts of Other Sciences, 56* (1), 331–346.
18. Génin, J. M. R., Bourrié, G., Trolard, F., Abdelmoula, M., Jaffrezic, A., Refait, P., ... & Herbillon, A. (1998). Thermodynamic equilibria in aqueous suspensions of synthetic and natural Fe (II)-Fe (III) green rusts: Occurrences of the mineral in hydromorphic soils. *Environmental Science & Technology, 32*(8), 1058–1068.
19. Gowan, J. E., & Wheeler, T. S. (1962). *Name Index of Organic Reactions*. Longmans, 1(1), 1–15.
20. Hamilton, T. P. and Schaefer, H. F., (1991). The prototypical zinc carbene and zinc carbyne molecules $ZnCH_2$ and HZnCH: triplet electronic ground states, J. Chem. Soc., Chem. Commun., 1 (9), 621–622.
21. Hewthorne, M. F. (1956). Peroxytrifluoroacetic Acid as Reagent for Determination of Carbonyl Function in Aldehydes and Ketones. *Analytical Chemistry, 28*(4), 540–541.
22. Horner, L., & Schmitt, E. (1978). Hydrogen transfer studies, 54.

Mechanism and limitations of Clemmensen reduction of acetophenone and related compounds. *Justus Liebig's Annalen der Chemie, 1978* (10), 1617–1633.

23. House, H. O. (1972). Modern synthetic reactions. *The Organic Chemistry Monograph Series, 1(1), 2–13.*
24. Klyszejko-Stefanowicz, L., Krajewska, W. M., & Lipinska, A. (1989). Histone occurrence, isolation, characterization and biosynthesis. *Histones and Other Basic Nuclear Proteins,* L. S. Hnilica, G. S. Stein, and J. L. Stein, eds. CRC Press, Boca Raton, Florida, 17–71.
25. Leonard, N. J., & Sentz, R. C. (1952). Rearrangement of α-Aminoketones during Clemmensen Reduction. VIII. The Fate of an Eight-membered Ring1, 2. *Journal of the American Chemical Society*, *74*(7), 1704–1709.
26. Lipp, P. (1922). Comparison of α-campholanic acid with Mahla-Tiemann's dihydro-campholenic acid and with isocampholic acid. *Reports of the German Chemical Society (A and B Series)*, *55* (6), 1883–1892.
27. Long, N. J. (1995). Organometallic Compounds for Nonlinear Optics—The Search for En-light-enment!. *Angewandte Chemie International Edition in English*, *34*(1), 21–38.
28. Luder, W. F., & Zuffanti, S. (1961). The Electronic Theory of Acids and Bases, 2nd. edn., New York, USA: Dover, 1961, vol.1, pp. 71–72.
29. Minabe, M, Yoshida, M., Fujimoto, M. and Suzuki, K. (1976). Clemmensen reduction of 2-acetylfluorene. Pathways for the formation of 2,3-di(fluorenyl)butane and its homologues, *J. Org. Chem.*, *41*(11), 1935–1938.
30. Nakabayashi, T., (1960). Studies on the mechanism of Clemmensen reduction. II. Evidence for the formation of an intermediate carbonium ion, *J. Am. Chem. Soc.*, *82*(15), 3906–3908.
31. Nakabayashi, T., (1960). Studies on the mechanism of Clemmensen reduction. I. The kinetics of Clemmensen reduction of p-hydroxyacetophenone, *J. Am. Chem. Soc.*, *82*(15), 3900–3906.
32. Quintero, T., Pernía, E., Guzmán, W. J., & Márquez, C. J. (2000). Synthesis of 3-carbethoxi-2-phenyl-4-quinolones. Potential antimitotics. *Acta cientifica venezolana*, *51*(3), 137–142.
33. Reusch, W. (1968). *Clemmensen Reduction*, in R. L. Augustine, Ed., Reduction, Techniques and Applications in Organic Synthesis, New York, USA: Marcel Dekker, 1968, pp. 186–188.

34. Risinger, G. E. and Thompson, J. A., (1963). Studies in the zinc reduction series: The reduction of certain 2-benzoylbenzoic acids and esters, *Rec. Trav. Chim. (Pays-Bas), 82*, 801–803.

35. Royals, E. E. (1954). *Advanced Organic Chemistry, 1*(1), 1–10.

36. Roylance, J. J., & Choi, K. S. (2016). Electrochemical reductive biomass conversion: direct conversion of 5-hydroxymethylfurfural (HMF) to 2, 5-hexanedione (HD) via reductive ring-opening. *Green Chemistry*, *18*(10), 2956–2960.

37. Ruzicka, L., Plattner, PA, & Wild, H. (1945). For information on the carbon ring (38th communication). On the dissociation constants of the cyanohydrins of some cyclic ketones. *Helvetica Chimica Acta, 28* (1), 613–617.

38. Sofer, Z., Jankovský, O., Libánská, A., Šimek, P., Nováček, M., Sedmidubský, D., ... & Pumera, M. (2015). Definitive proof of graphene hydrogenation by Clemmensen reduction: use of deuterium labeling. *Nanoscale*, *7*(23), 10535–10543.

39. Sorokin, A., Meunier, B., & Séris, J. L. (1995). Efficient oxidative dechlorination and aromatic ring cleavage of chlorinated phenols catalyzed by iron sulfophthalocyanine. *Science*, *268*(5214), 1163–1166.

40. Stashevsky, D. (1959). The mechanism of Clemmensen reduction. *Angewandte Chemie*, *71* (23), 726–736.

41. Stoll, M., & Rouvé, A. (1935). Many-membered heterocyclic compounds VIII. To the knowledge of the high-linked mono and poly-lcton rings. *Helvetica Chimica Acta*, *18* (1), 1087–1125.

42. Suh, J. J., Hong, Y. H., & Bae, M. (1991). Synthesis of 3-amino-1, 4-dihydropyridine derivativevia an intramolecular rearrangement of 1, 4-dihydropyridine-3-hydroxamate. *Archives of Pharmacal Research*, *14*(4), 319.

43. Surrey, A. R. (2013). *Name reactions in organic chemistry, vol. 1, pp. 1–12*. Elsevier.

44. Train, C., Gheorghe, R., Krstic, V., Chamoreau, L. M., Ovanesyan, N. S., Rikken, G. L., ... & Verdaguer, M. (2008). Strong magneto-chiral dichroism in enantiopure chiral ferromagnets. *Nature materials*, *7*(9), 729.

45. Vedejs, E. (2004). C lemmensen Reduction of Ketones in Anhydrous Organic Solvents. *Organic Reactions*, *22*, 401–422.

46. Weber, W. P., Felix, R. A., & Willard, A. K. (1969). Mass spectral rearrangements of siliconium ions. Migration of electronegative groups from carbon to silicon. *Journal of the American Chemical Society, 91*(23), 6544–6545.
47. White, J. H. (1960) *A Reference Book of Chemistry,* Section One, London, U.K.: University of London Press, 1960, vol.1, p. 66, Hydroxonium ion.

8

Carbonyl Group Reactions

CONTENTS

8.1. INTRODUCTION

Ketones and aldehydes organic compounds which integrate a *carbonyl functional group*, C=O. The carbon atom of this assembly has two residual bonds that may be engaged by hydrogen or aryl or alkyl substituents. The compound is an aldehyde if in any case one of these substituents is hydrogen. The compound is a ketone if neither is hydrogen, (Jencks, 1964; Burgi et al., 1974; Mikami et al., 2004).

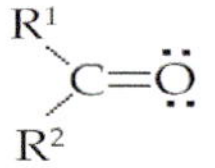

Aldehyde: R^1 and/or R^2 = H
Ketone: R^1 and R^2 = alkyl or aryl

Figure 8.1: The organic structure of carbonyl group.

[Source: https://www2.chemistry.msu.edu/faculty/reusch/VirtTxtJml/aldket1.htm]

A characteristic suffix is assigned to these classes by the IUPAC system of nomenclature, **one** to ketones and **al** to aldehydes. For instance, $H_2C=O$ is methanal, more frequently called formaldehyde. Since an aldehyde carbonyl group essentially always lie at the termination of a carbon chain, it is by defaulting location #1, and consequently describes the numbering course (Brown et al., 1939; Noyce & Jorgenson, 1962; Kawabata et al., 1991). A ketone carbonyl function may be situated anyplace within a ring or chain, and its location is assumed by a locator number. Chain numbering generally initiates from the end adjacent to the carbonyl group (Imamoto et al., 1989; Adams & Spero, 1991; Zamora et al., 2000). The carbonyl group is allocated location #1, in cyclic ketones, and this number is not quoted in the name, except more than one carbonyl group is existing. Cases of IUPAC names are given (in blue) in the subsequent diagram (Bender, 1960; Jørgensen, 2000; Zatsepin et al., 2005). Communal names are in red, and derivative names in black. In communal names, carbon atoms close to the carbonyl group are frequently elected by Greek letters. The atom close to the function is alpha, the following detached is beta and so on. Subsequently, ketones have two sets of adjacent atoms, one set is categorized α, β, etc., and the other α,' β', etc.

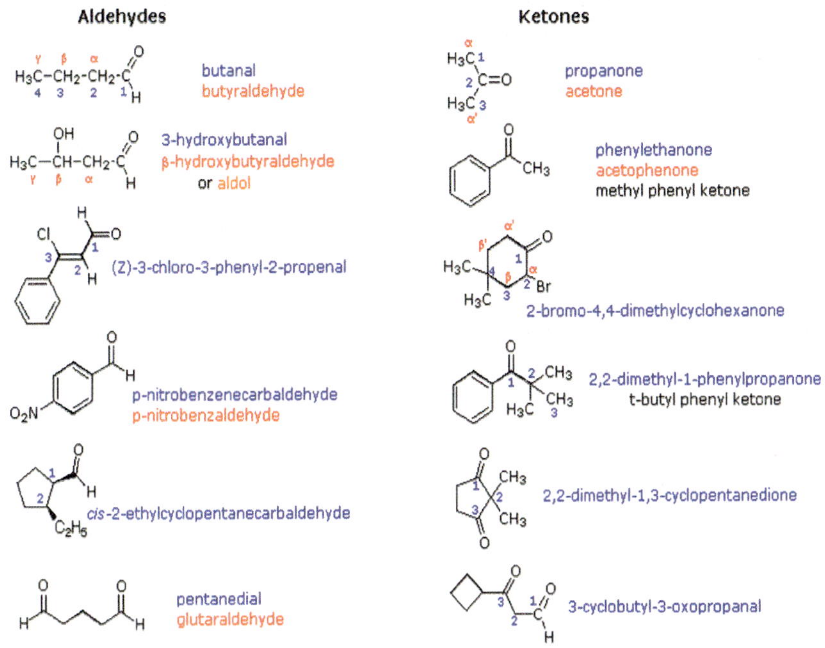

Figure 8.2: Different ketones and aldehydes containing a carbonyl group.

[Source: https://www2.chemistry.msu.edu/faculty/reusch/VirtTxtJml/aldket1.htm]

Very common ketones, such as phenylethanone and propanone (first two cases in the right column), do not need a locator number, as there is only one probable location for a ketone carbonyl function. Similarly, at the bottom left, locator numbers are absent for the simple dialdehyde, because aldehyde functions must inhabit the termination of carbon chains site (Koelsch & LeClaire, 1941; Johnson & Evans, 2000). The hydroxy propenal and butanal cases (left column, 2nd & 3rd from the top) and the oxopropanal case (bottom right) exemplify the nomenclature precedence of IUPAC suffixes. In all situations, the aldehyde function has a greater status than either an alkene, alcohol or ketone and gives the nomenclature suffix (Mukaiyama et al., 1974; Shinohara et al., 1980; Lee & Harris, 1989). The further functional groups are considered as substituents. Because in nomenclature suffix priority, ketones are just underneath aldehydes, the "oxo" substituent terminology is rarely desired. Simple substituents integrating a carbonyl group are frequently met (Trost & Massiot, 1977; Noyori et al., 1979). The general name for these groups is **acyl**. Three cases of acyl groups having precise names are revealed beneath.

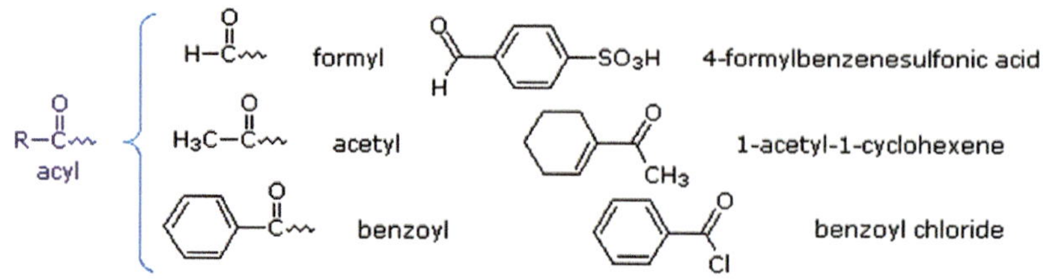

Figure 8.3: Compounds containing carbonyl group present in an acyl group.

[Source: https://www2.chemistry.msu.edu/faculty/reusch/VirtTxtJml/aldket1.htm]

8.2. THE OCCURRENCE OF ALDEHYDES AND KETONES

Ketones and aldehydes are extensive in nature, frequently united with other functional groups. Cases are revealed in the subsequent diagram (Ohwada et al., 1991; Wiberg et al., 1994; Kuroda et al., 2004). The compounds in the upper row are observed mainly in microorganisms or plants; those in the lowest row have animal roots. With the exclusion of the initial three compounds (upper row) these molecular structures are all known chiral. When these chiral compounds are observed in nature, they are typically enantiomerically unadulterated, though diverse causes may produce dissimilar enantiomers (Brown & Hess, 1969; Sakaitani & Ohfune, 1985; McMurry, 1989; Griesbeck et al., 1998). For instance, carvone is observed in spearmint oil, as its levorotatory (R)-enantiomer, while, caraway seeds comprise of the dextrorotatory (*S*)-enantiomer. Observe that the aldehyde function is frequently printed as –**CHO** in complex or condensed formulas (Kamernitzky & Akhrem, 1962; Corey & Boaz, 1985).

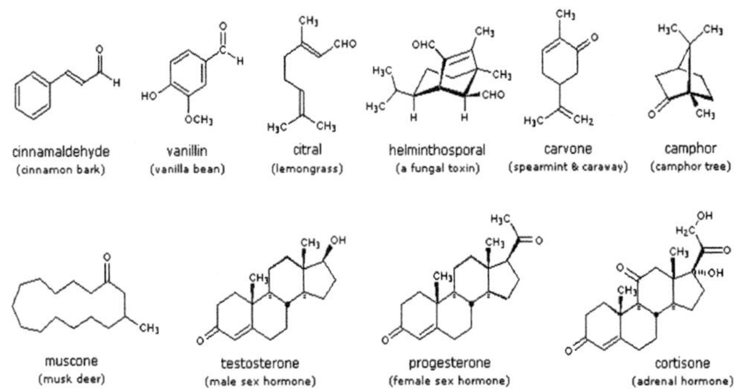

Figure 8.4: Common compounds containing carbonyl groups in aldehydes and ketones.

[Source: https://www2.chemistry.msu.edu/faculty/reusch/VirtTxtJml/aldket1.htm]

8.3. SYNTHETIC PREPARATION OF ALDEHYDES AND KETONES

Ketones and aldehydes are attained as yields from numerous reactions debated in former segments of this transcript. The subsequent diagram sums up the most significant of these.

To assess the former debate of any of these reaction types just click on the number (**1** to **5**) or expressive caption for the group. With the omission of Friedel-Crafts acylation, these approaches do not rise the complexity or size of molecules. In the subsequent segments of this chapter, we shall discover that one of the most suitable features of ketones and aldehydes is their reactivity to carbon nucleophiles and the resultant expansion of molecular structure that results. Briefly, ketones and aldehydes are significant intermediates for the synthesis or assembly of complex organic molecules.

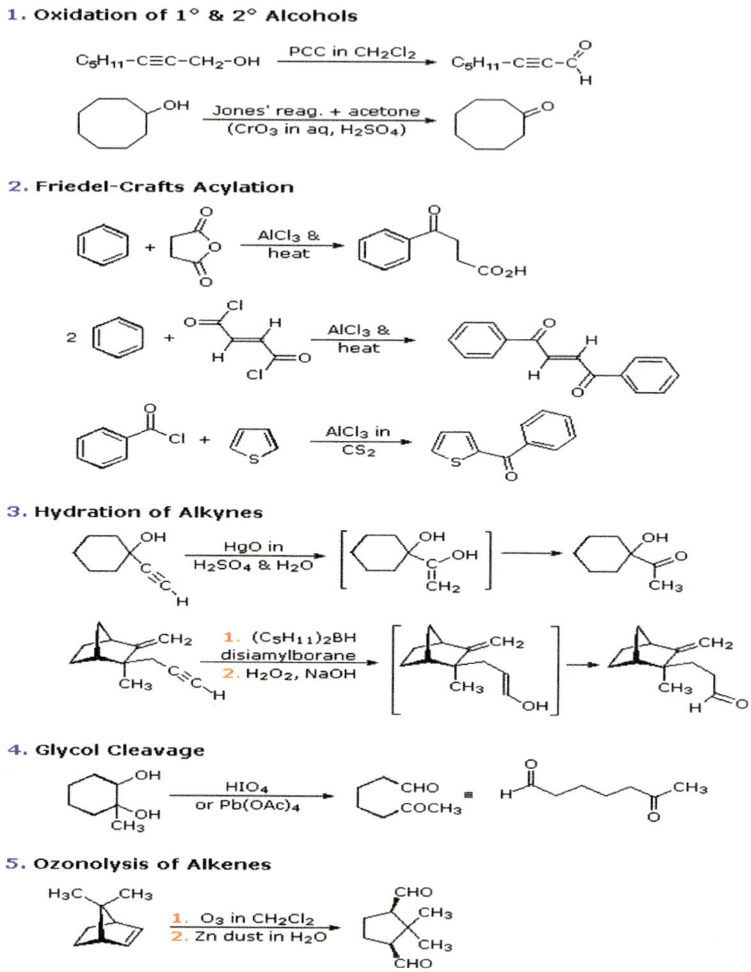

Figure 8.5: Reactions involved in the synthetic form of aldehydes and ketones.

[Source: https://www2.chemistry.msu.edu/faculty/reusch/VirtTxtJml/aldket1.htm]

8.4. PROPERTIES OF ALDEHYDES AND KETONES

An assessment of the reactivity and characteristics of ketones and aldehydes with those of the alkenes is necessary because both have a dual bond functional group. The carbonyl group is polar, because of the superior electronegativity of oxygen, and ketones and aldehydes have greater molecular dipole moments (D) than do alkenes. This polarity is illustrated by the resonance structures on the right, and the relative dipole moments of formaldehyde, other ketones and aldehydes approve the stabilizing effect that alkyl substituents have on carbocations. Therefore, we suppose, that ketones and aldehydes will have greater boiling points than alike sized alkenes.

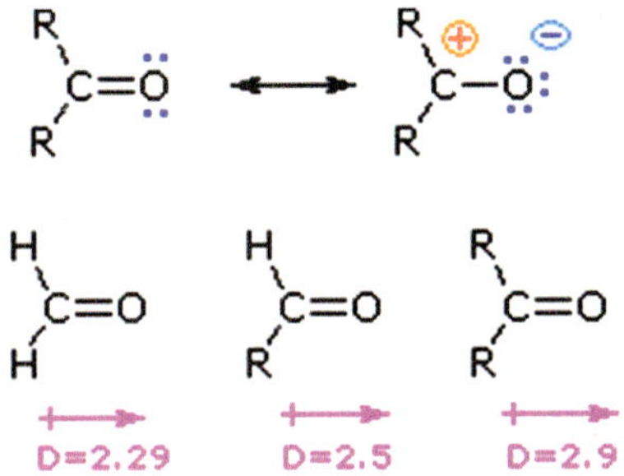

Figure 8.6: Dipole moments of different carbonyl compounds.

[Source: https://www2.chemistry.msu.edu/faculty/reusch/VirtTxtJml/aldket1.htm]

Moreover, ketones and aldehydes are made hydrogen-bond acceptors because of the existence of oxygen with its non-bonding electron pairs and should raise their solubility of water relative to hydrocarbons. Particular instances of these associations are given in the subsequent table.

Equated with the non-polar double bonds of alkenes, the polarity of the carbonyl group also has a deep impact on its chemical reactivity. Therefore, reversible addition of water to the carbonyl function is fast, whereas water addition to alkenes is infinitely slow in the nonappearance of a strong acid catalyst. Inquisitively, relative bond energies affect the thermodynamics of such addition reactions in a conflicting manner.

$(CH_3)_2C{=}CH_2$	56	-7.0°C	0.04 g/100
$CH_3CH_2CH_2CH{=}O$	72	76.0°C	7 g/100
$CH_3CH_2CH_2CH{=}CH_2$	70	30.0°C	0.03 g/100
$(CH_3)_2C{=}O$	58	56.5°C	infinite
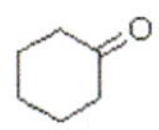	98	155.6°C	5 g/100
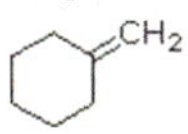	96	103.0°C	insoluble

Figure 8.7: Physical characteristics of different carbonyl compounds.

The C=C of alkenes has an average bond energy of 146 kcal/mole. Since a C–C σ-bond has a bond energy of 83 kcal/mole, the π-bond energy may be estimated at 63 kcal/mole (i.e., less than the energy of the sigma bond). The C=O bond energy of a carbonyl group, on the other hand, varies with its location, as follows:

$H_2C=O$	170 kcal/mole
$RCH=O$	175 kcal/mole
$R_2C=O$	180 kcal/mole

The C–O σ-bond is found to have average bond energy of 86 kcal/mole. Consequently, with the exception of formaldehyde, the carbonyl function of aldehydes and ketones has π-bond energy greater than that of the sigma-bond, in contrast to the pi-sigma relationship in C=C. This suggests that addition reactions to carbonyl groups should be thermodynamically disfavored, as is the case for the addition of water. All of this is summarized in the following diagram (ΔH° values are for the addition reaction).

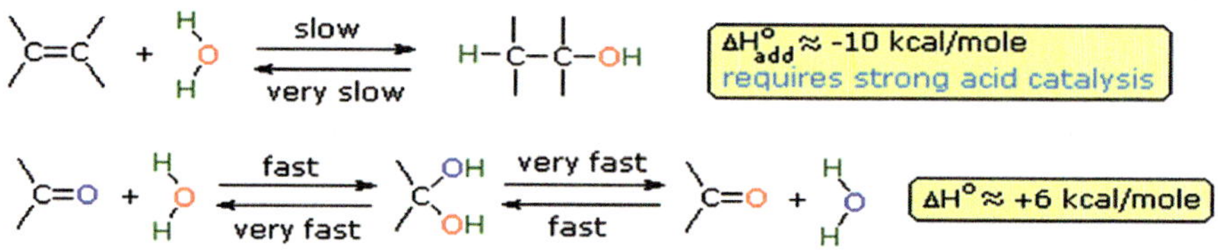

Figure 8.8: Reaction kinetics of alkene and carbonyl compound with the water molecule

[Source: https://www2.chemistry.msu.edu/faculty/reusch/VirtTxtJml/aldket1.htm]

Although the addition of water to an alkene is exothermic and gives a stable product (alcohol), the uncatalyzed reaction is extremely slow due to a high activation energy. The reverse reaction (dehydration of alcohol) is even slower, and because of the kinetic barrier, both reactions are practical only in the presence of a strong acid. The microscopically reversible mechanism for both reactions was described earlier.

In contrast, both the endothermic addition of water to a carbonyl function and the exothermic elimination of water from the resulting geminal-diol are fast. The inherent polarity of the carbonyl group, together with its increased basicity (compared with alkenes), lowers the transition state energy for both reactions, with a resulting increase in rate. Acids and bases catalyze both the addition and elimination of water. Proof that rapid and reversible addition of water to carbonyl compounds occurs is provided by experiments using isotopically labeled water. If a carbonyl reactant composed of ^{16}O (colored blue above) is treated with water incorporating the ^{18}O isotope (colored red above), a rapid exchange of the oxygen isotope occurs. This can only be explained by the addition-elimination mechanism shown here.

8.5. REACTIONS OF ALDEHYDES AND KETONES

8.5.1. Hydration and Hemiacetal Formation

It has been demonstrated (above) that water adds rapidly to the carbonyl function of aldehydes and ketones. In most cases, the resulting **hydrate** (a geminal-diol) is unstable relative to the reactants and cannot be isolated. Exceptions to this rule exist, one being formaldehyde (a gas in its pure monomeric state). Here the weaker pi-component of the carbonyl double bond, relative to other aldehydes or ketones, and the small size of the hydrogen substituents favor addition (Noyce & Pryor, 1955; Noyce & Reed, 1958). Thus, a solution of formaldehyde in water (formalin) is almost exclusively the hydrate or polymers of the hydrate. Similar reversible additions of alcohols to aldehydes and ketones take place. The equally unstable addition products are called **hemiacetals**.

$R_2C{=}O + R'OH\ R'O{-}(R_2)C{-}O{-}H$ (a hemiacetal)

8.5.2. Acetal Formation

Acetals are geminal-diether derivatives of aldehydes or ketones, formed by reaction with two equivalents of alcohol and elimination of water. Ketone derivatives of this kind were once called ketals, but modern usage has dropped that term. The following equation shows the overall stoichiometric change in acetal formation, but a dashed arrow is used because this conversion does not occur on simple mixing of the reactants.

$R_2C{=}O + 2\ R'OH\ R_2C(OR')_2 + H_2O$ (an acetal)

In order to achieve effective acetal formation, two additional features must be implemented. First, an acid catalyst must be used; and second, the water produced with the acetal must be removed from the reaction. The latter is important since acetal formation is reversible. Indeed, once pure acetals are obtained they may be hydrolyzed back to their starting components by treatment with aqueous acid. The mechanism shown here applies to both acetal formation and acetal hydrolysis by the principle of microscopic reversibility (Pinhas et al., 1981; Reetz, 1984; Prakash et al., 1989).

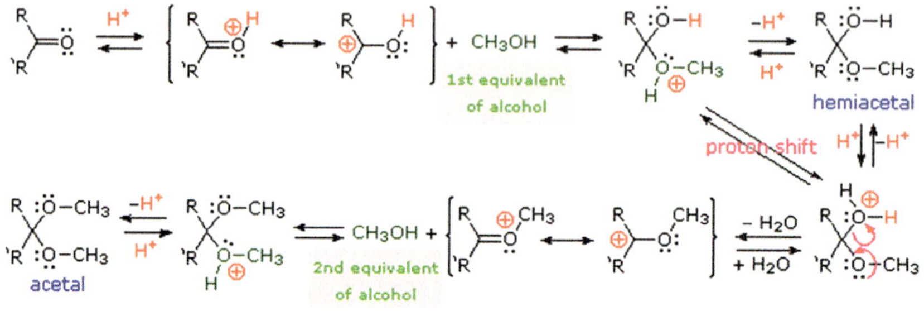

Figure 8.9: Acetal formation and acetal hydrolysis reactions involving carbonyl compounds.

[Source: https://www2.chemistry.msu.edu/faculty/reusch/VirtTxtJml/aldket1.htm]

Some examples of acetal formation are presented in the following diagram. As noted, p-toluenesulfonic acid ($pK_a = -2$) is often the catalyst for such reactions. Two equivalents of the alcohol reactant are needed, but these may be provided by one equivalent of a diol (example #2). Intramolecular involvement of a gamma or delta hydroxyl group (as in examples #3 and 4) may occur and is often more facile than the intermolecular reaction. Thiols (sulfur analogs of alcohols) give thioacetals (example #5). In this case, the carbonyl functions are relatively hindered, but by using excess ethanedithiol as the solvent and the Lewis acid BF_3 as a catalyst a good yield of the bis-thioacetal is obtained. Thioacetals are generally more difficult to hydrolyze than are acetals (Amici et al., 1989; Padwa & Austin, 1994).

The importance of acetals as carbonyl derivatives lies chiefly in their stability and lack of reactivity in neutral to strongly basic environments. As long as they are not treated by acids, especially aqueous acid, acetals exhibit all the lack of reactivity associated with ethers in general. Among the most useful and characteristic reactions of aldehydes and ketones is their reactivity toward strongly nucleophilic (and basic) Metallo-hydride, alkyl, and aryl reagents (to be discussed shortly). If the carbonyl functional group is converted to an acetal these powerful reagents have no effect; thus, acetals are excellent protective groups, when these irreversible addition reactions must be prevented (Juhl et al., 1991; Marson, 1992).

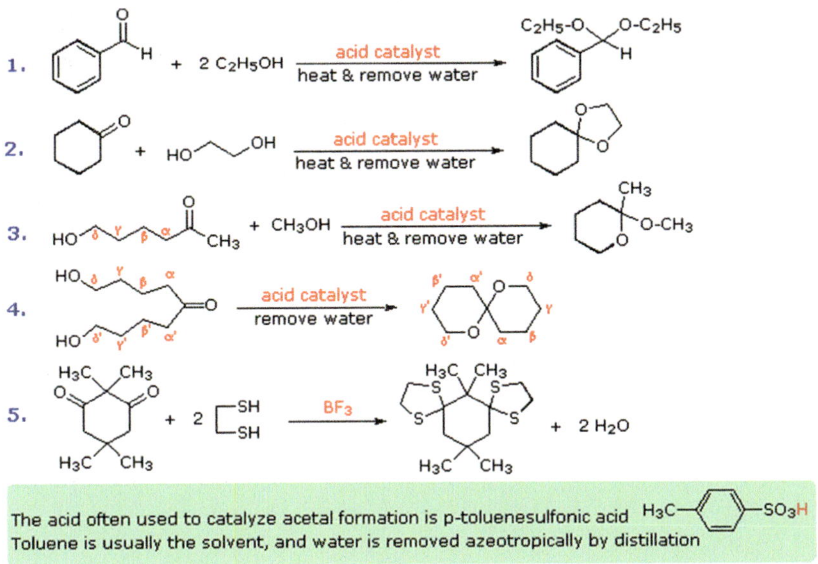

Figure 8.10: Examples of acetal formation.

[Source: https://www2.chemistry.msu.edu/faculty/reusch/VirtTxtJml/aldket1.htm]

8.5.3. Formation of Imines and Related Compounds

The reaction of ketones and aldehydes with 1°-amines or ammonia forms **imine derivatives** also-called as **Schiff bases**, (compounds with a C=N function). This reaction acts as a significant part in the formation of 2°-amines. Water is removed in the reaction, which is reversible and acid-catalyzed in the same manner as acetal synthesis.

$R_2C{=}O + R'NH_2\ R'NH{-}(R_2)C{-}O{-}H\ R_2C{=}NR' + H_2O$

An addition-removal mechanism for this reaction was planned, and an animation viewing this mechanism is triggered by the button.

Imines are occasionally problematic to purify and isolate due to their sensitivity to hydrolysis. Therefore, other reagents of the kind Y–NH_2 have been performed and observed to provide steady products ($R_2C{=}N{-}Y$) valuable in describing the ketones and aldehydes from which they are arranged. Some of these reagents are registered in the next table, together with the names and structures of their carbonyl reaction products. A stimulating feature of these carbonyl byproducts is that stereoisomers are conceivable when the R-groups of the carbonyl reactant are diverse. Therefore, a low-melting isomer, benzaldehyde produces two stereoisomeric oximes, having the hydroxyl group cis to the aldehyde hydrogen (called *syn*), and a higher melting isomer in which the hydrogen and hydroxyl group are trans (the anti isomer). At ambient temperature or below the configuration of the dual-bonded nitrogen atom is seemingly secure in one trigonal shape, dissimilar the quickly interconverting pyramidal shapes of the sp3 hybridized amines (Hanzlik & Leinwetter, 1978; Shibata et al., 2005).

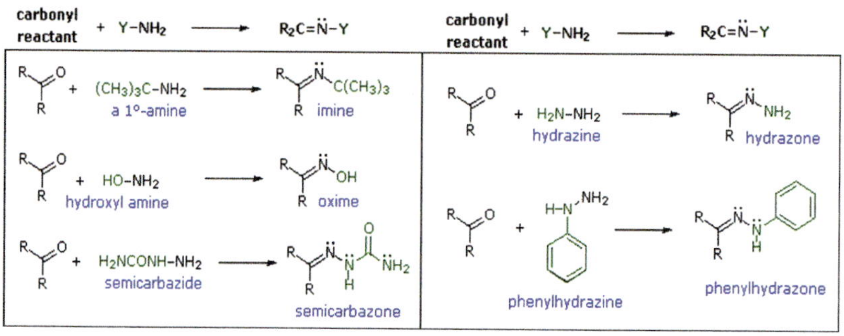

Figure 8.11: Formation of imines and miscellaneous compounds from carbonyl compounds.

[Source: https://www2.chemistry.msu.edu/faculty/reusch/VirtTxtJml/aldket1.htm]

These derivatives are usually crystalline solids and their preparation is easy with the exceptional case of unsubstituted hydrazones – although if the maternal aldehyde or ketone is a liquid. Subsequently, **melting points can be determined more quickly and precisely than boiling points,** these derivatives are suitable for judgment and identification of carbonyl compounds. The hydrazone derivatives and resulting reagent that are given when the aromatic ring of phenylhydrazine is replaced with nitro groups at the 2- & 4-positions are powerfully colored and easily identify. As semicarbazide has two amino groups ($–NH_2$) but one of them is a reactive amine. The next one is amide-like which is neutralized by the adjacent carbonyl group (Mehta & Muthusamy, 2002).

The formation rate of these imine-alike compounds is usually highest almost a pH of 5 and drips at lower and higher pH's. **This agrees with general acid catalysis in which the conjugate acid of the carbonyl reactant combines with a free amino group,** as shown in the above animation. There will be a desperately low concentration of the carbonyl conjugate acid at high pH, and utmost of the amine reactant will be occupied as its ammonium conjugate acid at low pH. There is no obligation of active removal of water (not shown as a product in the previous equations) for most of this derivatization reaction with the special case of imine synthesis itself. Equilibrium is not recognized rapidly, although the reactions are reversible and when they are formed, the products often precipitous from solution.

8.5.4. Enamine Formation

All tangled reagents of the type, the earlier reactions have **Y–NH$_2$**, i.e., reactions of 1°-amino group. The products known as enamines are obtained by the reaction of 2°-amines with most aldehydes and ketones. Two examples of these reactions are offered in the following diagram. It should be renowned that, these are acid-catalyzed reversible reactions in which water vanishes, such as acetal formation. Therefore, by acid-catalyzed hydrolysis enamines are simply transformed back to their carbonyl precursors. A procedure for enamine formation may be seen by persistent the "Show Mechanism" button.

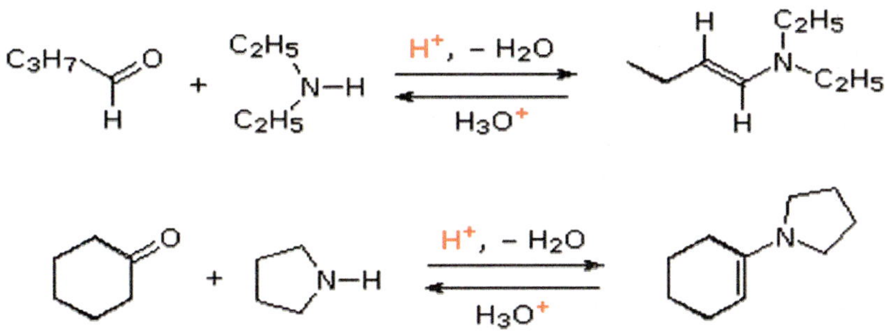

Figure 8.12: Enamine Formation reactions.

[Source: https://www2.chemistry.msu.edu/faculty/reusch/VirtTxtJml/aldket1.htm]

8.5.5. Cyanohydrin Formation

Hydrogen cyanide (HC≡N) is the latter example of reversible addition, which produces products called **cyanohydrins** by adding to aldehydes and many ketones.

RCH=O + H–C≡N RCH(OH)CN (a cyanohydrin)

Since hydrogen cyanide is an acid (pK_a = 9.25) by itself, the adding is not acid-catalyzed. Actually, cyanide anion, $C{\equiv}N^{(-)}$ must be existing for best results, it actually means that catalytic base must supply. The reaction of cyanohydrin formation is exothermic feebly and is preferred for unhindered cyclic, aldehydes, and methyl ketones. Two examples of reactions like this are shown below.

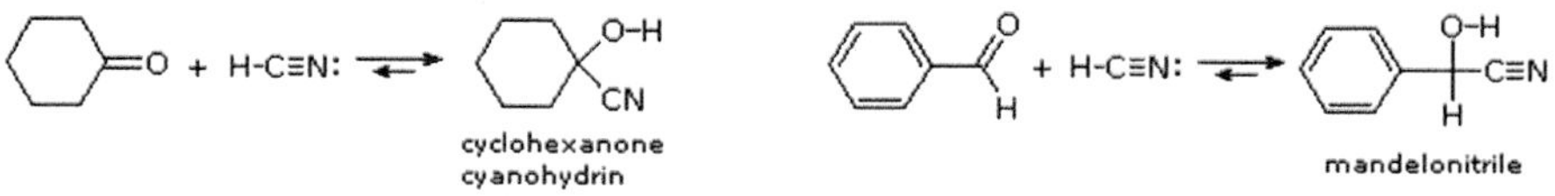

Figure 8.13: Cyanohydrin Formation reaction.

[Source: https://www2.chemistry.msu.edu/faculty/reusch/VirtTxtJml/aldket1.htm]

Mandelonitrile is the name of cyanohydrin from benzaldehyde. Millipede *Apheloria corrugata* uses the reversible formation of cyanohydrin in a notable resisting procedure. Releases mandelonitrile secrete this arthropod from an internal storing gland into an external cavity, where it is shattered down into benzaldehyde and hydrogen cyanide by using enzymes beforehand being scattered at an enemy.

8.6. REDUCTION REACTIONS

As mentioned above, The metal hydride reductions and organometallic flourishes to aldehydes and ketones, the carbonyl carbon's oxidation state is decreased by both, and maybe ordered as reductions.it it is observed that they advance by the strong nucleophilic dose of species at the electrophilic carbon. Other beneficial reductions of carbonyl compounds may occur by different procedures, either to alcohols or to hydrocarbons. For example, all reactions such as sodium or potassium in hydroxylic or amine solvents and, hydrogenation (Pt, Pd, Ni or Ru catalysts), reaction with diborane, and reduction by lithium have been described to convert carbonyl compounds into alcohols. Though, the complex metal hydrides offer cleaner products in high yield so are usually favored for such transformations.

Complete removal of the oxygen is the requirement of the reductive conversion of a carbonyl group to a methylene group and is called **deoxygenation**. The [H] symbol refers to indefinite reduction conditions which affect the wanted change is shown by simple equation here. Three very different methods of accomplishing this transformation will be defined here.

$R_2C=O + [H] R_2CH_2 + H_2O$

8.6.1. Wolff-Kishner Reduction

A hydrazone derivative is produced by the reaction of an aldehyde or ketone with extra hydrazine, which gives the matching hydrocarbon on heating with base. The required temperatures are usually obtained by utilizing high-boiling hydroxylic solvent, such as diethylene glycol. How this reduction may be utilized to transform cyclopentanone to cyclopentane is shown by the upcoming diagram. Another example, in which an aldehyde is correspondingly concentrated to a methyl group, also exemplifies again the use of an acetal defensive group. On pressing the "Show Mechanism" button it will display the useful transformation of the appliance which contains tautomerization of the initially formed hydrazone to an azo isomer. Some solidly basic conditions which we use in this reaction causes to stop its application near base sensitive compounds.

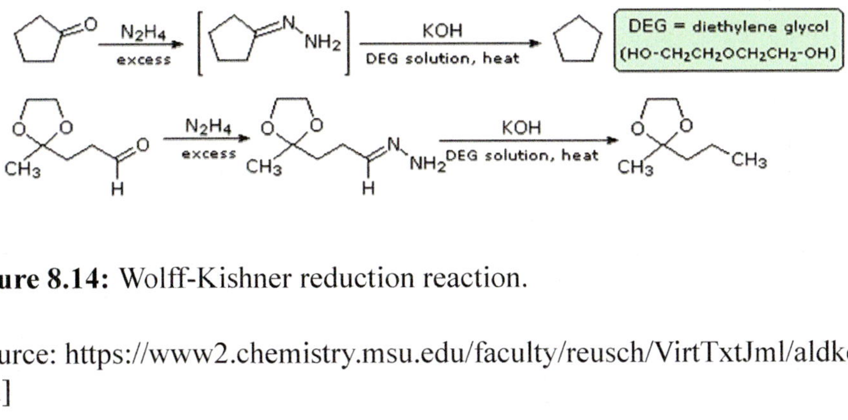

Figure 8.14: Wolff-Kishner reduction reaction.

[Source: https://www2.chemistry.msu.edu/faculty/reusch/VirtTxtJml/aldket1.htm]

8.6.2. Clemmensen Reduction

This substitute reduction includes heating a carbonyl compound with finely separated, combined zinc in a hydroxylic solvent (frequently an aqueous mixture) comprising of a mineral acid such as HCl. The mercury alloyed with the zinc does not contribute to the reaction, it acts only to deliver a fresh active metal surface. The primary case underneath displays a normal user of this reduction, the alteration of a Friedel-Crafts acylation product to an alkyl side-chain. Another case exemplifies the lability of functional substituents alpha to the carbonyl group. Substituents such as hydroxyl, halogens, and alkoxyl are condensed first, the resultant unsubstituted ketone or aldehyde is then condensed to the parent hydrocarbon. By clicking the "Show Mechanism" button, a probable mechanism for the Clemmensen reduction will be displayed.

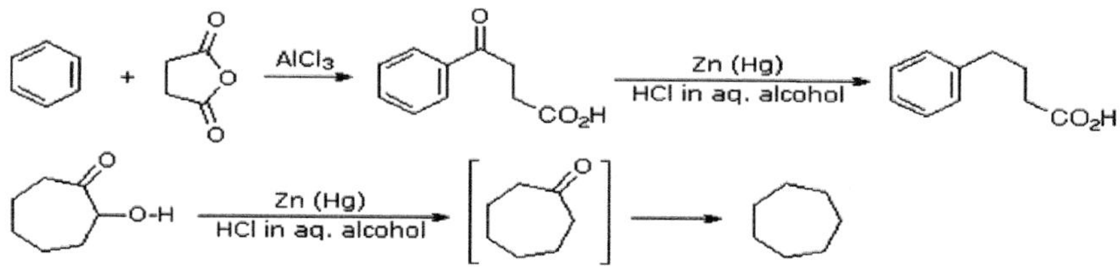

Figure 8.15: Clemmensen reduction reactions.

[Source: https://www2.chemistry.msu.edu/faculty/reusch/VirtTxtJml/aldket1.htm]

8.7. OXIDATION REACTIONS

The carbon atom of a carbonyl group has a comparatively high oxidation state. This is replicated in the detail that most of the reactions defined therefore far either reason no change in the oxidation state (e.g. imine creation and acetal) or result in a reduction (e.g., deoxygenations or organometallic additions). The most characteristic and normal oxidation reaction is the adaptation of aldehydes to carboxylic acids. In the shorthand equation revealed here the [O] sign denotes to undetermined oxidation circumstances which result in the anticipated alteration. Numerous dissimilar approaches to achieving this transformation will be defined here.

RCH=O + [O] RC(OH)=O

In debating the oxidations of 1° and 2°-alcohols, we distinguished that Jones' reagent converts aldehydes to carboxylic acids, apparently via the hydrate. Other reagents affect the same transformation, amongst them dilute bromine and aqueous potassium permanganate. Even the oxygen in the air will gradually oxidize aldehydes to peracids or acids, most probable by a radical mechanism. Beneficial examinations for aldehydes, **Benedict's test, Tollens' test** & **Fehling's test**, take benefit of this affluence of oxidation by the means $Cu^{(2+)}$ and $Ag^{(+)}$ as oxidizing agents (oxidants).

RCH=O + 2 [$Ag^{(+)}$ $OH^{(-)}$] RC(OH)=O + 2 Ag (metallic mirror) + H_2O

As in the above equation, when silver cation is the oxidant, it is reduced to metallic silver in the sequence of the reaction, and this bonds as an attractive mirror on the internal surface of the reaction container. The Benedict and Fehling tests use cupric cation as the oxidant. This profound blue reagent is reduced to cuprous oxide, which causes a red to yellow solid. All these cation oxidations must be shown under alkaline circumstances. The cations must be stabilized as complexed ions to evade precipitation of the insoluble metal hydroxides. Cupric ions are used as citrate or tartrate complexes and silver is applied as its ammonia complex, $Ag(NH_3)_2^{(+)}$.

Saturated ketones are usually inert to oxidation circumstances that change aldehydes to carboxylic acids. However, under energetic acid-catalyzed oxidations with chromic or nitric acids ketones may undergo carbon-carbon bond cleavage at the carbonyl group.

REFERENCES

1. Adams, J., & Spero, D. M. (1991). Rhodium (II) catalyzed reactions of diazo-carbonyl compounds. *Tetrahedron, 47*(10–11), 1765–1808.
2. Amici, A., Levine, R. L., Tsai, L., & Stadtman, E. R. (1989). Conversion of amino acid residues in proteins and amino acid homopolymers to carbonyl derivatives by metal-catalyzed oxidation reactions. *Journal of Biological Chemistry, 264*(6), 3341–3346.
3. Bender, M. L. (1960). Mechanisms of catalysis of nucleophilic reactions of carboxylic acid derivatives. *Chemical Reviews, 60*(1), 53–113.
4. Brown, H. C., & Hess, H. M. (1969). Selective reductions. XIII. Reaction of 2-cyclopentenones with representative complex hydrides. Aluminum hydride as a selective reagent for the reduction of the carbonyl group in 2-cyclopentenones. *The Journal of Organic Chemistry, 34*(7), 2206–2209.
5. Brown, H. C., Schlesinger, H. I., & Burg, A. B. (1939). Hydrides of boron. XI. The reaction of diborane with organic compounds containing a carbonyl group. *Journal of the American Chemical Society, 61*(3), 673–680.
6. Burgi, H. B., Dunitz, J. D., & Shefter, E. (1973). Geometrical reaction coordinates. II. Nucleophilic addition to a carbonyl group. *Journal of the American Chemical Society, 95*(15), 5065–5067.
7. Bürgi, H. B., Lehn, J. M., & Wipff, G. (1974). Ab initio study of nucleophilic addition to a carbonyl group. *Journal of the American Chemical Society, 96*(6), 1956–1957.
8. Corey, E. J., & Boaz, N. W. (1985). The reactions of combined organocuprate-chlorotrimethylsilane reagents with conjugated carbonyl compounds. *Tetrahedron letters, 26*(49), 6019–6022.
9. Dunitz, J. D., Lehn, J. M., & Wipff, G. (1974). Stereochemistry of reaction paths at carbonyl centres. *Tetrahedron, 30*(12), 1563–1572.
10. Griesbeck, A. G., Buhr, S., Fiege, M., Schmickler, H., & Lex, J. (1998). Stereoselectivity of Triplet Photocycloadditions: 1 Diene–Carbonyl Reactions and Solvent Effects. *The Journal of Organic Chemistry, 63*(12), 3847–3854.
11. Hanzlik, R. P., & Leinwetter, M. (1978). Reactions of epoxides and carbonyl compounds catalyzed by anhydrous copper sulfate. *The Journal of Organic Chemistry, 43*(3), 438–440.

12. Imamoto, T., Takiyama, N., Nakamura, K., Hatajima, T., & Kamiya, Y. (1989). Reactions of carbonyl compounds with Grignard reagents in the presence of cerium chloride. *Journal of the American Chemical Society*, *111*(12), 4392–4398.
13. Jencks, W. P. (1964). Mechanism and catalysis of simple carbonyl group reactions. *Prog. Phys. Org. Chem*, *2*, 63–128.
14. Johnson, J. S., & Evans, D. A. (2000). Chiral bis (oxazoline) copper (II) complexes: versatile catalysts for enantioselective cycloaddition, aldol, Michael, and carbonyl ene reactions. *Accounts of Chemical Research*, *33*(6), 325–335.
15. Jørgensen, K. A. (2000). Catalytic Asymmetric Hetero-Diels–Alder Reactions of Carbonyl Compounds and Imines. *Angewandte Chemie International Edition*, *39*(20), 3558–3588.
16. Juhl, K., Gathergood, N., & Jørgensen, K. A. (2001). Catalytic Asymmetric Direct Mannich Reactions of Carbonyl Compounds with α-Imino Esters. *Angewandte Chemie International Edition*, *40*(16), 2995–2997.
17. Kamernitzky, A. V., & Akhrem, A. A. (1962). The stereochemistry of reactions of nucleophilic addition to the carbonyl group of cyclic ketones. *Tetrahedron*, *18*(6), 705–750.
18. Kawabata, T., Yahiro, K., & Fuji, K. (1991). Memory of chirality: enantioselective alkylation reactions at an asymmetric carbon adjacent to a carbonyl group. *Journal of the American Chemical Society*, *113*(25), 9694–9696.
19. Koelsch, C. F., & LeClaire, C. D. (1941). The reactions and enolization of cyclic diketones. V.(1) Some carbonyl reactions. *The Journal of Organic Chemistry*, *6*(4), 516–533.
20. Kuroda, H., Hanaki, E., Izawa, H., Kano, M., & Itahashi, H. (2004). A convenient method for the preparation of α-vinylfurans by phosphine-initiated reactions of various substituted enynes bearing a carbonyl group with aldehydes. *Tetrahedron*, *60*(8), 1913–1920.
21. Lee, M., & Harris, C. B. (1989). Ultrafast studies of transition metal carbonyl reactions in the condensed phase. Solvation of coordinatively unsaturated pentacarbonyls. *Journal of the American Chemical Society*, *111*(24), 8963–8965.
22. Marson, C. M. (1992). Reactions of carbonyl compounds with (monohalo) methyleniminium salts (Vilsmeier reagents). *Tetrahedron*,

48(18), 3659–3726

23. McMurry, J. E. (1989). Carbonyl-coupling reactions using low-valent titanium. *Chemical Reviews, 89*(7), 1513–1524.
24. Mehta, G., & Muthusamy, S. (2002). Tandem cyclization–cycloaddition reactions of rhodium generated carbenoids from alpha-diazo carbonyl compounds. *Tetrahedron, 58*(47), 9477–9504.
25. Mikami, K., Itoh, Y., & Yamanaka, M. (2004). Fluorinated carbonyl and olefinic compounds: basic character and asymmetric catalytic reactions. *Chemical Reviews, 104*(1), 1–16.
26. Mukaiyama, T., Banno, K., & Narasaka, K. (1974). New cross-aldol reactions. Reactions of silyl enol ethers with carbonyl compounds activated by titanium tetrachloride. *Journal of the American Chemical Society, 96*(24), 7503–7509.
27. Noyce, D. S., & Jorgenson, M. J. (1962). Carbonyl Reactions. XIX. The Basicities of Substituted Chalcones. *Journal of the American Chemical Society, 84*(22), 4312–4319.
28. Noyce, D. S., & Pryor, W. A. (1955). Carbonyl Reactions. I. Kinetics and Mechanism of the Acid-catalyzed Aldol Condensation of Benzaldehyde and Acetophenone1. *Journal of the American Chemical Society, 77*(6), 1397–1401.
29. Noyce, D. S., & Reed, W. L. (1958). Carbonyl Reactions. VI. Evidence for Alternate Mechanisms for the Dehydration of β-Hydroxy Ketones1. *Journal of the American Chemical Society, 80*(20), 5539–5542.
30. Noyori, R., Tomino, I., & Tanimoto, Y. (1979). Virtually complete enantioface differentiation in carbonyl group reduction by a complex aluminum hydride reagent. *Journal of the American Chemical Society, 101*(11), 3129–3131.
31. Ohwada, T., Yamagata, N., & Shudo, K. (1991). Friedel-Crafts-type reactions involving di-and tricationic species. Onium-allyl dications and O, O-diprotonated aci-nitro species bearing a protonated carbonyl group. *Journal of the American Chemical Society, 113*(4), 1364–1373.
32. Padwa, A., & Austin, D. J. (1994). Ligand Effects on the Chemoselectivity of Transition Metal Catalyzed Reactions of α-Diazo Carbonyl Compounds. *Angewandte Chemie International Edition in English, 33*(18), 1797–1815.
33. Pinhas, A. R., Samuelson, A. G., Risemberg, R., Arnold, E. V., Clardy, J., & Carpenter, B. K. (1981). Experimental evidence for orbital

symmetry control of stereochemistry in some (methylenecyclopropane) iron-carbonyl reactions. *Journal of the American Chemical Society, 103*(7), 1668–1675.

34. Prakash, G. S., Krishnamurti, R., & Olah, G. A. (1989). Synthetic methods and reactions. 141. Fluoride-induced trifluoromethylation of carbonyl compounds with trifluoromethyltrimethylsilane (TMS-CF3). A trifluoromethide equivalent. *Journal of the American Chemical Society, 111*(1), 393–395.
35. Reetz, M. T. (1984). Chelation or Non-Chelation Control in Addition Reactions of Chiral α-and β-Alkoxy Carbonyl Compounds [New Synthetic Methods (44)]. *Angewandte Chemie International Edition in English, 23*(8), 556–569.
36. Sakaitani, M., & Ohfune, Y. (1985). Selective transformation of Nt-butoxycarbonyl group into n-alkoxy-carbonyl group via n-carboxylate ion equivalent. *Tetrahedron Letters, 26*(45), 5543–5546.
37. Shibata, N., Kohno, J., Takai, K., Ishimaru, T., Nakamura, S., Toru, T., & Kanemasa, S. (2005). Highly enantioselective catalytic fluorination and chlorination reactions of carbonyl compounds capable of two-point binding. *Angewandte Chemie International Edition, 44*(27), 4204–4207.
38. Shinohara, K., Wu, R. T., Jahan, N., Tanaka, M., Morinaga, N., Murakami, H., & Omura, H. (1980). Mutagenecity of the Browing Mixtures by Amino-Carbonyl Reactions on Salmonella typhimurium TA 100. *Agricultural and Biological Chemistry, 44*(3), 671–672.
39. Trost, B. M., & Massiot, G. S. (1977). New synthetic reactions. A chemoselective approach to cleavage. alpha. to a carbonyl group via. beta.-keto sulfides. Preparation of 1, 2-diketones. *Journal of the American Chemical Society, 99*(13), 4405–4412.
40. Wiberg, K. B., Morgan, K. M., & Maltz, H. (1994). Thermochemistry of carbonyl reactions. 6. A study of hydration equilibria. *Journal of the American Chemical Society, 116*(24), 11067–11077.
41. Zamora, R., Alaiz, M., & Hidalgo, F. J. (2000). Contribution of pyrrole formation and polymerization to the nonenzymatic browning produced by amino-carbonyl reactions. *Journal of Agricultural and Food Chemistry, 48*(8), 3152–3158.
42. Zatsepin, T. S., Stetsenko, D. A., Gait, M. J., & Oretskaya, T. S. (2005). Use of carbonyl group addition-elimination reactions for synthesis of nucleic acid conjugates. *Bioconjugate chemistry, 16*(3), 471–489.

INDEX

C

D

E

F

G

H

X

Z